# Land Degradation in Environmentally Sensitive Areas (ESA) : Assessment and Conservation

# Land Degradation in Environmentally Sensitive Areas (ESA) : Assessment and Conservation

Guest Editors

**Mario Al Sayah**
**Rita Der Sarkissian**
**Rachid Nedjaï**

MDPI

Basel • Beijing • Wuhan • Barcelona • Belgrade • Novi Sad • Cluj • Manchester

*Guest Editors*
Mario Al Sayah
Remote Sensing Center
National Council for
Scientific Research
Beirut
Lebanon

Rita Der Sarkissian
LAB'URBA
University of Gustave Eiffel
University of Paris Est Creteil
Ecole des Ingénieurs de la Ville
de Paris (EIVP)
Marne-la-Vallée
France

Rachid Nedjaï
Centre D'études et de
Développement des
Territoires et de
l'Environnement
Université d'Orléans
Orléans
France

*Editorial Office*
MDPI AG
Grosspeteranlage 5
4052 Basel, Switzerland

This is a reprint of the Special Issue, published open access by the journal *Land* (ISSN 2073-445X), freely accessible at: https://www.mdpi.com/journal/land/special_issues/J5F93TZ23O.

For citation purposes, cite each article independently as indicated on the article page online and as indicated below:

| Lastname, A.A.; Lastname, B.B. Article Title. *Journal Name* **Year**, *Volume Number*, Page Range. |
|---|

**ISBN 978-3-7258-4479-1 (Hbk)**
**ISBN 978-3-7258-4480-7 (PDF)**
**https://doi.org/10.3390/books978-3-7258-4480-7**

# Contents

# About the Editors

**Mario Al Sayah**

Mario Al Sayah holds a Ph.D. in Land Management, Geography, and Environmental Sciences, jointly awarded by the University of Orléans (France), the National Council for Scientific Research (Lebanon), and the Lebanese University. His expertise lies in land degradation and the development of related frameworks, particularly the concept of Land Degradation Neutrality (LDN). He has received two awards for his doctoral research and has been nominated for several others. Dr. Al Sayah specialized in Remote Sensing and GIS at the National Remote Sensing Center of Lebanon, followed by postgraduate studies at the University of Orléans and the École Nationale des Ponts et Chaussées in France. He serves as the principal investigator for numerous projects focused on land degradation. His academic contributions include teaching graduate and postgraduate courses in GIS and remote sensing, the development of master's level modules, as well as supervising numerous master's and doctoral theses. He is the author or co-author of over 40 publications, including peer-reviewed journal articles, conference proceedings, book chapters, and technical reports. He also serves as a reviewer for several international scientific journals. Dr. Al Sayah has extensive experience working in Mediterranean and European contexts. His research spans multiple scales and aligns with international and regional agendas, focusing on concepts such as LDN and Nature-based Solutions (NbS).

**Rita Der Sarkissian**

Rita Der Sarkissian is a lecturer and researcher at the École des Ingénieurs de la Ville de Paris (EIVP), affiliated with Université Gustave Eiffel. Her expertise lies in urban resilience, Disaster Risk Reduction (DRR), the degradation of natural resources, and geospatial sciences. She holds a Ph.D. in DRR and Geospatial Information Technologies, jointly awarded by the University of Orléans (France) and the CNRS Remote Sensing Center of Lebanon. Her doctoral research focused on leveraging geospatial data to enhance the resilience of critical infrastructure systems. Dr. Der Sarkissian's research spans key areas such as urban resilience, DRR, geospatial modeling, evacuation planning, critical infrastructure protection, land degradation, and the "Build Back Better" framework. Her work aims to strengthen the resilience of urban environments and essential services in the face of natural hazards, while also bridging the gap between geospatial technologies and their practical applications in disaster risk management. She is an active member of the Arab Science and Technology Advisory Group (STAG) and is recognized as a prominent expert in DRR across the Middle East and Arab states. Through her research, publications, and conference contributions, Dr. Der Sarkissian is a significant contributor to disaster risk reduction and geospatial sciences.

**Rachid Nedjaï**

Rachid Nedjaï is a full Professor in the University of Orléans, France. He started his career in applied geology with a focus on hydrogeology in Algeria. He then pursued a PhD on Algerian thermomineral springs, which was completed in 1987 at the University of Grenoble 1, France. He then held academic and professional roles in both public and private sectors, including a position as assistant lecturer and the leadership of a private hydrogeology firm. He then specialized in geographic information systems and environmental data simulation through a dual-competency program in computer science in 1992. Professor Nedjaï then held positions at the Alpine Geography Institute and the CNRS-affiliated Transport Economics Laboratory. In 1998, he followed a senior lectureship program in hydrology and geomatics at the University of Grenoble 1. He then pursued a habilitation in lake geochemistry and geomatics in 2010. A professorship in limnology and geomatics was taken up in 2013 at the University of Orléans. Since then, Professor Nedjaï has assumed numerous responsibilities, including program leadership and directorship of the UFR LLSH at the University of Orléans. Professor Nedjaï leads multiple research teams, namely on land degradation, urban heat islands, greenhouse gas emissions from ponds, and hydrogeochemical dynamics in the Loire Basin.

# Preface

The articles featured in this Special Issue, entitled "Land Degradation in Environmentally Sensitive Areas (ESA): Assessment and Conservation", cover a broad range of contexts and were conducted in Europe and Asia at different scales. The variation in study areas reflects the multifaceted and cross-scalar nature of land degradation. Notably, many of the studies stem from collaborations between universities and research institutions across various countries and continents, demonstrating the motivation of the international scientific community to address the issue of land degradation. Within this Special Issue, articles can be classified into the following five categories—Category A: "Agricultural practices and soil pollution"; Category B: "Ecosystem services and land use management"; Category C: "Wind erosion and desertification"; Category D: "Salinization and climate hazards"; and Category E: "Policy impacts on rural livelihoods". Articles 1, 2 and 3 belong to Category A. In the first article (Ding et al.), an econometric analysis using survey data was carried out to investigate the impact of farm size on the waste recycling behavior regarding pesticide packaging among farmers in China. In the second article (Andreeva et al.), the ecological impact of copper-containing fungicides on the accumulation and distribution of copper, manganese, chromium, and cobalt in the upper soil horizons of vineyards was investigated. In the third article (Si et al.), soil was sampled across six land use types, and a principal component analysis was used to derive the soil quality index, with recommendations of soil parameters that are essential for land degradation assessment. Articles 4 and 5 belong to Category B: In the fourth article (Richiedei et al.), a biophysical and monetary assessment of six ecosystem services was carried out using a set of indicators. Subsequently, a soil quality assessment methodology based on ecosystem value at sub-regional levels in Italy was developed. In article five (Chen et al.), the IVST and PLUS models were used to assess habitat degradation and its driving mechanisms in Urumqi, China for the period of 2000–2022. Future projections were then carried out up to 2035. Articles 6 and 7 belong to Category C: In the sixth article (Xiong et al.), an improvement of the RWEQ model, incorporating uncertainty analysis, scenario simulations, and environmental factor evaluations, enabled the dynamic modeling of both the supply and demand of windbreak and sand fixation ecosystem services in the Wuding River Basin, China. In the seventh article (Zhang et al.), wind erosion in the Tarim River Basin, China's largest inland river basin, was studied. A soil wind erosion model tailored to the terrain was developed to assess erosion-induced degradation in nine subbasins. Article 8 belongs to Category D: In this article (Lin et al.), a global-scale study on the salinity stress thresholds of maize fields was carried out. Using the EPIC model and eliminating environmental stress factors, daily salinity stress thresholds during corn growth stages were simulated. The results show that high-risk areas, where the disruption index exceeded 0.8, were mainly concentrated in arid and semi-arid regions such as Central Asia, northwestern China, southern South America, and the southern coast of Africa, while humid regions exhibited relatively lower risk. As the recurrence interval increased (e.g., over 10, 20, 50, and 100 years), both the geographic extent and severity of salinization hazards intensified, with countries such as Oman, Egypt, and Mongolia consistently showing the highest average salinization intensity (above 0.7). Article 9 belongs to Category E: In this article (Wang et al.), household surveys, regression analysis, and the sustainable livelihood frameworks were used to meticulously assess the impact of grain for green projects on farmer well-being in Zhangbei, China.

**Mario Al Sayah, Rita Der Sarkissian, and Rachid Nedjaï**
*Guest Editors*

*land*

*Article*

# Impact of Farm Size on Farmers' Recycling of Pesticide Packaging Waste: Evidence from Rural China

**Jingyi Ding, Kun Song, Kuan Zhang and Xin Deng ***

College of Economics, Sichuan Agricultural University, Chengdu 611130, China;
dingjingyi@stu.sicau.edu.cn (J.D.); songkun@stu.sicau.edu.cn (K.S.); 14869@sicau.edu.cn (K.Z.)
* Correspondence: dengxin@sicau.edu.cn

**Abstract:** Scale management has become an essential form of modern agricultural production. However, it is still unclear how farm size influences farmers' pesticide packaging waste recycling behavior (FPPWRB). Based on the data from the China Rural Revitalization Survey 2020, this study quantitatively explores the impact of farm size on FPPWRB. This study found that (1) the ratio for FPPWRB is low, with only about 41.7% of the sample farmers expressing participation in recycling. (2) The empirical results show that for every 1% increase in farm size, the probability of FPPWRB increases by 3.59%. (3) Farmers in urban suburbs or younger farmers are more inclined to FPPWRB. (4) Farm size can improve FPPWRB by enhancing farmers' environmental cognition levels. The research in this study provides insights into the improper disposal of pesticide packaging waste and offers references for formulating policies related to the resource utilization of pesticide packaging waste. Thus, the findings of this study can help provide a reference for the introduction of policies to manage pesticide packaging waste, which, in turn, can help promote sustainable agricultural development.

**Keywords:** farm size; pesticide packaging waste; farmer recycling behavior; rural China

## 1. Introduction

Pesticides play an important role in achieving the global goal of "zero hunger". However, they also pose environmental threats that must not be ignored. About 3.5 million tons of pesticides are applied globally each year [1]. Asia has the highest pesticide usage globally, accounting for 38%, while Europe has the lowest usage, accounting for only 13% [2]. Between 1990 and 2021, pesticide usage in Oceania saw the fastest growth, increasing by 206%, while the growth rates in the Americas, Africa, Asia, and Europe were 191%, 175%, 67%, and 1%, respectively [3]. Oceania applies a relatively low level of pesticides per hectare of farmland, which totals 1.66 kg/hectare. Still, when standardized by the agricultural production value (0.83 kg/USD1000), the pesticide application per capita is higher (1.30 kg/person) [4]. In contrast, the Americas have reached higher levels in all three indicators, with 3.01 kg/hectare, 1.49 kg/USD1000, and 1.23 kg/person, respectively [5]. The massive generation of pesticide packaging waste (PPW) accompanies the abuse of pesticides. Approximately 10 billion pieces of this waste are generated annually worldwide. However, there is limited research on the subsequent disposal of PPW and its potential threat to sustainable development [6].

The improper disposal of PPW hinders the comprehensive green transformation of agriculture and rural areas. Improper handling and disposal of this waste may bring hazards. Firstly, the presence of residual pesticides means improper disposal can exacerbate

 https://doi.org/10.3390/land14030465

the harmful effects of pesticide overuse [4,7]. Li et al. [8] found that the misuse of pesticides can lead to pesticide resistance in pests, creating a vicious cycle that threatens biodiversity; Zhao et al. [9] pointed out that 20 types of pesticides have been detected in the serum, urine, and cerebrospinal fluid of urban populations in China. Second, the haphazard discarding of plastic pesticide packaging greatly harms soil and water. Studies show that white plastic pollution and microplastic contamination are significant driving factors of soil compaction and water body damage [10,11]. Third, the burning of PPW pollutes the air. Open burning releases large amounts of carcinogens, such as dioxins, polycyclic aromatic hydrocarbons, and particulate matter, into the atmosphere, causing severe environmental pollution that threatens human health [5]. Therefore, promoting the recycling of PPW is the key to achieving sustainable development in agriculture and rural areas.

Differences in farm size are considered a key factor in understanding the differences in farm management performance [12]. According to economies of scale, expanding the scale of operations within a specific period makes better use of production facilities and resources, thereby reducing average costs and increasing operating profits. In farm management, studies have found that large-scale farms have higher production efficiency compared to small-scale farms [13]. At the same time, studies have shown that large-scale farms have lower transaction costs compared to small-scale farms [14]. Existing studies have verified that farm size positively affects agricultural production. However, how it does so, i.e., the quantitative relationship between farm size and how farmers recycle PPW, is still unclear.

According to statistics from the Ministry of Agriculture and Rural Affairs of China, approximately 3.5 billion pieces of PPW are generated annually in China. Therefore, this study aims to identify the causal relationship between farm size and farmers' recycling of PPW. Based on existing research, the marginal contribution of this study is as follows: (1) As the world's largest developing country, sustainable agriculture is the foundation of China's sustainable development [15]. This study is an empirical analysis based on rural big data in China, using the 2020 China Rural Revitalization Comprehensive Survey (CRRS 2020) data as the research sample. It aims to provide empirical evidence for sustainable farm management in other developing countries. (2) In discussing the relationship between farm size and farmers' pro-environmental behaviors, existing studies have downplayed or neglected endogeneity, thus failing to pinpoint the causal relationship between farm size and farmers' pro-environmental behaviors. This study introduces the average farm size of other farmers in the same village as an instrumental variable to address the endogeneity issue. It then examines the causal relationship between farm size and FPPWRB. (3) This study further examines the mechanism through which farm size influences FPPWRB, providing a theoretical foundation and empirical support for the design of targeted policies.

The scientific management of PPW plays a crucial role in ensuring food security in China. China needs to feed over 1.4 billion people annually, and ensuring food security requires the rational use of pesticides [15]. According to statistics, China uses over 1 million tons of pesticides annually, resulting in the generation of approximately 3.5 billion PPW items [16]. Previous research has primarily focused on how to scientifically and rationally use pesticides. For example, Qiao et al. [17], Liu and Huang [18], and Gong et al. [19] discussed the pesticide usage behavior of Chinese farmers and its influencing factors. However, there is limited research focusing on the FPPWRB. Therefore, the marginal contribution of this paper lies in the following points: (1) As the world's largest developing country, sustainable agriculture is the foundation of China's sustainable development [15]. Therefore, the scientific management of PPW in China is crucial for both China's and global sustainable development. Based on the large-scale survey data of Chinese farmers, discussing their FPPWRB helps to clarify the micro-mechanisms underlying recycling behavior. (2) Scaled operations are key to ensuring agricultural sustainability. This paper aims to examine the quantitative impact and mechanism between scaled operations and FPPWRB using econometric models, providing valuable insights into the development of policies on managing PPW. (3) In discussing the relationship between farm size and farmers' pro-environmental behaviors, existing studies have downplayed or neglected endogeneity, thus failing to pinpoint the causal relationship between farm size and farmers' pro-environmental behaviors. This study introduces the average farm size of other farmers in the same village as an instrumental variable to address the endogeneity issue.

The structure of the remaining sections of this paper is as follows: (1) Section 2 discusses the theoretical relationship between farm size and FPPWRB. (2) Section 3 discusses the data and research methods used in this study. (3) Section 4 examines the causal relationship between farm size and FPPWRB. (4) Section 5 summarizes the findings of this study and provides policy recommendations.

## 2. Theoretical Analysis and Research Hypotheses

### 2.1. *The Impact of Farm Size on FPPWRB*

The core concept of the economies of scale theory is that as the scale of production increases, the per-unit production cost decreases. Large-scale production can spread fixed costs and improve production efficiency. Subsequently, Sharma et al. [20] pointed out that the larger the production scale alongside the amount of prepaid capital, the more efficiently the entire production system operates. He et al. [21] and Wang et al. [22] found that land scale has a positive impact on straw return. Based on data from South Africa [23], Europe [24], and Uruguay [25], studies in the field of manufacturing indicate that the size of a firm is conducive to technological innovation.

Firm size promotes the adoption of technology [14,15]. A large firm can reduce the risks of market development and financing and is a manifestation of increased risk tolerance for technological innovation and the adoption of new technologies. The same is true in the agricultural sector, where small farms face more risks when adopting new technologies. For example, Li et al. [4] studied the relationship between agricultural technological change and farm size, finding that large-scale farms are more likely to adopt new technologies because they are more able to bear the risks and costs associated with technology adoption. Similarly, Deng et al. [26], Zheng et al. [27], and Yu et al. [28] found that large-scale farms are more proactive in adopting labor-saving mechanized technologies.

**Hypothesis 1.** *Farm size positively influences farmers' recycling of pesticide packaging waste.*

### 2.2. *Farm Size, Hazard Awareness, and FPPWRB*

Hazard awareness, which refers to people's awareness of the possible negative impact of a certain behavior or substance, significantly influences the recycling of PPW. Studies indicate that the greater the awareness of potential harm, the more likely individuals are to take preventive actions [20]. In the field of environmental protection, hazard awareness is widely regarded as one of the key factors promoting environmentally responsible behavior [29]. Deng et al. [30] believe that the public's awareness of environmental pollution hazards directly influences their willingness to engage in environmental protection behaviors. Similarly, Sikor et al. [5] found that there is a significant correlation between farmers' awareness of pesticide pollution hazards and their waste disposal behavior. In addition, Song and Ye [31] found that when farmers have a deeper understanding of the harmful effects of pesticide waste, they are more likely to adopt recycling and reuse measures. Therefore, hazard awareness is important in promoting the recycling and reuse of PPW. Farmers who recognize the harm generated by pesticide waste are more likely to adopt recycling measures to reduce environmental pollution.

Farm size has a positive impact on the awareness of the harmful effects of PPW. A large farm size often comes with more resources and better educational conditions, allowing farm owners to access more knowledge and information about the hazards of pesticides. The theory of economies of scale suggests that large-scale operations can improve the efficiency of information acquisition and technology learning [32]. Wang et al. [33] pointed out that large-scale farm owners are more likely to participate in agricultural training and environmental education programs, thereby increasing their awareness of the hazards of pesticides. Yan et al. [34] found that large-scale farms have significant advantages in information dissemination and technology application, making large-scale farmers more likely to understand and recognize the potential hazards of pesticides. Additionally, Zhang et al. [35] emphasized that large-scale farm owners, due to their operational scale, are more motivated and capable of acquiring knowledge related to environmental protection. Therefore, large-scale farm owners, with their resources and educational advantages, are better able to recognize the hazards of PPW, making them more sensitive and proactive in hazard awareness compared to small-scale farm owners.

**Hypothesis 2.** *Farm size positively influences the awareness of pesticide packaging hazards, further promoting the recycling and reuse of pesticide packaging waste.*

## 3. Data, Variables, and Model

### 3.1. Data

This study employs data from the China Rural Revitalization Survey, which was conducted by the Rural Development Institute, Chinese Academy of Social Sciences. According to the introduction on its official website, the survey has the following characteristics: (1) The survey covers a wide range of topics, including rural population and labor force, rural industrial structure, rural governance, and comprehensive rural reforms. (2) The survey covers a large geographic area. It includes 50 county-level units from both economically developed and underdeveloped provinces in China, with data collected from 300 village surveys and over 3800 farmer surveys. (3) The sample is representative. The sampling procedure is as follows: (a) sample provinces are randomly selected based on their economic development level, regional location, and agricultural development, covering eastern, central, western, and northeastern regions; (b) sample counties are randomly selected within each province based on per capita GDP at the county level using an equal interval random sampling method; (c) sample townships (or towns) and villages are randomly selected based on economic development levels; and (d) sample households from the household register are randomly selected provided by the village committees.

After excluding questionnaires with significant missing data and responses that deviated from reality, this study used a total of 2663 valid farmer surveys.

### 3.2. Variables

#### 3.2.1. Dependent Variable

The dependent variable in this study is "FPPWRB": farmers' pesticide packaging waste recycling behavior. In the CRRS (China Rural Revitalization Survey), farmers were asked about their methods for handling pesticide packaging waste. In this study, households that answered "recycle to a designated point" or "recycle to the agricultural supply market" are defined as households with FPPWRB (coded as 1); conversely, households that did not report recycling are defined as households without FPPWRB (coded as 0).

#### 3.2.2. Key Variable

The key variable in this study is farm size. Most studies use the actual total cultivated area to represent the scale of land management [36]. Therefore, this study uses the land area managed by the farm as the measure of farm size.

#### 3.2.3. Mediator Variables

The main mediator variable in this study is the awareness of the hazards of PPW. Zhang et al. [36] found that larger farms are more likely to adopt environmental protection measures due to their stronger resource integration capabilities and greater environmental awareness, providing theoretical support for using the awareness of the hazards of PPW as a mediator variable.

#### 3.2.4. Control Variables

This study introduces control variables in the empirical model to mitigate the impact of omitted variables on the estimation results. Drawing on the studies of Huang and Elahi [37], Wang et al. [38], and Chen et al. [39], the research includes the characteristics of the household's head (e.g., gender, age, marital status), household characteristics (e.g., level of organization, per capita income, number of household laborers), and community characteristics (e.g., village topography and distance from the village committee to the township government) as control variables. The variable definitions and descriptive statistics for this study are shown in Table 1.

**Table 1.** Descriptive statistical analysis.

| Variables | Definition | Mean | S.D. |
| --- | --- | --- | --- |
| FPPWRB | Whether pesticide packaging waste is recycled: 1 = Yes; 0 = No | 0.417 | 0.493 |
| farm size | Farm area under farm management (mu) | 25.586 | 80.887 |
| gender | Gender: 1 = Male; 0 = Female | 0.944 | 0.229 |
| age | Farmer's age (years) | 54.548 | 10.750 |
| marriage | Marital status: 1 = Married; 0 = Unmarried | 0.930 | 0.256 |
| education | Whether the farmer received education at the high school level or above: 1 = Yes; 0 = No | 0.145 | 0.352 |
| party | Whether the farmer is a member of the Communist Party of China: 1 = Yes; 0 = No | 0.221 | 0.415 |
| nonfarm work | Whether the farmer engages in nonfarm work: 1 = Yes; 0 = No | 0.093 | 0.291 |
| leader | Whether the farmer is a village official: 1 = Yes; 0 = No | 0.020 | 0.138 |
| farm cooperation | Whether the farmer joined an agricultural cooperative: 1 = Yes; 0 = No | 0.235 | 0.424 |
| income | Per capita income of the household | 9.361 | 1.103 |
| labor | Total number of laborers in the household | 2.919 | 1.284 |
| machine | Whether mechanized services are used for pesticide application: 1 = Yes; 0 = No | 0.201 | 0.401 |
| outsource | Whether the household outsources pesticide spraying services: 1 = Yes; 0 = No | 0.069 | 0.254 |
| Terrain 1 | Whether the village is located in a plain: 1 = Yes; 0 = No | 0.465 | 0.499 |
| Terrain 2 | Whether the village is located in a hilly area: 1 = Yes; 0 = No | 0.212 | 0.409 |
| recovery point | Whether the village has a pesticide packaging waste recycling point: 1 = Yes; 0 = No | 0.395 | 0.489 |
| distance | Distance from the village committee to the county government | 23.314 | 16.241 |

### 3.3. Methods

#### 3.3.1. Baseline Model

To examine the impact of farm size on FPPWRB, a probit regression model was constructed as follows:

$$pesticide\_packaging_i = \alpha_0 + \beta_1 Lnland\_area_i + \beta_2 Controls_i + \gamma_i + \varepsilon_i \quad (1)$$

In Equation (1), $i$ represents the household; $pesticide\_packaging$ denotes the binary dummy variable for the household's FPPWRB; $land\_area$ represents the land area operated by the household; *Controls* refers to a set of control variables; $\gamma_i$ indicates provincial fixed effects; and $\varepsilon_i$ is the random disturbance term.

#### 3.3.2. Mediation Effect Model

By using the awareness of pesticide packaging hazards as a mediating variable, a mediation effect model was constructed to explore the transmission mechanism of the land management scale's influence on FPPWRB. The specific equations are given as follows:

$$cognition_i = \alpha_0 + \beta_1 lnland\_area_i + \beta_2 Controls_i + \gamma_i + \varepsilon_i \quad (2)$$

$$pesticide\_packaging_i = \alpha_0 + \beta_1 cognition_i + \beta_2 lnland\_area_i + \beta_3 Controls_i + \gamma_i + \varepsilon_i \quad (3)$$

In Equations (2) and (3), $cognition$ represents the awareness of the hazards of pesticide packaging waste. Since $cognition$ is a type of ordinal data, Equation (2) is specified as an ordered probit model.

#### 3.3.3. Multicollinearity Test Method

To eliminate the influence of multicollinearity on empirical results, this study employed a variance inflation factor (VIF) test to assess whether the model suffers from multicollinearity. Following Chen [40], we constructed Equation (4) to calculate the variance inflation factor (VIF).

$$\mathrm{VIF} = \frac{1}{1 - \mathrm{R}^2} \quad (4)$$

## 4. Results

### 4.1. Results of Multicollinearity Test

Table 2 presents the test results. The VIF values of all variables, as well as the model's mean VIF, are less than two, indicating that the empirical estimates are not significantly affected by multicollinearity [41]. Additionally, the correlation coefficients between the core explanatory variables and control variables are all below 0.3, which preliminarily rules out multicollinearity among the independent variables and facilitates more accurate results in the subsequent regression analysis. However, it is worth noting that correlation analysis only measures the pairwise relationships between variables without accounting for the influence of other factors. Therefore, more accurate estimation results should be further addressed and adjusted in the subsequent multivariate regression analysis.

**Table 2.** Multicollinearity—VIF test.

| Variable | VIF | 1/VIF |
|---|---|---|
| Terrain 1 | 1.83 | 0.543755 |
| machine | 1.65 | 0.603058 |
| Terrain 2 | 1.54 | 0.647147 |
| outsource | 1.46 | 0.684195 |
| Ln (farm size) | 1.41 | 0.702131 |
| distance | 1.22 | 0.793182 |
| age | 1.18 | 0.844405 |
| labor | 1.14 | 0.873237 |
| nonfarm work | 1.14 | 0.876215 |
| recovery point | 1.12 | 0.885026 |
| party | 1.12 | 0.889887 |
| marriage | 1.09 | 0.895756 |
| education | 1.09 | 0.917843 |
| Ln(income) | 1.08 | 0.918074 |
| leader | 1.05 | 0.923301 |
| gender | 1.04 | 0.950771 |
| farm cooperation | 1.04 | 0.960651 |
| mean VIF | 1.25 | |

### 4.2. *Estimated Results for the Impact of Farm Size on FPPWRB*

We used a stepwise approach to add variables to mitigate the impact of omitted variables on the empirical results and examined the relationship between farm size and FPPWRB. Table 3 reports the estimation results.

Model (1) incorporates farm size Ln (farm size) as the only core explanatory variable for FPPWRB. Model (2) adds a series of farm-owner characteristics—sex, age, marriage, education, party affiliation, nonfarm work, and leader—as control variables on top of Model (1). Model (3) incorporates a set of farm characteristics—farm cooperation, Ln(income), labor, machine, and outsourcing—as control variables. Model (4) continues by introducing Terrain 1, Terrain 2, the recovery point, and distance, estimating a nonlinear probit model. Since the coefficients of Model (4) cannot directly explain the quantitative relationship between two variables, marginal effects were estimated; the results are presented in the fifth column of Table 3. In terms of model selection, the $\chi^2$ results support the validity of using a probit model, making the probit estimation appropriate.

From the regression results in columns (4) and (5), the estimated coefficient of Ln (farm size) is 0.1029, and this is positively significant at the 1% level. This indicates that the core explanatory variable (farm size) has a statistically significant positive effect on the dependent variable (FPPWRB) at the 1% level. As shown in column (5), the marginal effect of the land management scale on FPPWRB is 0.0359, and this effect is significant at the 1% level. In other words, the higher the farm size, the more likely FPPWRB is to increase by 3.59%. According to the regression results, farm size is significantly and positively associated with FPPWRB in all models. The empirical results of this study are consistent with the theoretical analysis presented earlier and the findings of Yan et al. [34], Zhang et al. [35], and Zhang et al. [42], implying that as farm size increases, farmers are more likely to recycle and reuse pesticide packaging waste. This finding aligns with the studies of Han et al. [43], Ren et al. [15], and Zheng and Luo [44], as a larger farm size implies that farmers have more resources and capacity to manage waste. Therefore, hypothesis 1 (i.e., farm size positively influences household willingness to participate in waste sorting) is confirmed.

**Table 3.** Regression results–probit model.

| | (1) | (2) | (3) | (4) | (5) |
|---|---|---|---|---|---|
| Ln (farm size) | 0.1227 *** | 0.1037 *** | 0.0928 *** | 0.1029 *** | 0.0359 *** |
| | (5.21) | (4.24) | (3.64) | (3.93) | (3.96) |
| gender | | 0.0735 | 0.0789 | 0.0807 | 0.0282 |
| | | (0.65) | (0.69) | (0.70) | (0.70) |
| age | | −0.0054 ** | −0.0048 * | −0.0057 ** | −0.0020 ** |
| | | (−2.12) | (−1.82) | (−2.15) | (−2.16) |
| marriage | | 0.0486 | 0.0276 | 0.0201 | 0.0070 |
| | | (0.48) | (0.27) | (0.19) | (0.19) |
| education | | −0.0443 | −0.0464 | −0.0710 | −0.0247 |
| | | (−0.59) | (−0.61) | (−0.93) | (−0.93) |
| party | | 0.0117 | 0.0131 | 0.0034 | 0.0012 |
| | | (0.18) | (0.20) | (0.05) | (0.05) |
| nonfarm work | | −0.2593 *** | −0.2631 *** | −0.2620 *** | −0.0913 *** |
| | | (−2.71) | (−2.74) | (−2.71) | (−2.72) |
| leader | | −0.1523 | −0.1554 | −0.0627 | −0.0219 |
| | | (−0.79) | (−0.81) | (−0.32) | (−0.32) |
| farm cooperation | | | −0.0193 | 0.0007 | 0.0002 |
| | | | (−0.31) | (0.01) | (0.01) |
| Ln (income) | | | 0.0131 | 0.0059 | 0.0021 |
| | | | (0.54) | (0.24) | (0.24) |
| labor | | | 0.0198 | 0.0193 | 0.0067 |
| | | | (0.91) | (0.88) | (0.88) |
| machine | | | 0.0931 | 0.0620 | 0.0216 |
| | | | (1.04) | (0.69) | (0.69) |
| outsource | | | 0.0215 | 0.0397 | 0.0139 |
| | | | (0.18) | (0.33) | (0.33) |
| Terrain 1 | | | | 0.2601 *** | 0.0907 *** |
| | | | | (3.40) | (3.42) |
| Terrain 2 | | | | 0.3418 *** | 0.1192 *** |
| | | | | (4.22) | (4.26) |
| recovery point | | | | 0.2545 *** | 0.0888 *** |
| | | | | (4.25) | (4.29) |
| distance | | | | −0.0036 ** | −0.0013 ** |
| | | | | (−1.98) | (−1.98) |
| _cons | −0.5929 *** | −0.3455 | −0.5650 * | −0.8621 ** | |
| | (−4.89) | (−1.51) | (−1.70) | (−2.51) | |
| prov effect | Yes | Yes | Yes | Yes | Yes |
| N | 2663 | 2663 | 2663 | 2663 | 2663 |
| Log likelihood | −1666.8552 | −1661.1485 | −1659.7635 | −1628.6105 | −1628.6105 |
| $\chi^2$ | 284.62 *** | 296.04 *** | 298.81 *** | 361.11 *** | 361.11 *** |

Note: *, **, and *** denote significance at the 10%, 5%, and 1% levels, respectively. The T-values are in parentheses. The fifth column reports the estimated marginal effects.

Another interesting finding from this study is that geographical location can influence FPPWRB. In Model (4), the coefficients for Terrain 1 and Terrain 2 are significantly positive (at the 1% statistical level), indicating that farmers in plain and hilly areas are more inclined to recycle pesticide packaging waste than those in mountainous regions. This finding suggests that establishing a universal resource recycling system requires increased attention to mountainous regions (geographically disadvantaged areas).

### 4.3. Endogeneity Test for Farm Size and FPPWRB

Endogeneity analysis verifies whether the relationship between the explanatory variable (farm size) and the dependent variable (FPPWRB) is accurately captured. If endogeneity is present, modeling results may lead to false causal relationships and misleading policy recommendations. Farm size serves as the explanatory variable, while FPPWRB is the dependent variable. FPPWRB may be influenced by other unobserved factors, which could simultaneously affect farm size. Such unobservable factors can include economic factors, policy factors, etc. Therefore, there may be an endogeneity issue between farm size and FPPWRB. If there are issues such as omitted variables or unobservable factors, endogeneity analysis can help identify them and explore how to incorporate them into the model. This can help eliminate model errors and more accurately measure the relationship.

Therefore, to determine whether variables or causal relationships are included or not in the model and to enhance the credibility of the research results, instrumental variables were used to conduct an endogeneity test on the model. Table 4 shows the endogeneity regression results. The instrumental variable for farm size is "the proportion of other large-scale households (i.e., households with land size greater than the sample mean) in the village (excluding the household itself)" [34,45]. In selecting the instrumental variable, the first criterion is validity, meaning that the instrumental variable should be correlated with the explanatory variable and capable of influencing its value. In this case, "the proportion of other large-scale households in the village (excluding the household itself)" reflects the farm size situation within a village, making it a potentially valid instrumental variable. Second, the instrumental variable is available. In some cases, obtaining raw data on farm size may not be easy. By using "the proportion of other large-scale households in the village (excluding the household itself)" as an instrumental variable, existing data sources or survey results can be leveraged, making it easier to obtain data and conduct the study. Additionally, using "the proportion of other large-scale households in the village (excluding the household itself)" as an instrumental variable provides a clearer explanation of the determinants of farm size. For example, if this instrumental variable is found to have a significant relationship with the explanatory variable, it can allow for further exploration of how other large-scale households influence farm size. The regression results in Table 4 show that the first-stage estimation of the instrumental variable test yields an F-value of 90.37 ($p$-value = 0.000), which is significant at the 1% level. This indicates that there is no weak instrument problem. In the second stage, the Wald test rejects the null hypothesis that the model has no endogenous variables at the 1% significance level, indicating that endogeneity is indeed a concern. From the estimation results using the instrumental variable method, the direction of the coefficient for farm size and FPPWRB is consistent with the baseline regression results. This suggests that even after addressing endogeneity with the instrumental variable method, farm size still has a significant positive effect on FPPWRB, further confirming H1.

**Table 4.** Endogeneity regression results.

| | Phase 1 | | Phase 2 |
|---|---|---|---|
| size_ratio | 1.7759 ***<br>(12.90) | | |
| Ln (farm size) | | | 0.1936 *<br>(1.81) |
| _cons | 0.8313 ***<br>(3.18) | | −1.0164 ***<br>(−2.63) |
| control variables | Yes | | Yes |
| prov effect | Yes | | Yes |
| F-value of Phase 1 | | 90.37 *** | |
| Wald chi2 | | 322.14 *** | |
| Wald endogeneity value | | 0.3809 | |
| N | 2663 | | 2663 |
| R-squared | 0.471 | | 0.125 |

Note: *, **, and *** denote significance at the 10%, 5%, and 1% levels, respectively. The T-values are in parentheses.

### 4.4. Robustness Check for the Impact of Farm Size on FPPWRB

To strengthen the reliability of the research conclusions and results, we conducted robustness checks by changing the measurement method of the core explanatory variable and employing alternative estimation models.

Given that parameter estimation using the probit model may be subject to variability due to different distributional assumptions, this study used the logit model to test for robustness. Logit and probit models are commonly used binary choice models for dealing with binary response variable problems. However, compared to the probit model, the logit model imposes less restrictive distributional assumptions on the explanatory variables, making it more effective in handling heteroskedasticity and outliers. In addition, the coefficients of the logit model can be interpreted as odds ratios, enhancing the intuitive interpretability of the model's results. Therefore, the robustness check using the logit model aims to confirm the reliability of the research findings and ensure that the results are not solely influenced by the choice of a specific model or data-handling approach. As shown in Table 5, after the probit model is replaced by the logit model, the coefficients of the core explanatory variable are still significantly positive, which verifies the fact that the robustness test of the replacement model is reasonable.

Additionally, to ensure the robustness of the core explanatory variable, we incorporated the "irrigable area" and "maximum plot size" as alternative measures. By replacing the original core explanatory variable (farm size) with these two new indicators, Models (2) and (3) were constructed. As shown in Table 5, regardless of whether the core explanatory variable was replaced by the "irrigable area" or "maximum plot size," the coefficients remained significantly positive. This further confirms the positive impact of farm size on FPPWRB, validating the effectiveness of the robustness check through variable substitution. Therefore, the robustness test conducted by replacing the estimation model and the measurement method of the core explanatory variable not only strengthened the credibility of the research conclusions but also ensured the robustness of the research results, showing that the conclusions remain consistent and reliable under different model frameworks and measurement indicators. The results of the robustness check are shown in Table 5.

**Table 5.** Robustness check.

| | (1) | (2) | (3) |
|---|---|---|---|
| | **Logit** | **Irrigable Area** | **Maximum Plot Size** |
| Ln (farm size) | 0.1685 *** | | |
| | (3.89) | | |
| Ln (farm size2) | | 0.0671 *** | |
| | | (3.06) | |
| Ln (farm size3) | | | 0.0889 ** |
| | | | (2.31) |
| _cons | −1.4562 ** | −0.7731 ** | −0.9415 *** |
| | (−2.55) | (−2.24) | (−2.66) |
| control variables | Yes | Yes | Yes |
| prov effect | Yes | Yes | Yes |
| N | 2663 | 2565 | 2545 |
| pseudo $R^2$ | 0.0999 | 0.0965 | 0.0974 |

Note: *, **, and *** denote significance at the 10%, 5%, and 1% levels, respectively. The T-values are in parentheses.

### 4.5. Heterogeneity Analysis for the Impact of Farm Size on FPPWRB

To further investigate the heterogeneity of the effect of farm size on FPPWRB, we conducted two key heterogeneity analyses.

First, to examine whether the effect of farm size on FPPWRB varies by location, households were categorized into urban and non-urban suburbs, and regressions were conducted separately for each group. From the heterogeneous regression results in Table 6, it can be observed that the effect of farm size on FPPWRB differs between urban and non-urban suburbs. Specifically, for farmers in urban suburbs, each unit increase in farm size results in a 0.2890 effect on FPPWRB, which is significantly greater than the 0.0723 effect observed in non-urban suburbs. The potential reasons for this discrepancy may include the following: (1) Urban suburbs are closer to markets, making it easier to access relevant technologies and beneficial information [46,47]. Therefore, farmers in urban suburbs are more likely to access technology and information related to waste management than those in non-urban suburbs, which may lead to greater motivation to participate in PPW recycling. (2) Urban suburbs are more likely to receive government support for the construction of recycling facilities. Therefore, from a regional perspective, the impact of farm size on FPPWRB may vary.

**Table 6.** Heterogeneity regression results.

| | (1) | (2) |
|---|---|---|
| | **Urban Suburbs** | **Non-Urban Suburbs** |
| Ln (farm size) | 0.2890 *** | 0.0723 ** |
| | (3.95) | (2.53) |
| _cons | −1.1736 | −1.0166 *** |
| | (−1.20) | (−2.67) |
| control variables | Yes | Yes |
| prov effect | Yes | Yes |
| N | 519 | 2144 |
| pseudo $R^2$ | 0.1324 | 0.1096 |

Note: *, **, and *** denote significance at the 10%, 5%, and 1% levels, respectively. The T-values are in parentheses.

Second, farmers were divided into older and younger generations based on a 40-year-age threshold, and separate regressions were conducted. The regression results in Table 7 show that the effect of farm size on FPPWRB is 0.1714 for younger farmers, compared to 0.0894 for older farmers. This indicates that each unit increase in farm size has a greater impact on FPPWRB among younger farmers than older farmers. This may be because younger farmers are more open to adopting new agricultural technologies and management practices. Additionally, they might place a greater emphasis on environmental protection, making them more inclined to recycle and reuse pesticide packaging waste.

**Table 7.** Heterogeneity regression results—generational differences.

| | (1) | (2) |
|---|---|---|
| | **Over 40 Years Old** | **Under 40 Years Old** |
| Ln (farm size) | 0.0894 *** | 0.1714 ** |
| | (3.20) | (2.08) |
| _cons | −0.6507 * | −0.4711 |
| | (−1.67) | (−0.39) |
| control variables | Yes | Yes |
| prov effect | Yes | Yes |
| N | 2411 | 252 |
| pseudo $R^2$ | 0.0980 | 0.1847 |

Note: *, **, and *** denote significance at the 10%, 5%, and 1% levels, respectively. The T-values are in parentheses.

### 4.6. Mediation Effect for the Impact of Farm Size on FPPWRB

To examine the impact of the hazard awareness of pesticide packaging on the relationship between farm size and FPPWRB, we further conducted a mediation effect analysis to explore the underlying mechanism.

As shown in Table 8, mediation effect analysis was conducted using three models to examine the relationships between farm size (Ln (farm size)), farmers' environmental awareness (cognition), and the impact of farm size on FPPWRB (pesticide_packaging). This analysis aimed to determine whether farmers' environmental awareness mediates the relationship between farm size and FPPWRB.

Model (1) examines the direct impact of farm size on FPPWRB. In this model, the coefficient of Ln (farm size) is 0.1029, which is statistically significant ($p < 0.01$). This indicates that a large farm size is associated with a higher probability of FPPWRB. The results for Model (1) indicate a positive relationship between farm size and FPPWRB, confirming the initial research hypothesis.

Model (2) analyzes the effect of farm size on farmers' environmental awareness. In this model, the coefficient of Ln (farm size) is 0.1698, which is also statistically significant. This indicates that a larger farm size is associated with stronger environmental awareness among farmers. This finding suggests that farm size not only directly affects FPPWRB but may also indirectly influence it by enhancing farmers' environmental awareness.

Finally, Model (3) incorporates both farm size and farmers' environmental awareness to assess their combined effect on FPPWRB. After including farmers' environmental awareness as a mediating variable, the effect of farm size on FPPWRB decreased from 0.1029 to 0.0847 but remained statistically significant. Meanwhile, the coefficient for farmers' environmental awareness is 0.2425, which is also highly significant. This suggests that farmers' environmental awareness indeed plays a mediating role in the relationship between farm size and FPPWRB, thus confirming H2.

In summary, the analysis results of this series of models provide empirical evidence for our hypothesis—namely, the fact that farm size indirectly promotes FPPWRB by enhancing farmers' environmental awareness. At the same time, farm size also has a direct positive effect. These findings are consistent with existing studies, which have found that larger farms are more likely to adopt environmentally friendly practices, partly due to their stronger resource integration capabilities and greater awareness of the environment. This analysis not only provides a new perspective on the impact of farm size on FPPWRB but also highlights the crucial importance of enhancing farmers' environmental awareness when promoting pro-environment actions.

**Table 8.** Regression results of mediation effect.

| | (1) | (2) | (3) |
|---|---|---|---|
| | **Pesticide_Packaging** | **Cognition** | **Pesticide_Packaging** |
| Ln (farm size) | 0.1029 *** | 0.1698 *** | 0.0847 *** |
| | (3.93) | (7.68) | (3.17) |
| cognition | | | 0.2425 *** |
| | | | (7.44) |
| _cons | −0.8621 ** | | −1.1927 *** |
| | (−2.51) | | (−3.41) |
| control variables | Yes | Yes | Yes |
| prov effect | Yes | Yes | Yes |
| N | 2663 | 2663 | 2663 |
| pseudo $R^2$ | 0.0998 | 0.0346 | 0.1152 |

Note: *, **, and *** denote significance at the 10%, 5%, and 1% levels, respectively. The T-values are in parentheses.

## 5. Conclusions

### *5.1. Findings*

Based on the micro-survey data from CRRS 2020, this study covers a sample of 2775 rural households. Using the probit model, we thoroughly examined the impact of farm size on FPPWRB and its underlying mechanisms. The main findings are given as follows:

(1) Farmers engage in limited recycling of PPW. Descriptive statistics show that only 41.7% of the sample farmers participated in pesticide packaging waste recycling. This indicates that PPW management has not been given adequate attention by farmers, presenting new challenges to the fragile agricultural production environment.
(2) There is a significant positive correlation between farm size and FPPWRB. The empirical results show that for every 1% increase in farm size, the probability of farmers participating in pesticide packaging waste recycling increases by 3.59%. Namely, larger farms are more likely to engage in recycling. This may be because these farms have more resources and incentives to implement environmental protection measures, thereby reducing the environmental impact of agricultural production.
(3) Farmers in urban suburbs or younger farmers are more inclined to FPPWRB. In the heterogeneity analysis, the coefficient of the farm size variable was 0.2980 for the urban suburbs group and 0.1714 for the younger group (under 40 years old), both of which are higher than those for the non-urban suburbs group and the middle-aged group (over 40 years old). Namely, farm size has a more positive impact on FPPWRB among younger farmers and those in urban suburbs, suggesting that these groups are more likely to adopt recycling measures because of their specific advantages.
(4) Farm size can improve FPPWRB by enhancing farmers' environmental cognition levels. The mediation effect estimation results show that farm size significantly and positively affects environmental awareness at the 1% level. Meanwhile, environmental awareness significantly and positively influences FPPWRB at the 1% level. Namely, farm size influences farmers' perception of the hazards associated with pesticide packaging waste, which, in turn, affects FPPWRB. Therefore, enhancing farmers' knowledge and awareness is a key approach to promoting FPPWRB.

### 5.2. Implications

First, the government and relevant authorities should strengthen public awareness campaigns and provide technical guidance to enhance farmers' awareness of PPW recycling, especially among small-scale farmers. Collaboration with local villagers can be established by offering gift incentives to encourage active participation in PPW recycling. At the same time, recycling bins can be set up at the village level, and staff should be organized to visit the village regularly for collection and processing. Encouraging land transfer and large-scale farming through policy guidance and economic incentives can help integrate resources and improve recycling and utilization efficiency.

Second, differentiated support measures should be developed based on the characteristics of different farmers. For farmers with strong technical skills and higher education levels, market information and technical support should be provided. For farmers in regions with poor natural conditions, financial and technical assistance should be offered. For example, in hilly areas—where the terrain is complex and transportation inconvenient—innovative PPW recycling strategies can be developed. Incentive-based methods, such as the "pesticide bottle exchange for gifts" program, can be used to encourage recycling.

Third, media campaigns and policy interpretation can strengthen farmers' awareness of environmental protection and economic benefits, continuously enhancing their overall understanding of PPW recycling and utilization. At the same time, this can improve the incentive mechanism to promote the importance of recycling and utilization from both environmental and economic perspectives.

### 5.3. Limitations

This study also has some limitations, and future research can further improve upon these aspects. (1) There may be a dynamic relationship between farm size and FPPWRB. Therefore, future research could construct panel data to explore the dynamic relationship between these two variables. (2) The environmental impact of FPPWRB has not been sufficiently explored. Therefore, future research could develop new datasets to assess the environmental impact of FPPWRB. (3) Based on the Chinese case, it was found that as farm size increased, farmers were more willing to engage in environmentally friendly actions. Therefore, future research could examine whether the conclusions of this study are applicable to other countries.

**Author Contributions:** Conceptualization, J.D., K.S. and X.D.; methodology, X.D.; software, X.D.; formal analysis, K.S. and J.D.; investigation, X.D.; resources, X.D.; writing—original draft preparation, J.D., K.S., K.Z. and X.D.; writing—review and editing, J.D., K.S., K.Z. and X.D.; visualization, J.D.; funding acquisition, X.D. All authors have read and agreed to the published version of the manuscript.

**Funding:** This study was funded by the Natural Science Foundation of Sichuan Province, China (Grant No. 2025ZNSFSC1148), and Sichuan Philosophy and Social Key Laboratory of Monitoring and Assessing for Rural Land Utilization (Grant No. NDZDSC2023003).

**Data Availability Statement:** Data is contained within the article.

**Acknowledgments:** We gratefully acknowledge the financial support received from the Natural Science Foundation of Sichuan Province, China (Grant No. 2025ZNSFSC1148), and Sichuan Philosophy and Social Key Laboratory of Monitoring and Assessing for Rural Land Utilization (Grant No. NDZDSC2023003). The authors also extend their gratitude to the anonymous reviewers and editors for their helpful review and critical comments.

**Conflicts of Interest:** The authors declare no conflicts of interest.

## References

1. Gullett, B.K.; Tabor, D.; Touati, A.; Kasai, J.; Fitz, N. Emissions from Open Burning of Used Agricultural Pesticide Containers. *J. Hazard. Mater.* **2012**, *221*, 236–241. [CrossRef] [PubMed]
2. Haji, J. Production Efficiency of Smallholders' Vegetable-Dominated Mixed Farming System in Eastern Ethiopia: A Non-Parametric Approach. *J. Afr. Econ.* **2007**, *16*, 1–27. [CrossRef]
3. Li, B.; Shen, Y. Effects of Land Transfer Quality on the Application of Organic Fertilizer by Large-Scale Farmers in China. *Land Use Policy* **2021**, *100*, 105124. [CrossRef]
4. Li, B.; Xu, C.; Zhu, Z.; Kong, F. How to Encourage Farmers to Recycle Pesticide Packaging Wastes: Subsidies Vs Social Norms. *J. Clean. Prod.* **2022**, *367*, 133016. [CrossRef]
5. Sikor, T.; Müller, D.; Stahl, J. Land Fragmentation and Cropland Abandonment in Albania: Implications for the Roles of State and Community in Post-Socialist Land Consolidation. *World Devel.* **2009**, *37*, 1411–1423. [CrossRef]
6. Feder, G.; O'Mara, G.T. Farm Size and the Diffusion of Green Revolution Technology. *Econ. Devel. Cult. Change* **1981**, *30*, 59–76. [CrossRef]
7. Garbounis, G.; Komilis, D. A Modeling Methodology to Predict the Generation of Wasted Plastic Pesticide Containers: An Application to Greece. *Waste Manag.* **2021**, *131*, 177–186. [CrossRef] [PubMed]
8. Li, M.; Wang, J.; Chen, K.; Wu, L. Willingness and Behaviors of Farmers' Green Disposal of Pesticide Packaging Waste in Henan, China: A Perceived Value Formation Mechanism Perspective. *Int. J. Environ. Res. Public Health* **2020**, *17*, 3753. [CrossRef] [PubMed]
9. Zhao, K.-X.; Zhang, M.-Y.; Yang, D.; Zhu, R.-S.; Zhang, Z.-F.; Hu, Y.-H.; Kannan, K. Screening of Pesticides in Serum, Urine and Cerebrospinal Fluid Collected from an Urban Population in China. *J. Hazard. Mater.* **2023**, *449*, 131002. [CrossRef]
10. Wang, W.; Ge, J.; Yu, X.; Li, H. Environmental Fate and Impacts of Microplastics in Soil Ecosystems: Progress and Perspective. *ScTEn* **2020**, *708*, 134841. [CrossRef] [PubMed]
11. Wang, F.; Wang, Q.; Adams, C.A.; Sun, Y.; Zhang, S. Effects of Microplastics on Soil Properties: Current Knowledge and Future Perspectives. *J. Hazard. Mater.* **2022**, *424*, 127531. [CrossRef] [PubMed]

12. Gao, J.; Gai, Q.; Liu, B.; Shi, Q. Farm Size and Pesticide Use: Evidence from Agricultural Production in China. *China Agric. Econ. Rev.* **2021**, *13*, 912–929. [CrossRef]
13. Lu, H.; Hu, L.; Zheng, W.; Yao, S.; Qian, L. Impact of Household Land Endowment and Environmental Cognition on the Willingness to Implement Straw Incorporation in China. *J. Clean. Prod.* **2020**, *262*, 121479. [CrossRef]
14. Townsend, R.F.; Kirsten, J.; Vink, N. Farm Size, Productivity and Returns to Scale in Agriculture Revisited: A Case Study of Wine Producers in South Africa. *Agr. Econ.* **1998**, *19*, 175–180. [CrossRef]
15. Ren, C.; Liu, S.; Van Grinsven, H.; Reis, S.; Jin, S.; Liu, H.; Gu, B. The Impact of Farm Size on Agricultural Sustainability. *J. Clean. Prod.* **2019**, *220*, 357–367. [CrossRef]
16. MARA. Opinions of the Ministry of Agriculture and Rural Affairs of the People's Republic of China on Implementing the Law of the People's Republic of China on the Prevention and Control of Environmental Pollution by Solid Waste. Available online: http://www.fgs.moa.gov.cn/flfg/202108/t20210830_6375173.htm (accessed on 18 February 2025).
17. Qiao, F.; Huang, J.; Zhang, L.; Rozelle, S. Pesticide Use and Farmers' Health in China's Rice Production. *China Agric. Econ. Rev.* **2012**, *4*, 468–484. [CrossRef]
18. Liu, E.M.; Huang, J. Risk Preferences and Pesticide Use by Cotton Farmers in China. *J. Dev. Econ.* **2013**, *103*, 202–215. [CrossRef]
19. Gong, Y.; Baylis, K.; Kozak, R.; Bull, G. Farmers' Risk Preferences and Pesticide Use Decisions: Evidence from Field Experiments in China. *Agric. Econ.* **2016**, *47*, 411–421. [CrossRef]
20. Sharma, A.; Kumar, V.; Shahzad, B.; Tanveer, M.; Sidhu, G.P.S.; Handa, N.; Kohli, S.K.; Yadav, P.; Bali, A.S.; Parihar, R.D. Worldwide Pesticide Usage and Its Impacts on Ecosystem. *SN Appl. Sci.* **2019**, *1*, 1446. [CrossRef]
21. He, J.; Zhou, W.; Guo, S.; Deng, X.; Song, J.; Xu, D. Effect of Land Transfer on Farmers' Willingness to Pay for Straw Return in Southwest China. *J. Clean. Prod.* **2022**, *369*, 133397. [CrossRef]
22. Wang, X.; Song, Y.; Huang, W. The Effects of Agricultural Machinery Services and Land Fragmentation on Farmers' Straw Returning Behavior. *Agribusiness* **2024**. [CrossRef]
23. Booyens, I. Are Small, Medium-and Micro-Sized Enterprises Engines of Innovation? The Reality in South Africa. *Sci. Public Policy* **2011**, *38*, 67–78. [CrossRef]
24. Vaona, A.; Pianta, M. Firm Size and Innovation in European Manufacturing. *Small Bus. Econ.* **2008**, *30*, 283–299. [CrossRef]
25. Aboal, D.; Garda, P.; Lanzilotta, B.; Perera, M. Innovation, Firm Size, Technology Intensity, and Employment Generation: Evidence from the Uruguayan Manufacturing Sector. *Emerg. Mark. Financ. Trade* **2015**, *51*, 3–26. [CrossRef]
26. Deng, X.; Yan, Z.; Xu, D.; Qi, Y. Land Registration, Adjustment Experience, and Agricultural Machinery Adoption: Empirical Analysis from Rural China. *Land* **2020**, *9*, 89. [CrossRef]
27. Zheng, H.; Ma, W.; Zhou, X. Renting-in Cropland, Machinery Use Intensity, and Land Productivity in Rural China. *Appl. Econ.* **2021**, *53*, 5503–5517. [CrossRef]
28. Yu, X.; Yin, X.; Liu, Y.; Li, D. Do Agricultural Machinery Services Facilitate Land Transfer? Evidence from Rice Farmers in Sichuan Province, China. *Land* **2021**, *10*, 466. [CrossRef]
29. Sheng, Y.; Zhao, S.; Nossal, K.; Zhang, D. Productivity and Farm Size in a Ustralian Agriculture: Reinvestigating the Returns to Scale. *Aust. J. Agr. Resour. Econ.* **2015**, *59*, 16–38. [CrossRef]
30. Deng, X.; Song, Y.; He, Q.; Xu, D.; Qi, Y. Does Internet Use Improve Farmers' Perception of Environmental Pollution? Evidence from Rural China. *Environ. Sci. Pollut. Res.* **2022**, *29*, 44832–44844. [CrossRef] [PubMed]
31. Song, W.; Ye, C. Impact of the Cultivated-Land-Management Scale on Fertilizer Reduction—Empirical Evidence from the Countryside of China. *Land* **2022**, *11*, 1184. [CrossRef]
32. Wang, M.; Feng, C. Technological Gap, Scale Economy, and China's Industrial Energy Demand. *J. Clean. Prod.* **2019**, *236*, 117618. [CrossRef]
33. Wang, Y.; Li, X.; Lu, D.; Yan, J. Evaluating the Impact of Land Fragmentation on the Cost of Agricultural Operation in the Southwest Mountainous Areas of China. *Land Use Policy* **2020**, *99*, 105099. [CrossRef]
34. Yan, L.; Zhao, X.; Zhang, D.; Deng, J.; Zhang, Y. Associated Factors of Pesticide Packaging Waste Recycling Behavior Based on the Theory of Planned Behavior in Chinese Fruit Farmers. *Sustainability* **2022**, *14*, 10937. [CrossRef]
35. Zhang, B.; Niu, W.; Ma, L.; Zuo, X.; Kong, X.; Chen, H.; Zhang, Y.; Chen, W.; Zhao, M.; Xia, X. A Company-Dominated Pattern of Land Consolidation to Solve Land Fragmentation Problem and Its Effectiveness Evaluation: A Case Study in a Hilly Region of Guangxi Autonomous Region, Southwest China. *Land Use Policy* **2019**, *88*, 104115. [CrossRef]
36. Zhang, J.; Zhang, R.; Zhang, K.; Xu, D.; Qi, Y.; Deng, X. Does Land Management Scale Influence Farmers' Utilization of Straw Resources? Empirical Evidence from Rural China. *Environ. Impact Assess. Rev.* **2025**, *112*, 107820. [CrossRef]
37. Huang, S.; Elahi, E. Farmers' Preferences for Recycling Pesticide Packaging Waste: An Implication of a Discrete Choice Experiment Method. *Sustainability* **2022**, *14*, 14245. [CrossRef]
38. Wang, Y.-J.; Wang, N.; Huang, G.Q. How Do Rural Households Accept Straw Returning in Northeast China? *Resour. Conserv. Recycl.* **2022**, *182*, 106287. [CrossRef]

39. Chen, X.; Xing, L.; Li, B.; Zhang, Y. Substitution or Complementary Effects: The Impact of Neighborhood Effects and Policy Interventions on Farmers' Pesticide Packaging Waste Recycling Behavior. *J. Clean. Prod.* **2024**, *482*, 144198. [CrossRef]
40. Chen, Q. *Advanced Econometrics and Stata Applications*; Higher Education Press: Beijing, China, 2014.
41. Abadi, B. The Determinants of Cucumber Farmers' Pesticide Use Behavior in Central Iran: Implications for the Pesticide Use Management. *J. Clean. Prod.* **2018**, *205*, 1069–1081. [CrossRef]
42. Zhang, Y.; Yin, Y.; Li, F.; Duan, W.; Xu, K.; Yin, C. Can the Outsourcing Improve the Technical Efficiency of Wheat Production with Fertilization and Pesticide Application? Evidence from China. *J. Clean. Prod.* **2023**, *422*, 138587. [CrossRef]
43. Han, C.; Schröder, M.; Witthaut, D.; Böttcher, P.C. Formation of Trade Networks by Economies of Scale and Product Differentiation. *J. Phys. Complex.* **2023**, *4*, 025006. [CrossRef]
44. Zheng, W.; Luo, B. Understanding Pollution Behavior among Farmers: Exploring the Influence of Social Networks and Political Identity on Reducing Straw Burning in China. *Energy Res. Social Sci.* **2022**, *90*, 102553. [CrossRef]
45. Xu, X.; Zhang, Z.; Kuang, Y.; Li, C.; Sun, M.; Zhang, L.; Chang, D. Waste Pesticide Bottles Disposal in Rural China: Policy Constraints and Smallholder Farmers' Behavior. *J. Clean. Prod.* **2021**, *316*, 128385. [CrossRef]
46. Chen, X.; Orom, H.; Hay, J.L.; Waters, E.A.; Schofield, E.; Li, Y.; Kiviniemi, M.T. Differences in Rural and Urban Health Information Access and Use. *J. Rural Health* **2019**, *35*, 405–417. [CrossRef] [PubMed]
47. Fong, M.W. Digital Divide between Urban and Rural Regions in China. *Electron. J. Inf. Syst. Dev. Ctries.* **2009**, *36*, 1–12. [CrossRef]

land

MDPI

Article

# How Landscapes and History Shape Copper in Vineyard Soils: Example of Fruška Gora Region, Serbia

Irina Andreeva [1], Valeria Gabechaya [1], Dmitriy Morev [1], Miljan Samardžić [2], Zoran Galić [3] and Alexis Yaroslavtsev [1,*]

[1] Ecology Department, Russian Timiryazev State Agrarian University, Timiryazevskaya St. 49, 127434 Moscow, Russia
[2] Nature 4.0 BC SRL, Via Della Chimica 7, 01100 Viterbo, Italy
[3] Study Program Forestry, Faculty of Agriculture, University of East Sarajevo, Vuka Karadžića 30, Lukavica, 71126 East Sarajevo, Republic of Srpska, Bosnia and Herzegovina
* Correspondence: yaroslavtsevam@gmail.com; Tel.: +79051170532

**Abstract:** Vineyards are distinctive agroecosystems heavily influenced by local natural factors and traditional management practices, with significant implications for the quality and quantity of grape production. This study investigated the ecological impact of copper-containing fungicides on the accumulation and distribution of copper, manganese, chromium, and cobalt in the upper soil horizons of vineyards of varying ages in the Fruška Gora region, Serbia. The results indicated a marked difference in total copper content across vineyards, with the oldest vineyard exhibiting levels 6.9 times above the regulatory limit. Factor analysis delineated a strong correlation between copper accumulation and vineyard age while also highlighting the influence of landscape morphology on the spatial distribution of heavy metals. The findings suggest that copper accumulation is primarily related to agricultural practices, particularly the duration of fungicide application, while the distribution of other heavy metals is more closely associated with topographic features. The novelty of our research lies in the fact that we have shown that the assessment of copper accumulation in soil in vineyard ecosystems should take into account not only viticultural practices but also the history of land use and the landscape characteristics of the area.

**Keywords:** vineyard; soil; copper; heavy metals; accumulation; slope; migration; lateral differentiation coefficient; basal respiration

## 1. Introduction

Vineyards represent an authentic type of plantation-garden agroecosystem. Their peculiarity lies in the strong dependence of the quantity and quality of the obtained products on the combination of natural factors unique to each location and traditional management methods, as well as the large number of environmental risks associated with the cultivation of grapes in a particular area [1].

Due to the susceptibility of grape crops to infection by various pathogenic fungi, traditionally a wide range of chemicals have been used to protect grapevines [2]. In particular, copper-containing fungicides are of worldwide importance for the protection of grapes and fruit crops against such diseases as oidium or true powdery mildew (*Uncinula necator* Burill.), downy mildew (*Plasmopara viticola* (Berkeley & Curtis), and gray rot of grapes (*Botrytis cinerea*), which cause the greatest damage to the wine industry. The history of copper-based products being used in vineyards goes back about 150 years, when the protective properties of this element were discovered. Relatively high toxicity to plant pathogens, low cost, low toxicity to warm-blooded animals and humans, chemical stability, and ability to remain on the plant surface for a long time determined the wide commercialization of copper-containing fungicides in both traditional plant protection systems and integrated ones [3]. These protectants have also found wide application in organic farming, where they are almost the only authorized and effective chemical means of controlling grapevine diseases [4].

In the middle of the 20th century, when the use of copper-containing fungicides in orchards and vineyards in the world reached its maximum, evidence of their intensive accumulation in soils began to appear, especially in historical wine regions [5–7]. From the reports published in recent years on copper concentrations in the upper horizons of vineyard soils on the European, Australian, and South American continents, it can be concluded that the metal content ranges from 100 to 1500 mg/kg with multiple exceedances of the established regulatory values, and in some cases reaches a record 4500 mg/kg [8–11].

Copper compounds are stable in ecosystem components and can move along the soil profile and contaminate surface and ground waters [8], affect soil biota [12], and disturb the processes of organic matter mineralization in aquatic and soil environments due to fungicidal and bactericidal action. Copper pollution, together with the imperfection of the applied agrotechnologies, the development of erosion processes, and the spatial heterogeneity of slope soils on which vineyards are mainly located has led to them often being severely chemically, physically, and biologically degraded [13–16].

The level of copper accumulation, availability, and migration intensity in vineyard soils depends on many factors: climatic (primarily temperature, precipitation, and intensity) [17–19], soil (pH, organic matter, iron oxides, manganese, aluminum, clay minerals, etc.) [20–22], type of land use, and related agrotechnological practices, including dosage and frequency of applied plant protection products [23–25]. In addition, there is a certain relationship between the copper content in soil and vineyard age.

Against the background of a large volume of scientific data from studies on ampelocenoses up to 25 years old, studies on old vines are rather rare. Evidence of phytotoxicity in the presence of excess copper in the soil of old vineyards has been demonstrated [26], along with the efficiency of copper retention by the root system of old vines from the entry of the element into the aboveground part of the plant [27], the coprecipitation of copper with a number of other ions and their subsequent crystallization in the rhizosphere [28], and the possibility of the formation of several types of copper-containing minerals in the copper-enriched soil [29,30].

Vineyards are often located on slopes of different steepness, length, and shape, with pronounced or smoothed microrelief. These conditions determine spatial heterogeneity of soils of ampelocenoses by the content of macro- and microelements and pollutants. Landscape features are closely intertwined with the impact of other factors, primarily local climatic and agrogenic conditions. The applied agrotechnological methods of growing grapes, which in each area and in each farm have their own specifics, can strengthen or weaken the manifestation of landscape characteristics.

There are very few studies in the scientific literature that have carried out a comparative analysis of the behavior of heavy metals in the soil of ampelocenoses, taking into account their origin and identifying the factors that have the greatest influence on their accumulation and migration pathways. Particularly interesting results can be obtained in those wine-growing areas where there are many combinations of simultaneous and diverse landscape and agronomic factors. This contrast of natural conditions and grape growing methods is characteristic of the Fruška Gora region of the Autonomous Province of Vojvodina in the Republic of Serbia. In a restricted area of this historical wine-growing region, there are both industrialized wineries with intensive grape-growing technologies and small private vineyards, some of which have been cultivated for a long time using archaic technologies. However, common to them both is the use of chemical means of vine protection, including copper-containing ones, against pathogenic fungi.

The aim of the present study was the ecological assessment of the aftereffect of copper-containing fungicides on the accumulation and spatial and intra-profile distribution of copper, as well as manganese, chromium, and cobalt, in the upper (0–30 cm) soil horizons of different-aged vineyards cultivated by industrial, organic, and archaic technologies in the slope landscape of the Fruška Gora region of the Autonomous Province of Vojvodina, Republic of Serbia. The choice of manganese, chromium, and cobalt for the study was based on the fact that the agrogenic input of these elements into the soils of ampelocenoses, unlike copper, is relatively small, which makes it possible to demonstrate the contrasting behavior of these elements in vineyard agroecosystems.

## 2. Materials and Methods

### 2.1. Study Area

The research was conducted in the vineyards of three private wineries located on the right bank of the Danube River in the Autonomous Province of Vojvodina, Republic of Serbia (Figure 1). The farms are located in the eastern part of the Fruška Gora mountain range in Lipovac, Cherat, and Chushilovo in a historical wine region 1–4 km from the town of Sremski Karlovci. According to some sources, the first vineyards on the slopes of Fruška Gora date back to the reign of Roman emperors [31].

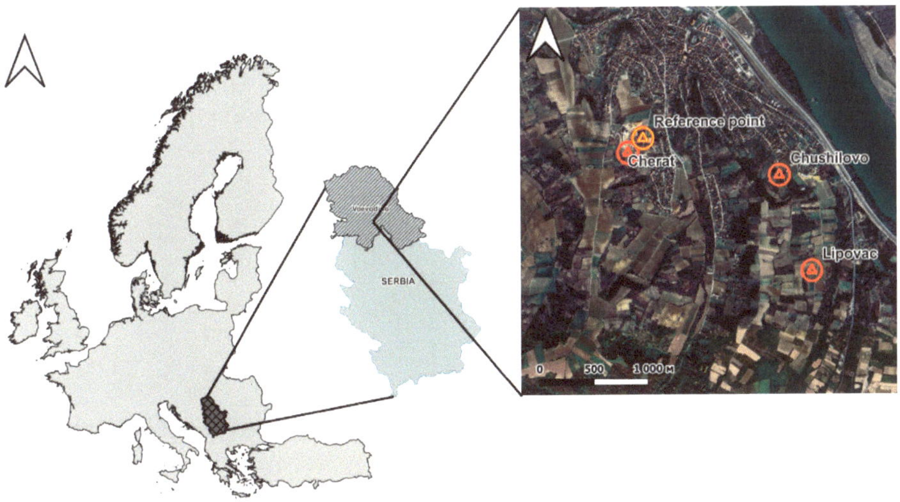


**Figure 1.** Location of the study region in the territory of the Republic of Serbia and the Autonomous Province of Vojvodina, with the sampling sites and reference point with a control soil.

The climate of the study area is temperate continental, humid, and mild. The average annual air temperature in the Fruška Gora area is 11.8 °C and the average annual precipitation is 764 mm, with 431 mm (56.4% of the average annual precipitation) falling during the growing season. Lack of moisture in the soil is observed in August and September and excess moisture - from December to April, with an average probability of occurrence of pluvial erosion [32].

The soil of the studied vineyards belongs to Eutric Cambisols according to the FAO (1988) and Haplic Cambisols Calcaric according to the WRB (2006). This soil was formed in the conditions of the slope landscape of the Middle Danube Foothills under broad-leaved forests. The profile thickness of the Eutric Cambisol at the studied sites is 115 cm. The traditional name of these soils in Serbia is *gajnjača*. They are brown, yellowish-brown or rusty reddish-brown, is the color being determined by the presence of hydrated iron oxide and clay minerals in the profile. These medium loamy soils are characterized by a favorable water and air regime and are traditionally used for orchards and vineyards.

The vineyards are located on slopes of different steepness and length with eastern exposure. Information on the peculiarities of the location of ampelocenoses in the landscape, age, and variety of the cultivated plantations is presented in Table 1.

**Table 1.** Characterization of the studied vineyards and the slopes on which they are located.

| Locality | Land-Use System | Trellis System | Cultivation Period, Years | Vine Variety | Exposition | Steepness, ° | Extent, m |
|---|---|---|---|---|---|---|---|
| Lipovac | Intensive | Vertical shoot position | 15 | Merlot | Eastern | 5–13 | 150 |
| Cherat | Archaic | Bush vine | More 200 | Slankamenka (autochthonous) | Eastern | 10 | 80 |
| Chushilovo | Organic | Vertical shoot position | More 100 | Merlot | Eastern | 4–8 | 27 |

The wineries under consideration practiced different grape-growing systems.

The farm in Lipovac uses an intensive industrial technology of cultivation of 15-year-old merlot grapes with a trellis-row system of bush management, cultivated row spacing, and application of a chemical plant protection system (Figure 2). The peculiarity of this vineyard is that it is located on a slope of complex shape with two terraces: the first was at the bottom of the concave upper part of the slope with a steepness of 13° (Lp2), and the second was at the bottom of the straight slope with a steepness of 5° (Lp3).

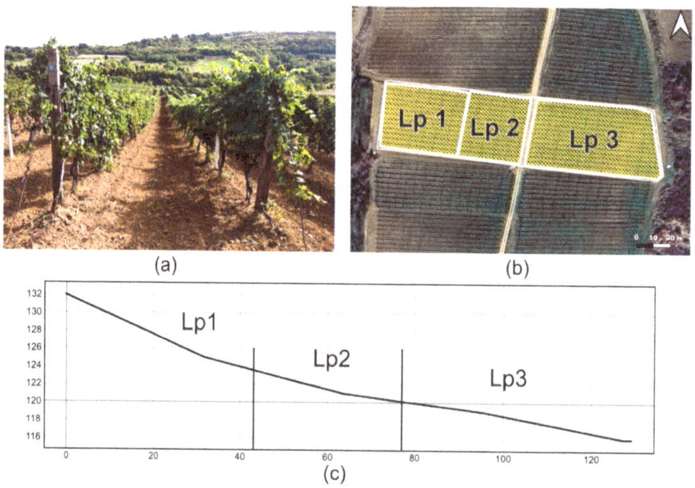


**Figure 2.** Vineyard in Lipovac: (**a**) general view; (**b**) location of sampling zones: upper Lp 1, middle Lp 2 and lower Lp 3 parts of the slope; (**c**) longitudinal profile of the slope of eastern exposure.

We classified the vineyard in Cherat as archaic, in accordance with the outdated system of cultivation of grapes in the form of bush vine–freestanding bushes of grape with sturdy 70–100 cm-high stems without supports evenly located on the whole slope (Figure 3). The advantage of this technology is that the bushes are evenly illuminated from all sides, which increases the assimilation activity of the leaves. On the other hand, the location of the bushes excludes the use of mechanized tillage and vineyard maintenance.

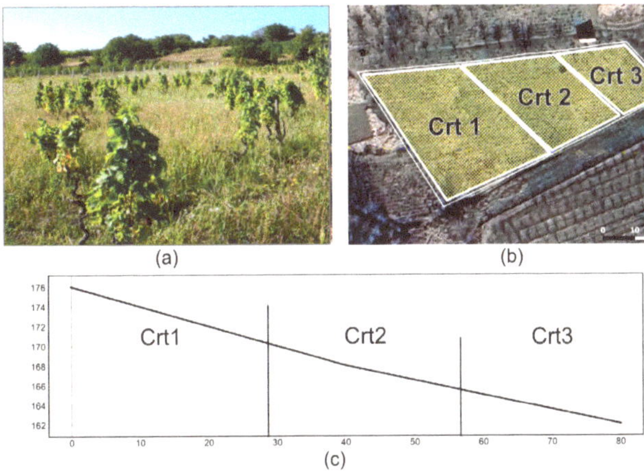


**Figure 3.** Vineyard in Cherat: (**a**) General view. (**b**) Location of sampling zones: upper Crt 1, middle Crt 2, and lower Crt 3 parts of the slope. (**c**) Scheme of longitudinal profile of the slope with eastern exposure.

This vineyard is of great scientific interest not only in terms of the cultivation system but also its duration: grapes have been grown on this site for more than 200 years. The age of the vines currently cultivated is 55 years.

The vineyard in Chushilovo is an abandoned vineyard on a slope of 4–8° and more than 100 years old on the eastern slope. It had not been maintained, tilled, fertilized, or treated with pesticides for the last 40 years (Figure 4).

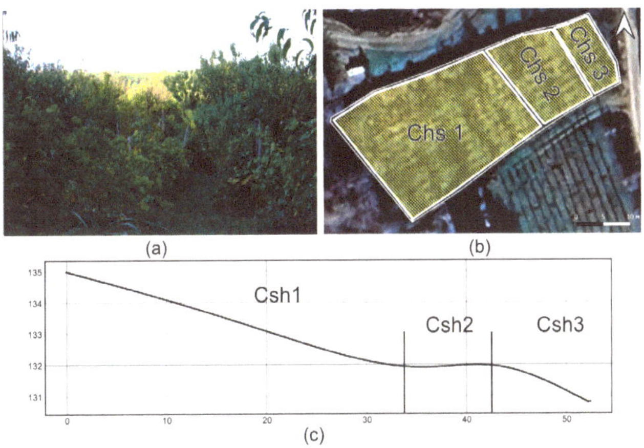


**Figure 4.** Vineyard in Chushilovo: (**a**) General view. (**b**) Location of sampling zones: upper Csh 1, middle Csh 2. and lower Csh 3 parts of the slope. (**c**) Scheme of longitudinal profile of the slope with eastern exposure.

We conditionally categorized this vineyard as an organic vineyard. There is herbaceous vegetation between the rows of vines.

### 2.2. Soil Sampling and Processing

Soil samples were collected using a soil auger at depths of 0–5, 5–15, and 15–30 cm in and between the rows of grapes within the contiguous elements of the upper, middle, and lower parts of the slopes of the three wineries in accordance with Russian Standard GOST R 58595-2019. Under the bush boll system, sampling was carried out directly inside the perimeter of the vine crown and within the free space between bushes. The "envelope" method was used to obtain a generalized picture of the soil property distribution at each site. This method involves taking five samples from a 5 m by 5 m plot (at the corners of the plot and in the middle of the plot). A 0.5 kg sample was taken from each point. The five subsamples were then combined and mixed, and one composite sample weighing 1.0 kg was retained from the total mass. All operations were carried out separately for samples from depths of 0–5, 5–15, and 15–30 cm. Mixed samples were collected from three sites within each vineyard as per the abovementioned procedure, placed in clean polyethylene bags, labeled, and transported to the laboratory.

Simultaneously with soil sampling from vineyards, control samples of background soil in the study area were collected (Figure 5). The reference site was approximately 0.5 ha in size and consisted of natural herbaceous vegetation dominated by cereals.

**Figure 5.** General view of the reference site where the reference soil samples were taken.

Reference soil samples were collected at depths of 0–20, 20–40, 40–60, 60–80, 80–100, 100–120, and 120–140 cm using a hand drill, similar to the above procedure.

The soil samples were air-dried at room temperature, plant roots, stones, and other inclusions removed, then disaggregated by hand in a mortar and sieved through a 1 mm sieve.

### 2.3. Soil Analysis

Soil reaction (pH) was determined by the potentiometric method in distilled water with a glass electrode using a SevenCompact pH meter S220 (Greifensee, Switzerland, Mettler Toledo) as per Russian Standard GOST 26423-85. The organic matter content was measured by the photometric method using a strong oxidizing agent ($K_2Cr_2O_7$) in the presence of $H_2SO_4$ (Walkley and Black method [33]) with a Leki UV2107 (Helsinki, Finland, MEDIORA OY) spectrophotometer at a wavelength of 600 nm.

For determining the total heavy metal content (Cu, Mn, Cr, Co), approximately 0.5 g of oven-dried (at 105 °C) soil was precisely weighed into a PFA vessel, and 7 mL aqua regia ($HNO_3/HCl$ = 1:3) was added. Before sample digestion, soil samples were ground in an agate ball mill to pass through a 250 μm sieve. All sample containers (tubes, vessels, volumetric flasks, etc.) were acid-washed, and clean acids were applied for sample digestion and multi-element standard dilution. Soil samples were digested in a Milestone ETHOS UP microwave oven following the recommendations of the company. Element concentrations in digested samples were determined by atomic absorption spectrometry (Agilent FS 240 AA, Santa Clara, CA, USA). Quality control was periodically carried out with reference materials GSO 11369-2019.

### 2.4. Lateral Differentiation Coefficient

To study the migration flows and pathways of the selected heavy metals along the slopes, we used the doctrine of elementary geochemical landscapes presented by Polynov (1956) [34] and further developed by Perelman (1975) [35] and Glazovskaya (1988) [36], whereby the nature of the links between elemental geochemical landscapes involved in geochemical conjugation is reflected in the redistribution of chemical elements and determines their accumulation or removal. Accordingly, lateral migration refers to the processes of movement of substances across the earth's surface from an autonomous elemental landscape to a subordinate one.

The quantitative assessment of the spatial (subhorizontal) heterogeneity of the heavy metal distribution along the slopes in our study was carried out on the basis of comparing the lateral differentiation coefficients (L) between the geochemically conjugated elements of the middle and lower parts of the slopes in relation to the upper part. This coefficient characterizes lateral migration of chemical substances in soils connected by unified moisture flows, which moves along the relief from top to bottom under the action of gravity [37]. The studies compared lateral differentiation coefficients between the metal concentration in the soil of the middle (T2) and lower (T3) parts of the slope relative to the upper (T1) part of the slope. The coefficient of lateral differentiation L of heavy metals in the soil of conjugate elements of trans-eluvial geochemical landscape was determined by the following formula:

$$L = \frac{Lx(T2 \text{ or } T3)}{Lx(T1)},$$

where Lx(T2 or T3) is the concentration of heavy metals in geochemically subordinate elements of the slope (middle part of T2 or lower part of T3) and Lx(T1) is the concentration of heavy metals in the upper part of the slope (T1).

When analyzing lateral migration of heavy metals along the slope, the following ranges of lateral differentiation coefficients were used: more than 1.7—high accumulation of element; 1.1–1.6—accumulation of medium strength; 0.6–0.9—insignificant removal of elements; less than 0.5—intensive removal of elements [37].

### 2.5. Basal Respiration

Basal respiration (BR) typically refers to soil respiration that is devoid of roots and driven solely by microbial activity. This index is extensively used to ascertain the physiological status of microorganisms [38,39]. Generally, an increase in the basal respiration rate suggests favorable soil microbiome conditions. However, evidence indicates that soil contamination can cause a pronounced increase in the soil respiration rate compared to optimal values [40].

Determination of basal respiration (BR) was conducted in accordance with CEN EN ISO 16072:2011: "Soil quality—Laboratory methods for determination of microbial soil respiration" [41]. Following the incubation of soil moistened with distilled water in vials for 24 h at 22 ± 0.5 °C, 10 mL of air was extracted using a syringe and analyzed for carbon dioxide content using a Chromatec-Crystall 5000 gas chromatograph (Yoshkar-Ola, Russia, Chromatec). BR rate is expressed in μg C-$CO_2$ $g^{-1}$ soil per hour, with fivefold repeatability.

### 2.6. Data Analysis

The data were processed using the R software environment for statistical computing and graphics. The FactoMiner package was used for factor analysis of the mixed data (FAMD) [42] and the FactoExtra package for visualization [43,44]. All variables except Cu, Mn, Cr, Co, and Corg were normally distributed according to the Shapiro–Wilk test ($p < 0.05$). Hence, a nonparametric test for group comparison was used (Kruskal–Wallis).

Mapping and spatial data were visualized with QGIS 3.34.5. Google satellite images were used for topography. To evaluate relief properties and prepare digital elevation models, SRTM data and a Garmin eTrex 10 GPS receiver were used.

## 3. Results

### 3.1. Agrochemical Properties and Basal Respiration of Soils Under Vineyards

Table 2 lists the basic agrochemical properties that allowed us to interpret the results of studies on the content and migration of the selected heavy metals in the soil of ampelocenoses. The pH value of water extract in the soil of all studied farms ranged from 8.5 to 8.9, which corresponds to an alkaline reaction of the medium. In the soil of the vineyard in Lipovac, the reaction of the medium did not change by horizons in the upper or middle part of the profile, while in the lower part it tended to increase in the horizon of 15–30 cm compared to the overlying ones. In the soil of the Cherat vineyard, pH changes based on depth, slope distribution, and position in and between rows were not significant. In the soil of the Chushilovo vineyard, the lowest pH values were found in the 0–5 cm horizon (8.0–8.2). pH values increased with depth up to 9.2, which was the maximum for all studied vineyards.

In contrast to pH, the organic matter content (Corg) in the studied vineyards fluctuated widely and varied in different parts of the slope. The lowest Corg in the soil of the vineyard in Lipovac was found in the upper part of the slope and ranged from 1.02% to 1.36%, depending on the horizon. In the soil of the middle and lower parts of the slope, the organic matter content in all horizons was significantly and reliably higher than in the upper part of the slope and ranged from 2.70% to 3.33%. At the same time, the change in Corg by soil horizon was unreliable both in and between rows.

In the Cherat vineyard, there was no mechanical tillage, so the differences between horizons in terms of organic matter content were more noticeable. For all parts of the slope, the highest accumulation of Corg was in the 0–5 cm horizon and consistently decreased by 1.5-fold to 2.4-fold to the 15–30 cm horizon. In general, the level of organic matter accumulation in the soil of the Cherat vineyard was significantly higher than in the soil of the Lipovac vineyard and was in the ranges of 2.34–7.96% and 3.17–7.17% for the soil under the bushes and between the bushes, respectively. In the 0–5 cm horizon under the bushes, Corg increased from 5.72% to 7.96%, whereas in the zone between the bushes, it varied from 6.86% to 7.17%, with no statistical significance.

The organic matter content in the soil of the vineyard in Chushilovo was also higher than in the soil of the vineyard in Lipovac, but less than in Cherat. There was a clear pattern in the distribution of organic matter in the upper part of the soil profile of the vineyard rows: the maximum accumulation of Corg was found in the 0–5 cm horizon (3.51–4.37%), followed by the 15–30 cm horizon (2.88–4.43%), with the least in the 5–15 cm horizon (2.35–2.88%). The same pattern was observed between the rows, but the differences between horizons in Corg accumulation were less noticeable.

**Table 2.** pH value, content of organic carbon (Corg), and basal respiration (BR) in soils of Fruška Gora vineyards with different land-use systems (means ± standard deviation).

| Vineyard (Land Use System) | Slope Part | Depth, cm | pH ($H_2O$) pH Unit | | Corg % | | BR µg $C\text{-}CO_2$ $g^{-1}$ Soil $hour^{-1}$ | |
|---|---|---|---|---|---|---|---|---|
| | | | Within Rows | Between Rows | Within Rows | Between Rows | Within Rows | Between Rows |
| Lipovac (Intensive) ($n$ = 45) | Lp 1 | 0–5 | 8.6 ± 0.1 | 8.6 ± 0.1 | 1.02 ± 0.15 | 0.88 ± 0.18 | 0.281 ± 0.018 | 0.270 ± 0.016 |
| | | 5–15 | 8.6 ± 0.1 | 8.6 ± 0.1 | 1.36 ± 0.27 | 1.10 ± 0.22 | 0.236 ± 0.031 | 0.221 ± 0.024 |
| | | 15–30 | 8.6 ± 0.1 | 8.6 ± 0.1 | 1.15 ± 0.23 | 1.07 ± 0.21 | 0.203 ± 0.019 | 0.118 ± 0.012 |
| | Lp 2 | 0–5 | 8.6 ± 0.1 | 8.6 ± 0.1 | 3.04 ± 0.46 | 3.00 ± 0.45 | 0.512 ± 0.054 | 0.393 ± 0.068 |
| | | 5–15 | 8.6 ± 0.1 | 8.6 ± 0.1 | 3.20 ± 0.48 | 2.95 ± 0.59 | 0.428 ± 0.027 | 0.353 ± 0.059 |
| | | 15–30 | 8.7 ± 0.1 | 8.7 ± 0.1 | 2.70 ± 0.54 | 2.53 ± 0.51 | 0.403 ± 0.020 | 0.324 ± 0.035 |
| | Lp 3 | 0–5 | 8.7 ± 0.1 | 8.6 ± 0.1 | 3.19 ± 0.48 | 3.12 ± 0.47 | 0.597 ± 0.034 | 0.513 ± 0.044 |
| | | 5–15 | 8.7 ± 0.1 | 8.7 ± 0.1 | 3.33 ± 0.50 | 3.04 ± 0.46 | 0.571 ± 0.018 | 0.462 ± 0.038 |
| | | 15–30 | 8.9 ± 0.2 | 8.7 ± 0.1 | 3.14 ± 0.47 | 3.08 ± 0.46 | 0.416 ± 0.025 | 0.343 ± 0.017 |
| Cherat (Archaic) ($n$ = 45) | Crt 1 | 0–5 | 8.5 ± 0.1 | 8.5 ± 0.1 | 5.72 ± 0.52 | 6.86 ± 0.71 | 3.344 ± 0.152 | 4.511 ± 0.289 |
| | | 5–15 | 8.8 ± 0.2 | 8.6 ± 0.1 | 4.41 ± 0.66 | 4.54 ± 0.68 | 2.491 ± 0.161 | 2.800 ± 0.116 |
| | | 15–30 | 8.7 ± 0.1 | 8.6 ± 0.1 | 2.35 ± 0.47 | 3.17 ± 0.48 | 1.024 ± 0.057 | 1.078 ± 0.048 |
| | Crt 2 | 0–5 | 8.6 ± 0.1 | 8.5 ± 0.1 | 5.51 ± 0.54 | 6.14 ± 0.59 | 3.570 ± 0.210 | 4.604 ± 0.154 |
| | | 5–15 | 8.8 ± 0.2 | 8.5 ± 0.1 | 3.57 ± 0.53 | 4.28 ± 0.64 | 1.923 ± 0.063 | 2.221 ± 0.138 |
| | | 15–30 | 8.8 ± 0.2 | 8.7 ± 0.2 | 2.34 ± 0.47 | 3.53 ± 0.53 | 0.924 ± 0.047 | 1.026 ± 0.082 |
| | Crt 3 | 0–5 | 8.5 ± 0.1 | 8.5 ± 0.1 | 7.96 ± 0.80 | 7.17 ± 0.69 | 7.520 ± 0.451 | 4.853 ± 0.337 |
| | | 5–15 | 8.7 ± 0.2 | 8.5 ± 0.1 | 5.67 ± 0.69 | 5.50 ± 0.53 | 2.677 ± 0.144 | 3.089 ± 0.238 |
| | | 15–30 | 8.7 ± 0.1 | 8.6 ± 0.1 | 3.81 ± 0.57 | 4.65 ± 0.70 | 1.516 ± 0.228 | 1.639 ± 0.121 |
| Chushilovo (Organic) ($n$ = 45) | Csh 1 | 0–5 | 8.0 ± 0.1 | 8.1 ± 0.1 | 4.07 ± 0.61 | 3.62 ± 0.54 | 5.405 ± 0.233 | 3.486 ± 0.195 |
| | | 5–15 | 9.2 ± 0.2 | 9.2 ± 0.2 | 2.56 ± 0.23 | 2.31 ± 0.33 | 2.328 ± 0.160 | 1.940 ± 0.117 |
| | | 15–30 | 8.2 ± 0.1 | 8.2 ± 0.1 | 3.77 ± 0.54 | 3.39 ± 0.53 | 1.775 ± 0.153 | 1.380 ± 0.085 |
| | Csh 2 | 0–5 | 8.0 ± 0.1 | 8.1 ± 0.1 | 4.37 ± 0.66 | 3.95 ± 0.59 | 5.218 ± 0.658 | 2.991 ± 0.139 |
| | | 5–15 | 9.2 ± 0.1 | 9.2 ± 0.2 | 2.88 ± 0.43 | 2.61 ± 0.39 | 2.546 ± 0.156 | 2.073 ± 0.236 |
| | | 15–30 | 8.2 ± 0.1 | 9.2 ± 0.1 | 3.86 ± 0.58 | 3.68 ± 0.58 | 2.803 ± 0.084 | 1.591 ± 0.177 |
| | Csh 3 | 0–5 | 8.1 ± 0.1 | 8.2 ± 0.1 | 3.51 ± 0.47 | 3.38 ± 0.51 | 6.067 ± 0.395 | 3.932 ± 0.319 |
| | | 5–15 | 8.4 ± 0.1 | 9.2 ± 0.2 | 2.35 ± 0.32 | 2.26 ± 0.24 | 1.511 ± 0.284 | 1.259 ± 0.247 |
| | | 15–30 | 9.2 ± 0.2 | 9.2 ± 0.2 | 4.43 ± 0.67 | 3.39 ± 0.49 | 1.448 ± 0.266 | 1.457 ± 0.113 |
| Reference ($n$ = 21) | - | 0–20 | 9.1 ± 0.2 | | 2.06 ± 0.33 | | 1.462 ± 0.214 | |
| | | 20–40 | 8.8 ± 0.1 | | 2.19 ± 0.41 | | 1.305 ± 0.107 | |
| | | 40–60 | 8.8 ± 0.1 | | 1.74 ± 0.25 | | 1.031 ± 0.053 | |
| | | 60–80 | 9.1 ± 0.2 | | 1.52 ± 0.17 | | 0.847 ± 0.115 | |
| | | 80–100 | 9.0 ± 0.2 | | 0.43 ± 0.07 | | 0.219 ± 0.038 | |
| | | 100–120 | 8.8 ± 0.1 | | 0.14 ± 0.03 | | 0.153 ± 0.012 | |
| | | 120–140 | 8.8 ± 0.2 | | 0.07 ± 0.01 | | 0.088 ± 0.004 | |

Changes in the basal respiration rate replicated the observed trends in the accumulation of organic matter: this indicator consistently decreased from the upper horizon

(0–5 cm) to the underlying horizon and significantly increased from the upper part of the slope to the lower part. Also, the basal respiration rate was higher in the soil of grape rows in the Lipovac vineyard than between the rows. The significantly higher values of this indicator for the vineyard in Cherat compared to the vineyard in Lipovac should be emphasized. In Lipovac, the BR varied between 0.118 and 0.597 μg $CO_2$-C $g^{-1}$ soil-$h^{-1}$, and in Cherat, the BR values were in the range of 1.024–7.520 μg $CO_2$-C $g^{-1}$ soil-$h^{-1}$, with predominant activity of soil microbiota in the uppermost soil horizon. Interestingly, in the vineyard in Cherat, the basal respiration rate in all considered soil horizons was higher between bushes than within the crown perimeter, except for the 0–5 cm horizon in the lower part of the slope, where the BR under bushes was 1.5 times higher than that between bushes, which corresponded to the nature of organic matter distribution. In terms of BR values, the soil under the 100-year-old abandoned vineyard in Chushilovo occupied an intermediate position between the 15-year-old vineyard in Lipovac and the 55-year-old vineyard in Cherat. The basal respiration rate, as in the soil of the other studied ampelocenoses, was highest in the 0–5 cm horizon and ranged from 5.218 to 6.067 μg $CO_2$-C $g^{-1}$ soil-$h^{-1}$ and from 2.991 to 3.932 μg $CO_2$-C $g^{-1}$ soil-$h^{-1}$ in and between rows, respectively. In the 5–15 cm horizon, BR fell sharply compared to the 0–5 cm horizon by a factor of 1.4 to 4, depending on the slope part.

The control soil sampled at the reference site was characterized by an alkaline reaction similar to the soil under vineyards. However, organic matter content in the control soil was lower than in the soil of the vineyards in Cherat and Chushilovo and in the soil of the middle and lower part of the slope in Lipovac, but higher than in the soils of the upper part of the vineyard slope in Lipovac, within the 0–30 cm soil layer. According to the basal respiration index, the control soil occupied an intermediate position between the soil of the vineyards in Cherat and Chushilovo and the vineyard in Lipovac.

### 3.2. *Total Heavy Metal Content in Vineyard Soil*

Total copper content in reference soil varied in the 0–20, 20–40, 40–60, and 60–80 cm horizons in a rather narrow range of 49.2–57.8 mg/kg, with a mean value of 52.1 mg/kg. In the 80–100 cm horizon of the reference soil, the element content consistently decreased to 120–140 cm, where it reached values of 20.0 and 17.0 mg/kg, respectively, with a mean value of 18.5 mg/kg (Figure 6).

Total copper content in the soil in and between vineyard rows in Lipovac was in the ranges of 20.3–57.9 and 20.5–43.2 mg/kg, respectively, varying both by soil horizon and part of slope (Table 3). The maximum values of copper content were found in grapevine rows in the 15–30 cm horizon of the upper and middle parts of the slope (54.2 and 57.9 mg/kg, respectively) and in the 0–5 cm horizon of the middle part of the slope (41.9 mg/kg). In the lower, more gentle part of the slope with a steepness of 5–6°, there was a noticeably lower copper accumulation in the surface soil horizons at the level of 20.3–22.7 mg/kg compared to the overlying parts of the slope. These values were close to the copper accumulation in the 0–5 and 5–15 cm horizons in the upper part of the slope profile and corresponded to the copper content in the parent rock for this soil type.

The calculated lateral differentiation coefficients confirmed that the main accumulation of copper of medium strength (L = 1.62 and 1.32 within and between rows, respectively) along the vineyard slope in Lipovac occurred in the middle part of the slope in the 0–5 cm horizon (Table 4).

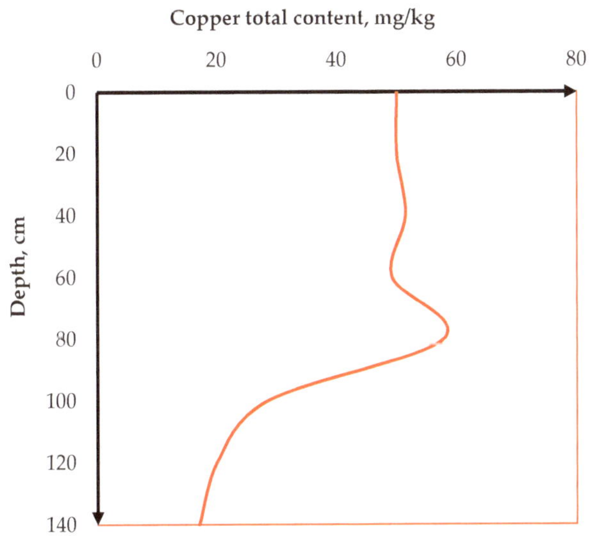


**Figure 6.** Copper distribution along the profile (0–140 cm) of the control soil from the reference site.

**Table 3.** Total content of heavy metals (mg/kg) in row and interrow soil of a 15-year-old vineyard in different parts of the slope in Lipovac (MAC—maximum allowed concentration, RC—remediation concentration for studied heavy metals in Serbia [45]).

| Slope Part | Depth, cm | Cu | | Mn | | Cr | | Co | |
|---|---|---|---|---|---|---|---|---|---|
| | | Within Rows | Between Rows | Within Rows | Between Rows | Within Rows | Between Rows | Within Rows | Between Rows |
| Lp 1 | 0–5 | 25.8 | 23.7 | 921 | 886 | 45.8 | 40.5 | 14.1 | 16.1 |
| | 5–15 | 24.8 | 23.4 | 866 | 743 | 36.7 | 38.2 | 15.7 | 16.8 |
| | 15–30 | 54.2 | 42.6 | 598 | 667 | 28.9 | 31.5 | 10.5 | 11.3 |
| Lp 2 | 0–5 | 41.9 | 31.4 | 750 | 682 | 36.6 | 24.2 | 13.9 | 12.0 |
| | 5–15 | 20.3 | 22.9 | 509 | 616 | 14.6 | 15.9 | 16.2 | 18.7 |
| | 15–30 | 57.9 | 43.2 | 748 | 829 | 38.6 | 41.5 | 15.7 | 16.0 |
| Lp 3 | 0–5 | 21.0 | 20.5 | 514 | 482 | 25.4 | 20.8 | 11.3 | 10.7 |
| | 5–15 | 20.3 | 21.6 | 687 | 648 | 18.6 | 13.0 | 13.9 | 14.1 |
| | 15–30 | 22.7 | 21.2 | 284 | 301 | 24.5 | 28.4 | 8.1 | 7.9 |
| MAC/RC | | 36/190 | | not specified | | 100/380 | | 9/240 | |

Total manganese content in different parts of the horizontal profile of the slope differed significantly (Table 3). It was concentrated in the 0–5 and 5–15 cm horizons of the upper part of the slope (886–921 and 743–866 mg/kg, respectively) and 0–5 and 15–30 cm horizons of the middle part of the slope (750 mg/kg and 748–829 mg/kg, respectively, in the grape row only) with lateral differentiation coefficients of 1.24–1.25 in the 15–30 cm horizon of the middle part of the slope (Table 4). As in the case of copper, the soil of the lower part of the slope accumulated 1.3–2.8 times less manganese compared to the upper and middle parts, especially in the 15–30 cm horizon (284–301 mg/kg).

**Table 4.** Lateral differentiation coefficients (L) of the transit landscape at the 15-year vineyard in Lipovac.

| Slope Part | Depth, cm | Cu | | Mn | | Cr | | Co | |
|---|---|---|---|---|---|---|---|---|---|
| | | Within Rows | Between Rows | Within Rows | Between Rows | Within Rows | Between Rows | Within Rows | Between Rows |
| Lp 2 | 0–5 | 1.62 | 1.32 | 0.81 | 0.77 | 0.80 | 0.60 | 0.99 | 0.75 |
| | 5–15 | 0.82 | 0.98 | 0.59 | 0.83 | 0.40 | 0.42 | 1.03 | 1.11 |
| | 15–30 | 1.07 | 1.01 | 1.25 | 1.24 | 1.34 | 1.32 | 1.50 | 1.42 |
| Lp 3 | 0–5 | 0.81 | 0.86 | 0.56 | 0.54 | 0.55 | 0.51 | 0.80 | 0.66 |
| | 5–15 | 0.82 | 0.92 | 0.79 | 0.87 | 0.51 | 0.34 | 0.89 | 0.84 |
| | 15–30 | 0.42 | 0.50 | 0.47 | 0.45 | 0.85 | 0.90 | 0.77 | 0.70 |

The highest amount of chromium was found in the 0–5 cm horizon of the upper part of the slope (40.5–45.8 mg/kg) and the 15–30 cm horizon of the middle part of the slope (38.6–41.5 mg/kg). In other horizons, including the lower part of the slope, the average content of the element was 24.8 mg/kg. A similar trend was observed for cobalt content, except that this element was also accumulated in the 5–15 cm horizon in the upper and middle parts of the slope. The gross cobalt content in these horizons ranged from 15.7 to 18.7 mg/kg, whereas in the lower part of the slope, it ranged from 7.9 to 14.1 mg/kg. For chromium and cobalt, as well as for manganese, lateral differentiation coefficients (1.32–1.34 and 1.42–1.50, respectively), indicating the accumulation of these elements of medium strength, were established in the 15–30 cm horizon of the middle part of the slope.

The total copper content in the soil of the upper (15–30 cm) and middle (0–5 and 15–30 cm) slope parts exceeded the maximum allowable concentration (MAC) of this element in Serbia [45] by 1.2–1.6 times. The soil of the Lipovac vineyard was also found to exceed the cobalt limit in almost all horizons and parts of the slope. The total chromium content was well below the MAC value.

Total copper content in the soil of the 55-year-old vineyard in Cherat varied widely: from 41.0 to 247.5 mg/kg (Table 5).

**Table 5.** Total content of heavy metals (mg/kg) in soil under and between grape bushes in different parts of the slope at the 55-year-old vineyard in Cherat (MAC—maximum allowed concentration, RC—remediation concentration for studied heavy metals in Serbia [45]).

| Slope Part | Depth, cm | Cu | | Mn | | Cr | | Co | |
|---|---|---|---|---|---|---|---|---|---|
| | | Under Bushes | Between Bushes | Under Bushes | Between Bushes | Under Bushes | Between Bushes | Under Bushes | Between Bushes |
| Crt 1 | 0–5 | 109.0 | 84.5 | 682 | 571 | 33.2 | 31.7 | 5.2 | 4.6 |
| | 5–15 | 95.1 | 73.7 | 656 | 636 | 34.4 | 33.1 | 7.5 | 11.2 |
| | 15–30 | 61.9 | 41.0 | 605 | 624 | 33.8 | 36.6 | 11.7 | 13.6 |
| Crt 2 | 0–5 | 132.7 | 127.6 | 650 | 624 | 29.0 | 28.4 | 8.5 | 9.4 |
| | 5–15 | 72.3 | 73.9 | 748 | 606 | 42.0 | 29.7 | 12.8 | 10.6 |
| | 15–30 | 59.4 | 52.8 | 732 | 755 | 35.3 | 33.2 | 14.3 | 12.1 |
| Crt 3 | 0–5 | 128.3 | 136.6 | 647 | 662 | 27.3 | 31.0 | 10.7 | 7.9 |
| | 5–15 | 146.1 | 189.4 | 661 | 741 | 26.3 | 28.4 | 11.8 | 16.6 |
| | 15–30 | 247.5 | 228.5 | 768 | 861 | 37.2 | 35.6 | 18.1 | 20.3 |
| MAC/RC | | 36/190 | | not specified | | 100/380 | | 9/240 | |

Maximum values of total copper content were found within the crown perimeter of grape bushes in the 0–5 cm soil horizon in the upper (109.0 mg/kg) and middle (132.7 mg/kg) parts of the slope, as well as in all horizons of the lower part of the slope (128.3–247.5 mg/kg). As a consequence, very high lateral differentiation coefficients were observed in the soil of the lower part of the slope both under (1.18–4.00) and between grape bushes (1.62–5.57), which indicates intensive migration of copper from the upper to the lower part of the slope, with pronounced accumulation in the 15–30 cm horizon (Table 6).

**Table 6.** Lateral differentiation coefficients (L) of the transit landscape at 55-year-old vineyard in Cherat.

| Slope Part | Depth, cm | Cu | | Mn | | Cr | | Co | |
|---|---|---|---|---|---|---|---|---|---|
| | | Under Bushes | Between Bushes | Under Bushes | Between Bushes | Under Bushes | Between Bushes | Under Bushes | Between Bushes |
| Crt 2 | 0–5 | 1.22 | 1.51 | 0.95 | 1.09 | 0.87 | 0.90 | 1.63 | 2.04 |
| | 5–15 | 0.76 | 1.00 | 1.14 | 0.95 | 1.22 | 0.90 | 1.71 | 0.95 |
| | 15–30 | 0.96 | 1.29 | 1.21 | 1.21 | 1.04 | 0.91 | 1.22 | 0.89 |
| Crt 3 | 0–5 | 1.18 | 1.62 | 0.95 | 1.16 | 0.82 | 0.98 | 2.06 | 1.72 |
| | 5–15 | 1.54 | 2.57 | 1.01 | 1.17 | 0.76 | 0.86 | 1.57 | 1.48 |
| | 15–30 | 4.00 | 5.57 | 1.27 | 1.38 | 1.10 | 0.97 | 1.55 | 1.49 |

The values of total copper content in the soil of the lower part of the slope at the 55-year-old Cherat vineyard exceeded the MAC value by 3.6–3.8, 4.1–5.3, and 6.3–6.9 times in the 0–5, 5–15, and 15–30 cm horizons, respectively. In addition, at the lower part of the slope in the 15–30 cm horizon, the copper content exceeded not only the MAC but also the remediation concentration (190 mg/kg).

Total manganese content in the soil of the vineyard in Cherat was in the range of 571–861 mg/kg, chromium 26.3–42.0 mg/kg, and cobalt 4.6–20.3 mg/kg (Table 5). As shown by lateral differentiation coefficients (Table 6), manganese accumulation was found in the middle and lower parts of the slope (L = 1.21–1.38) in the 15–30 cm horizon, whereas cobalt was actively accumulated in all studied horizons of the lower part of the slope (L = 1.48–2.06) and the surface horizon of the middle part of the slope (L = 1.63–2.04).

Total copper content in the soil of the abandoned vineyard in Chushilovo varied significantly depending on the horizon and slope element (Table 7). In the soil of the upper part of the slope, copper content in the soil in and between rows of the vineyard decreased 2.5 times from the surface horizon to the 15–30 cm horizon, whereas in the middle part of the slope, the highest accumulation of the element in the soil of the rows was observed in the 5–15 and 15–30 cm horizons and amounted to 115.4 and 114.1 mg/kg, respectively. In the lower part of the slope, similar to the trend noted above for the 15-year-old vineyard in Lipovac, the lowest accumulation of copper was found in all studied horizons, which ranged from 32.9 to 53.7 mg/kg. Thus, the distribution of copper in the upper soil horizons of the vineyard in Chushilovo along the slope was uneven. The total quantity of the element increased from the upper part of the slope to the middle part and then significantly decreased.

**Table 7.** Total content of heavy metals (mg/kg) in the soil in and between rows of a 100-year-old abandoned vineyard in different parts of the slope in Chushilovo (MAC—maximum allowed concentration, RC—remediation concentration for studied heavy metals in Serbia [45]).

| Slope Part | Depth, cm | Cu | | Mn | | Cr | | Co | |
|---|---|---|---|---|---|---|---|---|---|
| | | Within Rows | Between Rows | Within Rows | Between Rows | Within Rows | Between Rows | Within Rows | Between Rows |
| Csh 1 | 0–5 | 90.2 | 85.5 | 578 | 740 | 59.3 | 56.8 | 9.8 | 12.2 |
| | 5–15 | 78.7 | 67.8 | 662 | 738 | 38.9 | 48.4 | 10.6 | 13.2 |
| | 15–30 | 35.5 | 34.5 | 728 | 543 | 18.4 | 41.2 | 14.2 | 15.3 |
| Csh 2 | 0–5 | 96.4 | 86.6 | 559 | 654 | 38.9 | 31.6 | 5.3 | 9.7 |
| | 5–15 | 115.4 | 107.8 | 716 | 666 | 48.1 | 50.1 | 12.6 | 12.2 |
| | 15–30 | 114.1 | 27.9 | 692 | 784 | 43.5 | 68.7 | 15.8 | 16.8 |
| Csh 3 | 0–5 | 53.7 | 42.5 | 531 | 564 | 33.2 | 33.5 | 8.1 | 13.2 |
| | 5–15 | 45.6 | 37.9 | 609 | 688 | 31.8 | 29.7 | 10.3 | 14.7 |
| | 15–30 | 41.1 | 32.9 | 694 | 792 | 24.0 | 21.1 | 11.4 | 17.6 |
| MAC/RC | | 36/190 | | not specified | | 100/380 | | 9/240 | |

According to Table 8, the highest lateral differentiation coefficients with very high copper accumulation (L = 3.21) and medium strength accumulation (L = 1.47–1.59) were characterized by horizons 15–30 cm and 5–15 cm, respectively, in the middle part of the slope.

**Table 8.** Lateral differentiation coefficients (L) of the transit landscape at the 100-year-old abandoned vineyard in locality Chushilovo.

| Slope Part | Depth, cm | Cu | | Mn | | Cr | | Co | |
|---|---|---|---|---|---|---|---|---|---|
| | | In Rows | Between Rows | In Rows | Between Rows | In Rows | Between Rows | In Rows | Between Rows |
| Csh 2 | 0–5 | 1.07 | 1.01 | 0.97 | 0.88 | 0.66 | 0.56 | 0.54 | 0.80 |
| | 5–15 | 1.47 | 1.59 | 1.08 | 0.90 | 1.24 | 1.04 | 1.19 | 0.92 |
| | 15–30 | 3.21 | 0.81 | 0.95 | 1.44 | 2.36 | 1.67 | 1.11 | 1.10 |
| Csh 3 | 0–5 | 0.60 | 0.50 | 0.92 | 0.76 | 0.56 | 0.59 | 0.83 | 1.08 |
| | 5–15 | 0.58 | 0.56 | 0.92 | 0.93 | 0.82 | 0.61 | 0.97 | 1.11 |
| | 15–30 | 1.16 | 0.95 | 0.95 | 1.46 | 1.30 | 0.51 | 0.80 | 1.15 |

Despite the absence of treatments with chemical plant protection agents for grapes during the last 40 years, the copper content in the soil within rows at the middle part of the vineyard slope in Chushilovo exceeded the MAC level by 2.7 and 3.2 times, respectively, in the 0–5, 5–15, and 15–30 cm horizons. The results obtained indicate the presence of aftereffects of copper-containing pesticide application decades after the vineyard's useful life.

Total manganese content in the soil varied in the range of 531–728 mg/kg without any patterns confirmed statistically. Medium-strength accumulation of the element was observed only in the 15–30 cm horizons in the middle and lower part of the slope (L = 1.44 and 1.46, respectively).

Total chromium content in the soil of the vineyard under consideration varied in a wide range from 18.4 to 68.7 mg/kg, along both the depth and horizontal profiles of the slope (Table 7). In the 0–5 cm horizon, there was a significant decrease in chromium accumulation from the upper part of the slope to the middle and lower parts of the slope by 52–79% and by 70–80% in and between vineyard rows, respectively. At the same time, the opposite trend was established for the 5–15 and 15–30 cm horizons, in which chromium accumulation from the upper to the middle part of the profile, on the contrary, increased by 1.2–2.4 times, both in and between rows. A high level of chromium accumulation, as in the case of copper, was observed in the 15–30 cm horizon in and between rows (L = 2.36 and 1.67, respectively) in the middle part of the slope (Table 8).

Total cobalt content in the soil of the abandoned vineyard ranged from 5.3 to 17.6 mg/kg, which approximately corresponded to the level of accumulation of this element in the soil of vineyards in Lipovac and Cherat (Table 7). In all parts of the slope, cobalt content significantly increased 1.4–3 times from the surface horizon to a depth of 15–30 cm, and in rows was lower in rows than between. Based on the calculated lateral differentiation coefficients, no significant accumulation of cobalt was found in the considered horizons and along the slope profile (Table 8).

In terms of copper content, the profile of the control soil sampled from the reference site can be divided into two contrasting zones (Table 9). From the surface to a depth of 80 cm, the total content ranged from 48.1 to 58.1 mg/kg, which was 1.3 to 1.6 times higher than the MAC value. In soil layers from 80 to 140 cm, copper content ranged from 18.7 to 27.2 mg/kg and did not exceed the MAC value. As for the other heavy metals studied, their vertical distribution in the soil was relatively uniform, with a tendency for the total content to increase slightly from the upper to the lower part of the soil profile. The cobalt content was slightly above the MAC value in all horizons.

**Table 9.** Total content of heavy metals (mg/kg) in the control soil from reference site (MAC—maximum allowed concentration, RC—remediation concentration for studied heavy metals in Serbia [45]).

| Depth, cm | Cu | Mn | Cr | Co |
|---|---|---|---|---|
| 0–20 | 48.1 | 580 | 41.9 | 10.8 |
| 20–40 | 49.5 | 535 | 39.9 | 10.0 |
| 40–60 | 48.3 | 594 | 42.0 | 10.9 |
| 60–80 | 58.1 | 600 | 47.8 | 11.0 |
| 80–100 | 27.2 | 595 | 44.7 | 10.9 |
| 100–120 | 20.7 | 596 | 47.3 | 11.0 |
| 120–140 | 18.7 | 616 | 50.4 | 12.1 |
| MAC/RC | 36/190 | not specified | 100/380 | 9/240 |

### 3.3. Factor Analysis of Heavy Metal Distribution in Vineyard Soils

Since the dataset contained qualitative and quantitative variables, factor analysis of the mixed data (FAMD) was conducted to gain a general understanding of the relationships between the variables. FAMD is a principal component method dedicated to analyzing a dataset containing both quantitative and qualitative variables. It acts as principal component analysis for quantitative variables and as multiple correspondence analysis for qualitative variables.

The three principal components explained 53.14% of the total variance, i.e., more than all other principal components combined. The first two components are the most interesting for analysis and make the greatest contribution to the explanation of the total variance (Figure 7A,B). The main contribution to the first component comes from two strongly related variables: soil organic content (Corg) and vineyard age (Age), as well as slope shape (SlopeForm) and total copper content (Cu). The second component is mainly represented by variables describing relief: "SlopeForm", "Steepness", and "SlopePart". To a lesser extent, heavy metals (Mn, Cr, Cu) contribute to it. The third component is primarily determined by basal microbial respiration, slope length, sampling depth, and pH.

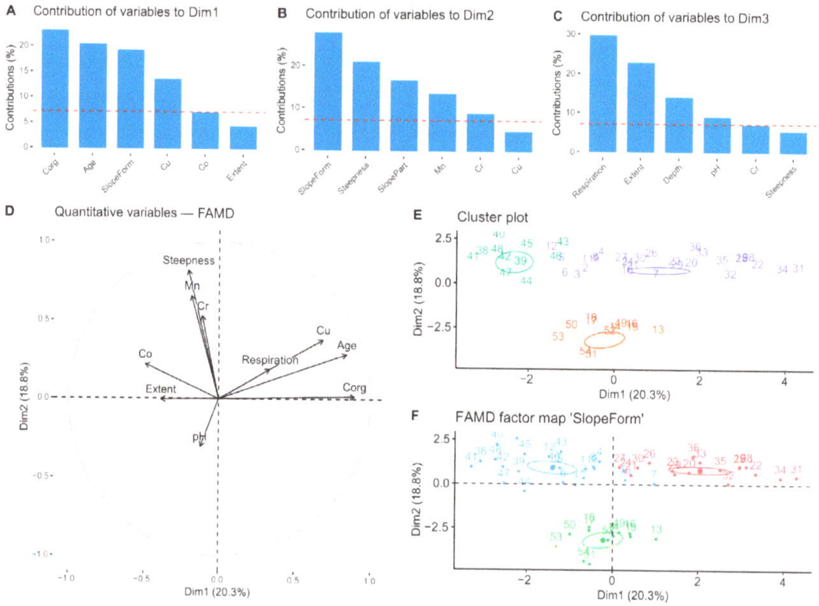

**Figure 7.** Factor analysis of mixed data (FAMD) of vineyard soil and landscape parameters. Variables' contribution to principal components shown for PC1 (**A**), PC2 (**B**), and PC3 (**C**). Factor map for quantitative variables (**D**). Agglomerative hierarchical clustering of results from factor analysis. Observations are represented by points in the plot, while ellipses represent 0.95 confidence intervals (**E**). Qualitative variable groupings for slope form (**F**).

The most important qualitative variable is slope shape. Comparison of the group of observations based on this qualitative factor with the groups of observations identified by agglomerative hierarchical clustering (Figure 7E,F) showed that the cluster analysis identified the same number of groups, with the majority of observations overlapping between them. Among the quantitative variables, two main groups were highlighted: total copper content, vineyard age, basal respiration, and Corg, and slope steepness, total copper content, and manganese content, oriented by the first and second principal components, respectively (Figure 7D).

Thus, it can be assumed that the distribution of copper and other heavy metals in the soils of the studied vineyards depends on different factors. Copper was more strongly related to parameters determined by agrogenic activity (duration of vineyard cultivation and organic matter content), while other heavy metals were more strongly related to relief parameters: slope steepness for chromium and manganese and slope extent for cobalt.

The introduction of heavy metal lateral differentiation coefficients into the FAMD would have led to multicollinearity, so in the next step, Spearman rank correlation coefficients were calculated for the qualitative variables, presented in Figure 8 as a correlation matrix. Variables in the matrix are grouped in hierarchical clustering order, based on which our previous assumption about the difference of copper migration from other studied heavy metals due to their location in different clusters was partially confirmed. The strongest correlation was observed between the total copper content and the age of the vineyard or duration of its cultivation, while a weaker correlation was noted between the total copper content and the land-use system (Figure 8).

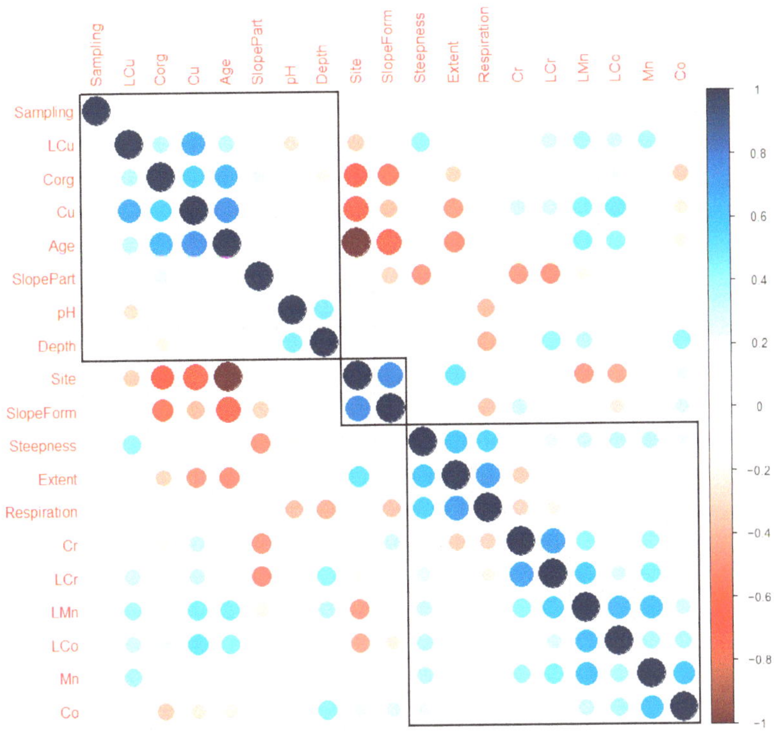


**Figure 8.** Spearman's rank correlation matrix for land-use system, slope parameters, agrochemical properties, basal respiration, total heavy metal content, and lateral differentiation coefficients in vineyard soil.

The Kruskal–Wallis method was chosen for pairwise comparison of median copper values in the vineyards because of the small number of samples in each group and the lack of normal distribution in the data (Figure 9). The comparisons showed that the median values of copper content in soil differed significantly among the vineyards ($p < 0.01$) All vineyards were characterized by an absence of significant difference ($p > 0.05$) between the copper content of soils in and between rows. The level of copper in the soil of the vineyards in Cherat and Chushilovo exceeded the maximum permissible values established in the Republic of Serbia.

The organic carbon content of soils of all surveyed vineyards at depths of 0–30 cm was compared by a similar method (Figure 10). Comparison of medians showed that the values of soil organic carbon content differed significantly for all vineyards and at all depths except for 15–30 cm. For this depth, a significant difference was observed only between the samples from Lipovac and Chushilovo. The maximum organic carbon content in soils was observed in the vineyard in Cherat and the minimum in Lipovac.

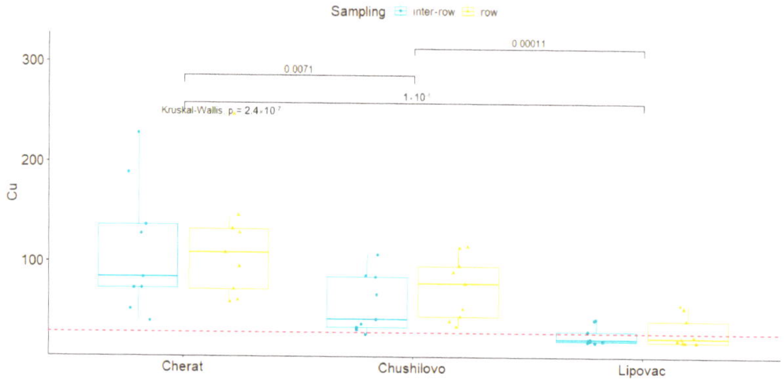


**Figure 9.** Total copper content (mg $kg^{-1}$) in vineyard soil in Lipovac, Cherat, and Chushilovo in according to Kruskal–Wallis criteria ($p < 0.01$). The horizontal dashed red line means the maximum allowed concentration of copper in soil established in the Republic of Serbia (36 mg $kg^{-1}$).

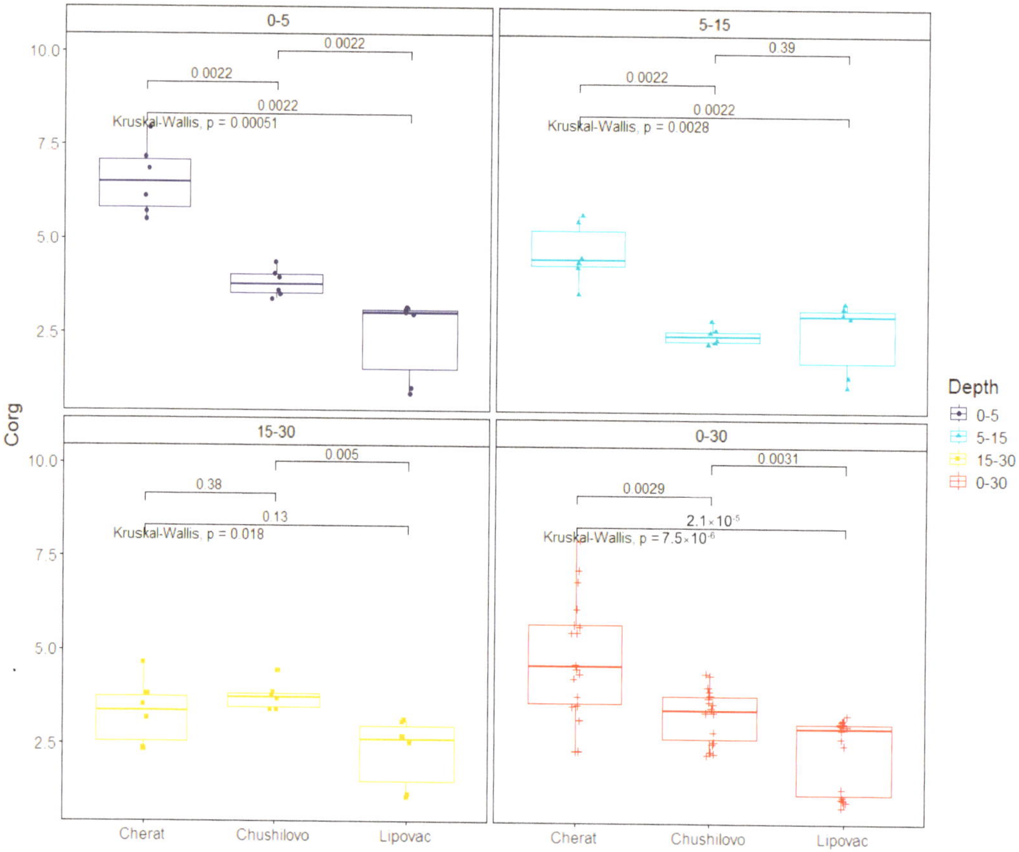


**Figure 10.** Organic matter content (Corg, %) in the surface layers (0–5, 5–15, 15–30 and overall 0–30 cm) of the vineyard soil in according to Kruskal–Wallis criteria ($p < 0.001$).

A general trend of downslope accumulation of copper in the soils of the studied vineyards was found (Figure 11). However, the distribution of copper and other heavy metals in the surface soil layer in each studied plot depended on the morphological characteristics of the slope (steepness, shape, length) and the agrotechnical methods applied.

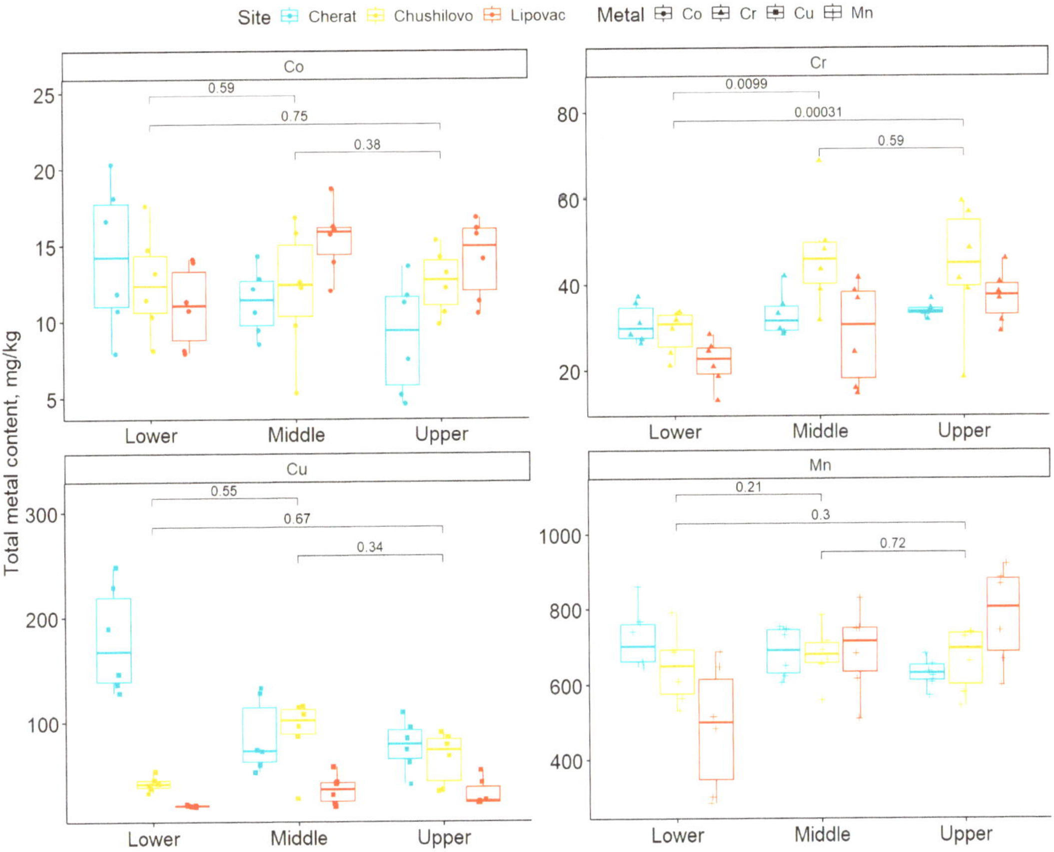


**Figure 11.** Slope distribution of total heavy metal content in the upper soil horizons (0–30 cm) of the vineyards according to the Kruskal–Wallis criterion: cobalt ($p = 0.045$), copper ($p = 0.017$), chromium ($p = 0.034$), and manganese ($p = 0.029$).

Pairwise comparison of median heavy metal content for individual slope elements for all farms using the Kruskal–Wallis method showed significant differences between individual elevation elements only for chromium (Figure 11). The chromium content in the upper part of the slope was higher than in the lower slope for all three farms.

A general trend of downslope accumulation of copper in the soils of the studied vineyards was found, but the difference in median values by slope was statistically significant only for the Cherat slope. A similar distribution characteristic only for Cherat was observed for cobalt as well. However, the distribution of copper and other heavy metals in the surface soil layer in each studied plot depended on the morphological characteristics of the slope (steepness, shape, length) and the agrotechnical methods applied.

## 4. Discussion

Statistical analysis supports our hypothesis that the overall level of copper accumulation is primarily determined by the history of land use in the study plots, including the features of the agrotechnology applied. When grapes are grown continuously on a certain plot for decades with the use of copper-containing fungicides, there is an accumulation of copper in the upper horizons of the soil profile, which is confirmed by the results of other authors. Brunetto et al. (2016) provided data that copper and zinc accumulated in the upper 2.5 cm soil layer of orchards in Canada, predominantly with a long (18 years) cultivation history, with consequent high fungicide loads of copper- and zinc-containing compounds [7]. A similar conclusion was made when comparing copper accumulation in the soil of 15- and 4-year-old vineyards in the state of Santa Catarina (Brazil), and in older vineyards, a greater proportion of copper was found in the most plant-available (soluble) fraction, associated with iron and manganese oxides and organic matter, with a simultaneous decrease in the proportion of the element in the residual fraction unavailable to plants at depths of 0–5, 5–10, and 10–20 cm. In New Zealand, young vineyards had lower levels of copper in soil than aged vineyards, prompting the authors to recognize the need for ongoing ecotoxicological monitoring in farms established more than 40 years ago, as well as those in which vineyards were established on the site of former orchards [46]. The results of soil monitoring of vine plantations more and less than 20 years old in Slovenia revealed copper content of 72 and 17.5 mg/kg, respectively, in the upper 20 cm layer, while in the background soil, the content was 0.8 mg/kg. A similar trend was observed in 20–40 and 40–60 cm soil horizons, but with lower absolute values for copper accumulation [47].

The agrogenic origin of the copper in the soil of the vineyards studied is also supported by the data on the distribution of copper in the profile of a reference soil sampled from a natural meadow ecosystem near the Cherat vineyard (Figure 6). A similar pattern of copper accumulation in soil samples taken as background soil was found by Ninkov et al. (2012) for a 50-year-old abandoned vineyard, where the total copper content in 0–60 cm horizons was 48.6–57.8 mg/kg [48]. It seems probable that this plot was previously used for grape cultivation, due to the fact that the slopes of Fruška Gora represent a historical center of viticulture and winemaking in southern Pannonia.

This total copper content was taken as the geochemical background for the territory where the studied vineyards were located. This value is close to the background value of 19.8 mg/kg for soils near the town of Sremski Karlovci, which was established by Ninkov et al. (2012) when conducting similar studies [48]. We suggest that the copper enrichment of the soil profile to a depth of 80 cm relative to the parent rock may be related to periodic deep ploughing during vineyard replanting, which facilitates the distribution of copper from the surface soil layers to the underlying ones.

Of all the vineyards studied, the highest total copper content was found in the soil of the archaic vineyard in Cherat, where it reached a value of 247.5 mg/kg, which is 6.9 and 13.4 times higher than the maximum allowed content and the geochemical background of copper, respectively (Figure 9). According to the owner of the vineyard in Cherat, grapes have been cultivated there for about 200 years, and although the current vineyard is 55 years old, it was replanted immediately after the previous vineyard was removed.

Copper content in the soil of Chushilovo vineyard was also 2.7–3.2 and 1.5–6 times higher than the MAC and background value, respectively (Figure 9). According to the cadastral statement available to the owner, the land-use history of this plot dates back to 1886, i.e., around the same time when Pierre-Marie-Alexis Millardet discovered the protective properties of Bordeaux liquid against mildew. Interestingly, the owner of the vineyard in Chushilovo has not applied any agrotechnical measures for the last 40 years.

However, the complete absence of copper-containing pesticide treatments during this period has not eliminated the problem of copper accumulation in the surface soil horizons at levels above the maximum allowed content. Similar results were found for abandoned vineyards in the Czech Republic [49] and in Galicia, Spain [50].

Overall, our results in the abandoned vineyard highlight the problem of land-use change in former vineyards, especially those that are being converted to other crops [51]. For example, since 2000, many old vineyards in France have ceased to exist for economic reasons, and most of these lands have been planted with wheat. At the same time, increased concentrations of copper in the soil as a result of long-term use of fungicides causes phytotoxicity and reduced yields of this crop [52]. Soil and ecotoxicological studies conducted on a large area of old Australian vine plantations vacated over the last 20 years have demonstrated the potential hazard of high copper concentrations to a range of cereals, vegetable crops, and perennial grasses [53]. Our studies on fallow soils of the northern Black Sea coast, previously used in viticulture with the use of a chemical plant protection system for about 40–60 years, revealed low indices of microbial biomass and respiratory activity of soil. In addition, ecophysiological indices exceeding the optimal values indirectly indicated unfavorable conditions for the functioning of the soil microbial community in long-term fallow conditions, even against the background of general accumulation of organic matter in the soil [54]. The possibility of such negative effects should certainly be taken into account when returning fallow soils under former vine plantations to agricultural turnover. At the same time, soils contaminated with copper can be relatively safely used for sowing of technical or energy crops, such as hemp [55].

The lowest concentrations of copper were found in the soil of the (youngest) 15-year-old industrial vineyard in Lipovac (Figure 9). Nevertheless, the accumulation level of the element was 2–3 times higher than the background level and 1.5–1.6 times higher in the 15–30 cm horizon of the rows than the maximum allowed content.

The correlation matrix confirmed the relationship between total copper content, organic matter content, slope part, and slope shape, indicating the involvement of organic matter together with copper in erosion-induced slope migration processes (Figure 10). A similar conclusion was reached in the work of Ha Pham et al. (2022) [56]. This can be explained by the fact that most of the copper is retained by soil organic matter due to its strong sorption and complexation properties [22,57–59].

The accumulation of organic matter in the soils of the studied vineyards showed a similar trend to that described above for the total copper content. The maximum Corg value was observed in the lower part of the vineyard slope in Cherat, the minimum in the soil of the vineyard in Lipovac (Figure 10). As can be seen in the figure, the differences in Corg content between the different vineyards were most pronounced at the 0–5 cm horizon, with the differences smoothing out with depth.

Surface runoff in vineyards with slopes of 5–10 degrees or more often leads to increased topsoil removal. The amount of erosion-induced soil losses depends on many factors: the frequency and seasonal distribution of rainfall, the morphological characteristics of the slope, the duration of vine cultivation, the granulometric composition and organic matter content, the position of the rows in relation to the slope, the presence of interrow cover crops and the intensity of their mechanical cultivation. Even the expression of some of the above factors can lead to the development of erosion processes, which are more common in vineyards than in other agroecosystems [28,60–62].

The migration of organic matter in the soil of the vineyard in Cherat was not as intensive as in Lipovac. This is probably explained by the absence of grape rows, due to the archaic bush vine trellis system used, and the uniform distribution of bushes along the slope, the preservation of natural meadow vegetation in the space between the bushes, and the complete absence of tillage. Approximately the same can be said for the abandoned vineyard in Chushilovo. The processes of mineralization of organic matter in the untreated soil of the vineyards in Cherat and Chushilovo were not as fast as in the cultivated soil of the vineyard in Lipovac. As a result, conditions were created for the accumulation of organic matter, the level of which was 1.2–5.6 and 1.1–4.1 times higher in the soil of the archaic and organic vineyards, respectively, than in the soil of the intensive type of vineyard (Figure 10).

Copper, and in some cases the other heavy metals studied, were also involved in the migration processes observed on the slopes of the vineyards (Figure 11). Soil particles from the surface layer, in which copper accumulates mainly because of the application of copper-containing fungicides, move from upper towards lower parts of the slope under the gravity effect. It has been shown that the total copper content in the soil at the foot of the slope significantly exceeds the content of its mobile forms, which indirectly indicates that the element does not move down the slope with easily mobile substances, but together with soil particles in the form of compounds with high chemical stability [14,63].

The longitudinal slope profile in Lipovac had three distinct mesorelief zones (Figure 2). The first zone corresponded to the upper part of the slope and represented a concave part with high steepness (13°). The middle part was essentially the leveled bottom of the steep part of the slope, followed by the lower gentle part of the slope with a steepness of about 4–5°. Due to enhanced migration processes in the upper, steepest part of the slope, a redistribution of copper from the upper horizons (0–5 and 5–15 cm) of the upper part of the slope to the surface horizon and the 15–30 cm horizon of the middle part of the slope was observed, as evidenced by the lateral differentiation coefficients above 1. Radial migration of copper to underlying soil horizons is also possible, but interpretation of these migration pathways of copper and other heavy metals has been made difficult by soil mixing as a result of tillage and its movement within and between rows. This contributed to the dispersion of copper in both the radial and lateral landscape structure. It is obvious that the lower part of the vineyard slope in Lipovac was not covered by the migration processes occurring in the upper and middle parts of the slope due to its low steepness and physical isolation (a technological main road passed between the middle and lower zones). As a result, the average level of copper accumulation in all investigated horizons of the lower part of the slope (21.2 mg/kg) was close to the background concentration of the element. Taking into account the essentially geogenic origin of the copper in the lower part of the slope, as well as the fact that limited doses of copper-containing fungicides were used in this vineyard (one preventive treatment before bud burst per year with average rainfall during the growing season), it can be assumed that the 2–2.5 times higher copper concentrations in the middle part of the slope compared to the surface horizons of the upper part and all horizons of the lower part are explained exclusively by the processes of agrogenic copper redistribution with runoff along the mesorelief elements.

The observed tendency for copper accumulation to decrease in the lower part of the vineyard slope in Lipovac was also true for manganese, chromium, and cobalt (Figure 11). The accumulation of cobalt increased similarly to copper from the upper to the middle part of the slope, while the opposite pattern was observed for manganese and chromium. A contradictory character in the distribution of these elements along the slope was also found by Ha Pham et al. (2022) for two vineyards of different ages in Hungary [56]. The different character of the spatial heterogeneity in the accumulation of heavy metals in vineyard soils can be related, in addition to the terrain morphology, to the granulometric composition of the soil and the distribution of individual fractions along the slope, the specificity of the bonding of the elements with manganese and iron oxides, organic matter, crystal structure of clay minerals, etc.

In contrast to the slope of the vineyard in Lipovac, the copper accumulation in soil of the vineyard in Cherat increased from the upper and middle parts of the slope to the lower part, with lateral differentiation coefficients ranging from 1.18 to 5.57. The absence of annual tillage not only revealed a pronounced lateral migration of copper along the slope zones but also the possibility of its radial movement deep into the soil profile, which was particularly noticeable in the lower part of the slope, where the copper content in the 15–30 cm horizon was 1.2–1.9 times higher than in the 0–5 cm horizon. On the other hand, in the upper and middle parts of the slope, the copper content in the surface horizon was higher than in the underlying horizon, indicating the predominance of lateral over radial migration of copper. Permanent vegetation of the slope could not stop or even slow the copper migration in the soil of the Cherat vineyard (Figure 11). The uniform distribution of bushes along the slope contributed to the coverage of the entire slope area by copper-containing pesticide treatments, which resulted in the involvement of soil areas between bushes in copper transport and the balancing of copper accumulation in the middle and lower parts of the slope, both within the perimeter of the grape crowns and between bushes. It appears that the long history of land use, combined with the high steepness and leveling of the slope, were the determining factors in the copper transport along the slope. In addition, the effect of the increased frequency of extreme rainfall events recorded in recent years in the Autonomous Province of Vojvodina on the intensity of slope runoff cannot be ignored [64].

In contrast to copper, no reliable dependencies were found in the distribution of manganese, chromium, and cobalt along the slope in the soil of the Cherat vineyard, except for a statistically significant sequential increase in the content of chromium and cobalt from the 0–5 cm horizon to the 15–30 cm horizon. The same trend was observed for manganese in the middle and lower parts of the slope. Thus, for these elements, radial migration prevailed over lateral, probably due to their active removal by the herbaceous vegetation.

As noted above, the abandoned vineyard in Chushilovo has not been treated with pesticides for the last 40 years, so the level of accumulation of copper and other metals in the upper soil horizons in different parts of the slope of this vineyard can be used to assess the behavior of agrogenic copper accumulated here many years ago. The slope under the vineyard in Chushilovo is short and has a pronounced mesorelief, based on which it can be divided into three zones: a relatively uniform upper part with a steepness of about 8°, a small depression in the middle part, and a gently leveled slope in the lower part (Figure 4). Similar to the slope distribution of copper in the industrial vineyard in Lipovac, the highest accumulation of the element was found in the rows of grape in the middle part of the slope within the depression, with lateral differentiation coefficients of 1.07–3.21. The lower part of the slope was characterized by the lowest copper concentration, which ranged from 32.9–53.7 mg/kg, and where the removal of this element prevailed based on lateral differentiation coefficients (0.50–1.16). In all the marked zones of the slope, the copper content of the soil in the rows of the abandoned vineyard was higher than that between the rows.

Among the other metals analyzed, a statistically significant difference was found for chromium by slope part (Figure 11). In the lower part of the slope, as in the case of copper, 1.2–1.8 times less metal was accumulated than in the overlying parts, which was also typical for the vineyard in Lipovac. In addition, a significant increase in cobalt content from the 0–5 cm horizon to the 15–30 cm horizon was found in all parts of the slope, both in rows and between rows, which was a general trend for all vineyards studied.

The total manganese content did not depend on slope characteristics, land-use system, or agrochemical indicators. Only a close relationship with cobalt and chromium in the soil was found. A similar relationship for these elements was found in a study by Ha Pham et al (2022) [56]. The authors suggested that this may indicate the association of Co and Cr with the reactive surface sites of Mn oxides, which was not observed for copper in either their or our studies.

## 5. Conclusions

Vineyard agroecosystems are of great scientific and practical interest because of the known problem of high copper accumulation in the upper soil horizons associated with the long-term use of copper-containing fungicides.

It is generally considered that copper of agrogenic origin in vineyard soils is not mobile and is concentrated in surface soil horizons, and therefore poses no threat to the environment or product quality.

However, our study on the example of vineyards of different ages on the slopes of the Fruška Gora mountain range in the Autonomous Province of Vojvodina, Republic of Serbia, has shown that under certain conditions, the migration of copper in vineyard soils is possible and its total content can vary significantly depending on the land-use systems in place, the meso- and microrelief parameters, and the duration of grape cultivation in the area.

The analysis of the data obtained using the principal component method allowed us to establish that copper was more strongly related to parameters determined by agrogenic activity (duration of vine cultivation and agrotechnologies used).

The soil of the 15-year-old vineyard had the lowest total copper content of all the vineyards studied, ranging from 20.3 to 57.9 mg/kg. In the soil of the vineyard more than 200 years old, the level of copper accumulation reached 247.5 mg/kg, which is 6.9 times higher than the maximum allowed concentration of this element established in the Republic of Serbia and 1.3 times higher than the remediation concentration.

The influence of the agrogenic factor was closely related to the content of organic matter in the soil, with had a strong correlation which copper. We assume that the migration of copper along the slope, which was increased by the location of the vineyards on the slope and the mechanical cultivation between rows in the absence of vegetation in the intensive vineyard, is related to its involvement in the migration of organic matter due to erosion processes occurring on the slopes.

Statistically significant copper accumulation in the vineyards was found in the lower or middle part of the slopes, depending on the terrain morphology. In general, slope shape was the most important qualitative variable, contributing to significant spatial variation in total copper content.

In contrast to copper, the accumulation and spatial variation of manganese, chromium, and cobalt were mainly determined by relief parameters—slope steepness for chromium and manganese and slope extent for cobalt—indicating their predominantly geogenic origin. In contrast to copper, radial migration into the soil profile predominated over lateral migration for these metals.

Our results highlight the problem of copper in vineyard soils in a more complex way, linking the level of its accumulation not only to the number of copper-containing fungicides used but also to the landscape characteristics of the vineyard location and to specific agrotechnological methods, which can vary greatly among wineries of different size and using different management models.

**Author Contributions:** Conceptualization, I.A.; methodology, I.A. and Z.G.; software, A.Y.; validation, V.G., D.M. and A.Y.; formal analysis, V.G. and D.M.; investigation, I.A., V.G. and D.M.; resources, I.A., A.Y. and M.S.; data curation, V.G. and A.Y.; writing—original draft preparation, I.A.; writing—review and editing, Z.G., A.Y. and M.S.; visualization, V.G. and D.M.; supervision, I.A.; project administration, I.A. and M.S.; funding acquisition, A.Y. All authors have read and agreed to the published version of the manuscript.

**Funding:** This research was funded by the Ministry of Science and Higher Education of the Russian Federation in accordance with agreement 075-15-2020-905, dated 16 November 2020, on providing a grant in the form of subsidies from the federal budget of the Russian Federation. The grant was provided for state support for the establishment and development of a world-class research center—Agricultural Technologies of the Future.

**Data Availability Statement:** The original contributions presented in this study are included in the article. Further inquiries can be directed to the corresponding author.

**Acknowledgments:** The authors would like to thank colleagues at the Institute of Lowland Forestry and Environment (University of Novi Sad, Serbia) for their assistance in selecting the study sites, as well as winegrowers from Sremski Karlovci (Viovodina) for agreeing to provide access to their vineyards for the research and for sharing information on land-use history and agronomic practices.

**Conflicts of Interest:** Author Miljan Samardžić was employed by the company Nature 4.0 BC SRL. The remaining authors declare that the research was conducted in the absence of any commercial or financial relationships that could be construed as a potential conflict of interest.

## References

1. Visconti, F.; López, R.; Olego, M.Á. The Health of Vineyard Soils: Towards a Sustainable Viticulture. *Horticulturae* **2024**, *10*, 154. [CrossRef]
2. Cataldo, E.; Fucile, M.; Mattii, G.B. A Review: Soil Management, Sustainable Strategies and Approaches to Improve the Quality of Modern Viticulture. *Agronomy* **2021**, *11*, 2359. [CrossRef]
3. Lamichhane, J.R.; Osdaghi, E.; Behlau, F.; Köhl, J.; Jones, J.B.; Aubertot, J.-N. Thirteen decades of antimicrobial copper compounds applied in agriculture. A review. *Agron. Sustain. Dev.* **2018**, *38*, 28. [CrossRef]
4. Tamm, L.; Pertot, I.; Gubler, W.D. Organic grape disease management. In *Plant Diseases and Their Management in Organic Agriculture*; Finckh, M.R., van Bruggen, A.H.C., Tamm, L., Eds.; APS Press: Eagan, MN, USA, 2015; pp. 335–350.
5. Ballabio, C.; Panagos, P.; Lugato, E.; Huang, J.; Orgiazzi, A.; Jones, A.; Fernandez-Ugalde, O.; Borrelli, P.; Montanarella, L. Copper distribution in European topsoils: An assessment based on LUCAS soil survey. *Sci. Total Environ.* **2018**, *636*, 282–298. [CrossRef] [PubMed]
6. Lorenzoni, P.; Valboa, G.; Papini, R.; Paone, R.; Aramini, G.; Colloca, C.; Corea, A.M. Soil Copper and Zinc Accumulation and Bioavailability Under a Long Term Vineyard Cultivation in South Italy. *Ital. J. Agron.* **2007**, *2*, 31. [CrossRef]
7. Brunetto, G.; de Melo, G.W.B.; Terzano, R.; Del Buono, D.; Astolfi, S.; Tomasi, N.; Pii, Y.; Mimmo, T.; Cesco, S. Copper accumulation in vineyard soils: Rhizosphere processes and agronomic practices to limit its toxicity. *Chemosphere* **2016**, *162*, 293–307. [CrossRef] [PubMed]
8. Komárek, M.; Čadková, E.; Chrastný, V.; Bordas, F.; Bollinger, J.-C. Contamination of vineyard soils with fungicides: A review of environmental and toxicological aspects. *Environ. Int.* **2010**, *36*, 138–151. [CrossRef] [PubMed]
9. Mackie, K.A.; Müller, T.; Kandeler, E. Remediation of copper in vineyards—A mini review. *Environ. Pollut.* **2012**, *167*, 16–26. [CrossRef]
10. Flores Velez, L.; Ducaroir, J.; Jaunet, A.M.; Robert, M. Study of the distribution of copper in an acidic vineyard soil by three different methods. *Eur. J. Soil Sci.* **1996**, *47*, 523–532. [CrossRef]
11. Kovačič, G.R.; Lesnik, M.; Vršič, S. An overview of the copper situation and usage in viticulture. *Bulg. J. Agric. Sci.* **2013**, *19*, 50–59.
12. Merrington, G.; Rogers, S.L.; Van Zwieten, L. The potential impact of long-term copper fungicide usage on soil microbial biomass and microbial activity in an avocado orchard. *Aust. J. Soil Res.* **2002**, *40*, 749–759. [CrossRef]
13. Andrés, P.; Doblas-Miranda, E.; Sanchez, A.; Mattana, S.; Font, F. Physical, Chemical, and Biological Indicators of Soil Quality in Mediterranean Vineyards under Contrasting Farming Schemes. *Agronomy* **2022**, *12*, 2643. [CrossRef]
14. Imfeld, G.; Guyot, B.; Wiegert, C.; Payraudeau, S. Soil Management Drives Copper and Zinc Export in Runoff from Vineyard Plots. *Water Air Soil Pollut.* **2023**, *234*, 357. [CrossRef]

15. Nóvoa-Muñoz, J.C.; Queijeiro, J.M.; Blanco-Ward, D.; Alvarez-Olleros, C.; Martínez-Cortizas, A.; García-Rodeja, E. Total copper content and its distribution in acid vineyards soils developed from granitic rocks. *Sci Total Environ.* **2007**, *378*, 23–27. [CrossRef] [PubMed]
16. Fernández-Calviño, D.; Pateiro-Moure, M.; López-Periago, E.; Arias-Estévez, M.; Nóvoa-Muñoz, J.C. Copper distribution and acid–base mobilization in vineyard soils and sediments from Galicia (NW Spain). *Eur. J. Soil Sci.* **2008**, *59*, 315–326. [CrossRef]
17. Deluisa, A.; Giandon, P.; Aichner, M.; Bortolami, P.; Bruna, L.; Lupetti, A.; Nardelli, F.; Stringari, G. Copper pollution in Italian vineyard soils. *Commun. Soil Sci. Plant Anal.* **1996**, *27*, 1537–1548. [CrossRef]
18. Brun, L.; Maillet, J.; Richarte, J.; Herrmann, P.A.; Remy, J. Relationships between extractable copper, soil properties and copper uptake by wild plants in vineyard soils. *Environ. Pollut.* **1998**, *102*, 151–161. [CrossRef]
19. Pietrzak, U.; McPhail, D.C. Copper accumulation, distribution and fractionation in vineyard soils of Victoria, Australia. *Geoderma* **2004**, *122*, 151–166. [CrossRef]
20. Duplay, J.; Semhi, K.; Errais, E.; Imfeld, G.; Babcsanyi, I.; Perrone, T. Copper, zinc, lead and cadmium bioavailability and retention in vineyard soils (Rouffach, France): The impact of cultural practices. *Geoderma* **2014**, *230–231*, 318–328. [CrossRef]
21. Trentin, E.; Cesco, S.; Pii, Y.; Valentinuzzi, F.; Celletti, S.; Feil, S.; Yorlady, M.; Zuluaga, A.; Ademar, P.; Ferreira, A.; et al. Plant species and pH dependent responses to copper toxicity. *Environ. Exp. Bot.* **2022**, *196*, 104791. [CrossRef]
22. Parat, C.; Chaussod, R.; Lévéque, J.; Dousset, S.; Andreux, F. The relationship between copper accumulated in vineyard calcareous soils and soil organic matter and iron. *Eur. J. Soil Sci.* **2002**, *53*, 663–669. [CrossRef]
23. La Torre, A.; Iovino, V.; Caradonia, F. Copper in plant protection: Current situation and prospects. *Phytopathol. Mediterr.* **2018**, *57*, 201–236. [CrossRef]
24. Colautti, A.; Civilini, M.; Contin, M.; Celotti, E.; Iacumin, L. Organic vs. conventional: Impact of cultivation treatments on the soil microbiota in the vineyard. *Front. Microbiol.* **2023**, *14*, 1242267. [CrossRef] [PubMed]
25. Hendgen, M.; Döring, J.; Stöhrer, V.; Schulze, F.; Lehnart, R.; Kauer, R. Spatial Differentiation of Physical and Chemical Soil Parameters under Integrated, Organic, and Biodynamic Viticulture. *Plants* **2020**, *9*, 1361. [CrossRef]
26. Kopittke, P.M.; Blamey, F.P.C.; Menzies, N.W. Toxicities of soluble Al, Cu, and La include ruptures to rhizodermal and root cortical cells of cowpea. *Plant Soil* **2008**, *303*, 217–227. [CrossRef]
27. Tiecher, T.L.; Tiecher, T.; Cereta, C.A.; Ferreira, P.A.A.; Nicoloso, F.T.; Soriani, H.H.; De Conti, L.; Kulmann, M.S.S.; Schneider, R.O.; Brunetto, G. Tolerance and translocation of heavy metals in young grapevine (*Vitis vinifera*) grown in sandy acidic soil with interaction of high doses of copper and zinc. *Sci. Hortic.* **2017**, *222*, 203–212. [CrossRef]
28. Bortoluzzi, E.C.; Korchagin, J.; Moterle, D.F.; Santos, D.R.; Caner, L. Accumulation and precipitation of Cu and Zn in a centenarian vineyard. *Soil Sci. Soc. Am. J.* **2019**, *83*, 492–502. [CrossRef]
29. Husson, O. Redox potential (Eh) and pH as drivers of soil/plant/microorganism systems: A transdisciplinary overview pointing to integrative opportunities for agronomy. *Plant Soil* **2012**, *362*, 389–417. [CrossRef]
30. Majzlan, J.; Zittlau, A.H.; Grevel, K.-D.; Schliesser, J.; Wood-field, B.F.; Dachs, E.; Števko, M.; Chovan, M.; Plášil, J.; Sejkora, J.; et al. Thermodynamic properties and phase equilibria of the secondary copper minerals libethenite, olivenite, pseudomalachite, kröhnkite, cyanochroite, and devilline. *Can. Mineral.* **2015**, *53*, 937–960. [CrossRef]
31. Ivaniševic, D.; Jakšic, D.; Korac, N. *Vinogradarski Atlas*; Statistical Office of the Republic of Serbia: Belgrade, Serbia, 2015; p. 414. ISBN -979-86-6161-138-4.
32. Babić, V.; Krstić, M. Climate characteristics of the sessile oak forest belt on Fruška Gora. *Shumarstvo* **2014**, *3–4*, 49–62. (In Serbian)
33. Walkley, A.; Black, I.A. An examination of the Degtjareff method for determining soil organic matter, and a proposed modification of the chromic acid titration method. *Soil Sci.* **1934**, *37*, 29–38. [CrossRef]
34. Polynov, B.B. *Geochemical Landscapes*; Izd. of the USSR Academy of Sciences: Moscow, USSR, 1956.
35. Perelman, A.I. *Geochemistry of Landscape*; Higher School: Moscow, USSR, 1975; p. 394.
36. Glazovskaya, M.A. *Geochemistry of Natural and Anthropogenic Landscapes of the USSR*; Vysh. Shk.: Moscow, USSR, 1988; p. 329.
37. Avessalomova, I.A. *Geochemical Indicators in the Study of Landscapes: Manual*; Izd-Vo Mosk. Un-Ta: Moscow, USSR, 1987; p. 108.
38. Zibilske, L.M. Carbon Mineralization. In *Methods of Soil Analysis: Part 2 Microbiological and Biochemical Properties*; John Wiley & Sons: Hoboken, NJ, USA, 2018; Chapter 38. [CrossRef]
39. Alef, K.; Nannipieri, P. *Methods in Applied Soil Microbiology and Biochemistry*; Academic Press: Cambridge, MA, USA, 1995; p. 608.
40. Vershinin, A.A.; Petrov, A.M.; Karimullin, L.K.; Ignatiev, Y.A. Influence of oil pollution on the ecological and biological state of various types of soils. *Bull. Kazan Technol. Univ.* **2012**, *8*, 207–211.
41. *DIN EN ISO 16072-2011*; Soil Quality—Laboratory Methods for Determination of Microbial Soil Respiration (ISO 16072:2002). European Committee for Standardization: Brussels, Belgium, 2011.
42. Le, S.; Josse, J.; Husson, F. FactoMineR: An R package for multivariate analysis. *J. Stat. Softw.* **2008**, *25*, 1–18. [CrossRef]

43. Kassambara, A.; Mundt, F. *Package 'Factoextra': Extract and Visualize the Results of Multivariate Data Analyses*; The R Foundation: Ames, IA, USA, 2017; Volume 76.
44. R Core Team. R: A Language and Environment for Statistical Computing. R Foundation for Statistical Computing, Vienna, Austria. 2024. Available online: https://www.R-project.org (accessed on 10 November 2024).
45. Uredba o Graničnim Vrednostima Zagađujućih, Štetnih I Opasnih Materija U Zemljištu "Sl. Glasnik RS", br. 30/2018 i 64/2019. Available online: https://www.paragraf.rs/propisi/uredba-granicnim-vrednostima-zagadjujucih-stetnih-opasnih-materija-zemljistu.html (accessed on 10 November 2024).
46. Morgan, R.K.; Taylor, E. Copper Accumulation in Vineyard Soils in New Zealand. *Environ. Sci.* **2004**, *1*, 139–167. [CrossRef]
47. Pavlovic, M. A review of agribusiness copper use effects on environment. *Bulg. J. Agric. Sci.* **2011**, *17*, 491–500.
48. Ninkov, J.; Paprić, Đ.; Sekulić, P.; Zeremski-Škorić, T.; Milić, S.; Vasin, J.; Kurjački, I. Copper content of vineyard soils at Sremski Karlovci (Vojvodina Province, Serbia) as affected by the use of copper-based fungicides. *Int. J. Environ. Anal. Chem.* **2012**, *92*, 592–600. [CrossRef]
49. Komárek, M.; Száková, J.; Rohošková, M.; Javorská, H.; Chrastný, V.; Balík, J. Copper contamination of vineyard soils from small wine producers: A case study from the Czech Republic. *Geoderma* **2008**, *147*, 16–22. [CrossRef]
50. Vázquez-Blanco, R.; Nóvoa-Muñoz, J.C.; Arias-Estévez, M.; Fernández-Calviño, D.; Pérez-Rodríguez, P. Changes in Cu accumulation and fractionation along soil depth in acid soils of vineyards and abandoned vineyards (now forests). *Agric. Ecosyst. Environ.* **2022**, *339*, 108146. [CrossRef]
51. Wightwick, A.M.; Reichman, S.M.; Menzies, N.W.; Allinson, G. Industry wide risk assessment: A case study of Cu in Australian vineyard soils. *Water Air Soil Pollut.* **2013**, *224*, 1702. [CrossRef]
52. Michaud, A.; Bravin, M.; Galleguillos, M.; Hinsinger, P. Copper uptake and phytotoxicity as assessed in situ for durum wheat (*Triticum turgidum durum* L.) cultivated in Cu-contaminated, former vineyard soils. *Plant Soil* **2007**, *298*, 99–111. [CrossRef]
53. Wightwick, A.M.; Mollah, M.R.; Partington, D.L.; Allinson, G. Copper fungicide residues in Australian vineyard soils. *J. Agric. Food Chem.* **2008**, *56*, 2457–2464. [CrossRef]
54. Gabechaya, V.; Andreeva, I.; Morev, D.; Yaroslavtsev, A.M.; Neaman, A.; Vasenev, I.I. Exploring the Influence of Diverse Viticultural Systems on Soil Health Metrics in the Northern Black Sea Region. *Soil Syst.* **2023**, *7*, 73. [CrossRef]
55. Quagliata, G.; Celletti, S.; Coppa, E.; Mimmo, T.; Cesco, S.; Astolfi, S. Potential Use of Copper-Contaminated Soils for Hemp (*Cannabis sativa* L.) Cultivation. *Environments* **2021**, *8*, 111. [CrossRef]
56. Ha Pham, N.; Babcsanyi, I.; Balling, P.; Farsang, A. Accumulation patterns and health risk assessment of potentially toxic elements in the topsoil of two sloping vineyards (Tokaj-Hegyalja, Hungary). *J. Soils Sediments* **2022**, *22*, 2671–2689. [CrossRef]
57. Fernández-Calviño, D.; Nóvoa-Muñoz, J.C.; Díaz-Raviña, M.; Arias-Estévez, M. Copper accumulation and fractionation in vineyard soils from temperate humid zone (NW Iberian Peninsula). *Geoderma* **2009**, *153*, 119–129. [CrossRef]
58. Steinmetz, Z.; Kenngott, K.; Azeroual, M.; Schäfer, R.; Schaumann, G. Fractionation of copper and uranium in organic and conventional vineyard soils and adjacent stream sediments studied by sequential extraction. *J. Soils Sediments* **2017**, *17*, 1092–1100. [CrossRef]
59. Strawn, D.G.; Baker, L.L. Molecular characterization of copper in soils using X-ray absorption spectroscopy. *Environ. Pollut.* **2009**, *157*, 2813–2821. [CrossRef] [PubMed]
60. Korchagin, J.; Moterle, D.F.; Escosteguy, P.A.V.; Bortoluzzi, E.C. Distribution of copper and zinc fractions in a Regosol profile under centenary vineyard. *Environ. Earth Sci.* **2020**, *79*, 439. [CrossRef]
61. Prosdocimi, M.; Cerdà, A.; Tarolli, P. Soil water erosion on Mediterranean vineyards: A review. *Catena* **2016**, *141*, 1–21. [CrossRef]
62. Ha Nhung, P.T.; Viet, N.Q. Assessing the Impact of Erosion and Farming Practices on the Spatial Distribution of Topsoil Characteristics in a Sloping Vineyard Using an Open-source QGIS Software. *VNU J. Sci. Earth Environ. Sci.* **2023**, *39*, 91–101. [CrossRef]
63. Tonello, M.; Moterle, D.; Tiecher, T.; Merten, G.; Petry, C.; Bortoluzzi, E. Copper transfer from vineyard watershed: Mineralogy and copper forms. *J. Sediment. Environ.* **2023**, *8*, 283–296. [CrossRef]
64. Malinović-Milićević, S.; Dragutin, T.; Mihailović, D.T.; Radovanović, M.M.; Drešković, N. Extreme precipitation indices in Voivodina region (Serbia). *J. Geogr. Inst. Cvijic.* **2018**, *68*, 1–15. [CrossRef]

land

MDPI

Article

# Soil Quality Assessment and Influencing Factors of Different Land Use Types in Red Bed Desertification Regions: A Case Study of Nanxiong, China

Fengxia Si [1], Binghui Chen [1,2,*], Bojun Wang [1], Wenjun Li [1], Chunlin Zhu [1], Jiafang Fu [3], Bo Yu [4] and Guoliang Xu [5]

1 School of Earth Sciences and Engineering, Sun Yat-Sen University, Zhuhai 519082, China; sifx@mail2.sysu.edu.cn (F.S.); wangbj29@mail2.sysu.edu.cn (B.W.); liwj276@mail2.sysu.edu.cn (W.L.); zhuchlin5@mail2.sysu.edu.cn (C.Z.)

2 Guangdong Provincial Key Laboratory of Geological Processes & Mineral Resources Survey, Zhuhai 519082, China

3 Natural Resources Bureau of Nanxiong City, Nanxiong 512400, China; fujiafang@163.com

4 Guangdong Institute of Environmental Geological Exploration, Guangzhou 510080, China; bobby.yu@163.com

5 School of Geography and Remote Sensing, Guangzhou University, Guangzhou 510006, China; xugl@gzhu.edu.cn

* Correspondence: eescbh@mail.sysu.edu.cn

**Abstract:** Soil environmental issues in the red bed region are increasingly conspicuous, underscoring the critical importance of assessing soil quality for the region's sustainable development and ecosystem security. This study examines six distinct land use types of soils—agricultural land (AL), woodland (WL), shrubland (SL), grassland (GL), bare rock land (BRL), and red bed erosion land (REL)—in the Nanxiong Basin of northern Guangdong Province. This area typifies red bed desertification in South China. Principal component analysis (PCA) was employed to establish a minimum data set (MDS) for calculating the soil quality index (SQI), evaluating soil quality, analyzing influencing factors, and providing suggestions for ecological restoration in desertification areas. The study findings indicate that a minimal data set comprising soil organic matter (SOM), pH, available phosphorus (AP), exchangeable calcium ($Ca^{2+}$), and available copper (A-Cu) is most suitable for evaluating soil quality in the red bed desertification areas of the humid region in South China. Additionally, we emphasize that exchangeable salt ions and available trace elements should be pivotal considerations in assessing soil quality within desertification areas. Regarding comprehensive soil quality indicators across various land use types, the red bed erosion soils exhibited the lowest quality, followed by those in bare rock areas and forest land. Within the minimal data set, $Ca^{2+}$ and pH contributed the most to overall soil quality, underscoring the significance of parent rock mineral composition in the red bed desertification areas. Moreover, the combined effects of SOM, A-Cu, and AP on soil quality indicate that anthropogenic land management and use, including fertilization methods and vegetation types, are crucial factors influencing soil quality. Our research holds significant implications for the scientific assessment, application, and enhancement of soil quality in desertification areas.

**Keywords:** soil quality assessment; land use type; minimum data set; red bed desertification areas

## 1. Introduction

Over the last few decades, salinization and land desertification have grown into serious environmental issues that have an impact on the ability of China's national economy to develop sustainably. Consequently, the state has made substantial efforts to combat karst rocky desertification, soil erosion on the Loess Plateau, and salinization in the North China Plain [1]. Unfortunately, not enough attention has been paid to land deterioration in the humid red beds of South China thus far. A variety of red continental sedimentary rocks, such as argillaceous, sandstone, siltstone, and conglomerate, make up the red beds. They cover an area of about 8.26 × $10^5$ $km^2$, accounting for 8.61% of the total land area, with 60% distributed in southern China [2].

The soil structure in the red bed area is poor, with rapid weathering and erosion processes. Although the soil formation rate is fast, organic matter accumulation is slow [3]. The parent material is loose, lacking clay and cementation between soil particles. The hot and rainy climate characteristics of southern China exacerbate this situation, causing the soil to be easily washed away by surface runoff. This further reduces vegetation coverage, diminishes the soil's water conservation function, and leads to severe soil erosion. This phenomenon and process is known as "red beds desertification" [4,5]. The deterioration of soil quality in the red bed desertification area will lead to a decline in the stability of the regional ecosystem and even the loss of agricultural use value. This significantly impedes the ability of the local ecology and socioeconomic sector to develop sustainably. Thus, a scientific assessment of the soil quality in the red bed desertification area is required.

Soil quality is broadly defined as "the capacity of soils to sustain biological productivity, preserve environmental quality, and promote plant and animal health within ecosystems and land-use boundaries" [6]. Since the concept of soil quality was introduced, numerous studies on soil quality assessment and its monitoring tools, including soil quality cards [7], test box methods [8], index methods [9], and multivariate indicator kriging methods, have been rapidly conducted in various countries [10]. Among these, the soil quality index method has become a more commonly used approach in recent years. Researchers have often utilized the soil quality index method to assess the quality of soils within a single land-use type, such as agricultural systems [11], forest systems [12], grassland systems [13], and wetland systems [14].

The soil quality assessment system includes physical, biological, and chemical indicators. Chemical indicators often have significant advantages because they are quantifiable and closely related to plant nutrition, which can indirectly reflect soil physical conditions and microbial quality [15]. Abraham et al. argue that pH determines the nutrient availability and physical conditions of the soil, thereby controlling the diversity of microorganisms in the soil [16]. Additionally, cation exchange capacity (CEC) is a sensitive indicator for determining soil nutrient retention capacity, fertility, and long-term productivity [17]. Nitrogen, phosphorus, and potassium are considered essential soil nutrients because they limit soil productivity by affecting various soil properties, plant growth, and soil microbial activity [18–21]. Soil organic matter has a multifaceted positive impact on the health and quality of (agricultural) ecosystems, and is therefore recognized as a highly desirable and complex soil variable [22,23]. It is evident from many previous studies that chemical indicators play a vital role in the comprehensive assessment of soil quality.

The extensive selection of measurement metrics often results in significant time and financial costs for sampling and measurement [24]. Therefore, it is essential to reduce the indicators to the smallest possible data set for assessing soil quality [10]. Initially, the selection of the minimum data set (MDS) was typically based on expert judgment [6]. However, more recently, methods such as principal component analysis (PCA), redundancy analysis (RDA) [25], and multiple regression have become more common for data dimensionality reduction.

Although the challenges posed by land desertification in the humid areas of South China have not received equal attention at the national policy level, in recent years, many scholars have begun to recognize the importance of assessing the quality of red bed soils and have conducted extensive research on this topic. These studies have, to varying degrees, addressed gaps in previous land management strategies in red bed desert areas. For example, they have analyzed the soil quality of red soil tillage layers in southern China using clustering and principal component analysis (PCA) [26]. Additionally, they have calculated the soil quality index and soil degradation index for different land use modes in the red bed region of southern China through PCA, investigating the influencing factors [27]. Yan et al. also explored the spatial differentiation of soil moisture and organic matter in the red bed region and identified their influencing factors [1,28,29].

Studies on soil quality in the red beds have primarily focused on single land use types or individual fixed soil attributes, with few addressing the differences in soil quality across various land use types in the red bed region or identifying the main factors affecting soil quality in red bed desertification areas. Therefore, this paper selected six different land use types in a typical red bed desertification area in the Nanxiong Basin, a humid zone in South China, and assessed soil quality by establishing a minimum data set. The objectives of this study were to (i) establish an objective MDS for assessing soil quality in the humid red bed desertification area, (ii) calculate the soil quality index (SQI) and assess soil quality across different land use types in the study area, and (iii) determine the individual contributions of the selected MDS components to soil quality in the study area and analyze their influencing factors. We hope that this study will provide ecological restoration suggestions and improvement directions for the red bed desertification areas in humid regions.

## 2. Materials and Methods

### 2.1. Overview of the Study Area

The study area is located in Nanxiong City (113°55′−114°44′ E, 24°56′−25°25′ N) in northeastern Guangdong Province, China (Figure 1). Greater heights in the northwest and lower elevations in the southeast define the area's general terrain. The central part of the area comprises red-layered hills formed from red terrestrial debris accumulated during the Cretaceous and Paleocene periods, creating the long and narrow Nanxiong Basin. The study area has a subtropical monsoon climate, with the northeast monsoon prevailing in winter, and the southwest and southeast monsoons in summer. Winters are short, summers are long, and the average annual temperature is approximately 20.6 °C. Annual rainfall ranges from 900 to 2100 mm, with the period from May to October receiving over 60% of the yearly total. The elevation of the basin varies from 79 to 568 m above sea level. Due to its inland location, low latitude, and distance from the sea, the area is less affected by typhoons. The natural soil in the study area is a calcareous purple soil developed in purplish-red sand shale or argillaceous rock, which is rich in iron ions [30]. This purple soil accumulates organic matter slowly, has poor water retention and drought tolerance, and exhibits high soil temperatures, all of which impose certain limitations on vegetation growth and recovery [4].

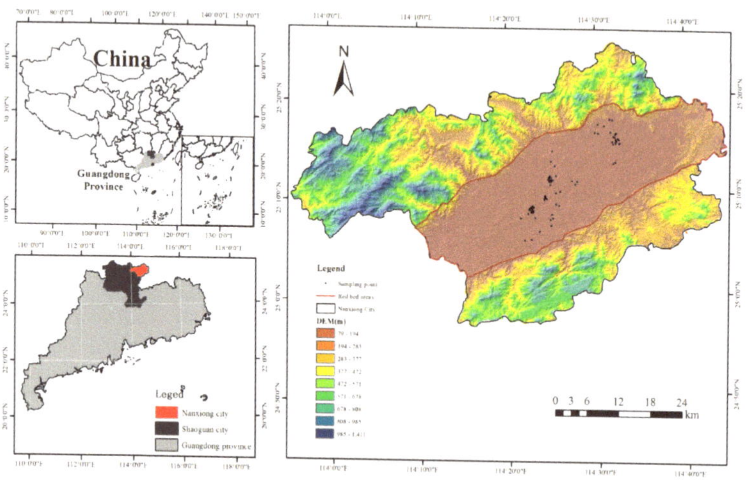


**Figure 1.** Distribution of the study area and sampling sites.

*2.2. Soil Sample Collection*

Six land types have been identified in the Nanxiong Basin, Guangdong Province, a typical red bed desertification area: agricultural land (AL), woodland (WL), shrubland (SL), grassland (GL), bare rock land (BRL), and red bed erosion land (REL), based on the actual land use at the time of sampling. A total of 87 sample plots, each measuring 20 m × 20 m, were selected. Within each plot, three sampling points were established using a random distribution method. A 5 cm diameter soil auger was used to retrieve soil columns (0–10 cm depth) at each sampling location. Due to the high diversity of crops in agricultural land (including vegetables, Nicotiana tabacum, and Chinese herbal medicine) and different tree species in woodland (such as *Pinus massoniana*, *Acacia confusa*, *Leucaena leucocephala*, and mixed forest), additional samples were collected to capture this variability. Specifically, 18 additional soil samples were collected from 6 agricultural land plots, and 12 additional samples were collected from 4 woodland plots. In total, 291 soil samples were gathered. The soil samples, weighing roughly 1.0 kg apiece, were packed into polyethylene bags and sent to the laboratory. Samples were cleaned in the lab to get rid of debris like plant litter and roots, and after that, they were dried and put through a 2 mm mesh screen to measure the physical and chemical characteristics of the soil. The six land types and their characteristics are shown in Table 1.

During the sampling stage, vegetation surveys were conducted concurrently in each sample plot. The typical vegetation types were recorded, and the degree of desertification in the basin was classified into four grades based on the grading method proposed by Shen et al., as detailed in Table 2 below [31].

**Table 1.** Field photographs and characteristics of the six land use types.

| Land Use Type | Field Photographs | Characteristics |
|---|---|---|
| Agricultural land (AL) |  | Plots where *Citrus sinensis*, *Nicotiana tabacum*, and *Chinese medicinal herbs* are cultivated after manual fertilization and irrigation. |
| Woodland (WL) |  | Primarily includes *Pinus massoniana*, *Acacia confusa*, and *Leucaena leucocephala*. |
| Shrubland (SL) |  | Plots with low shrubs such as *Vitex negundo*, *Maclura cochinchinensis*, and *Melia azedarach*. |

**Table 1.** *Cont.*

| Land Use Type | Field Photographs | Characteristics |
|---|---|---|
| Grassland (GL) |  | The vegetation is diverse, with some areas being abandoned agricultural land where grasses such as *Agave sisalana*, *Setaria viridis*, and *Caryopteris incana* naturally grow. |
| Bare rock land (BRL) |  | This area is in the initial stage of natural succession, with only small patches of vegetation cover, predominantly consisting of xerophytic shrubs and grasses. |
| Red bed erosion land (REL) |  | It is an area of severe red bed desertification, characterized by extensive bare ground, sparse vegetation, and, in extreme desertification zones, no plant growth at all. To carry out ecological restoration and prevent soil erosion in the red bed desertification area, experimental soil improvement has been conducted in some red bed erosion areas with extensive bare ground. The primary measure implemented has been the application of organic fertilizer. |

**Table 2.** Regional classification of red bed desertification in the study area.

| | Evaluate Factors | | |
|---|---|---|---|
| | **Vegetation Coverage** | **The Exposed Area Occupies the Total Area** | **Comprehensive Performance Characteristics** |
| Light desertification | 50%~70% | ≤10% | Erosion gullies are either developed or absent, with the topsoil layer missing. The vegetation is limited, consisting mostly of drought-tolerant shrubs and trees. |
| Moderate desertification | 30%~50% | 10%~25% | Erosion gullies are well developed, but the slopes are generally gentle, with only a small number of shrubs and grasses growing. |
| Severe desertification | 10%~30% | 25%~50% | Erosion gullies are widely distributed, and surface vegetation is scarce, consisting mainly of xerophytic shrubs and grasses. |
| Extreme desertification | ≤10% | ≥50% | The area is characterized by dense erosion gullies, extensive bare land surfaces, sparse vegetation mainly consisting of xerophytes, and an absence of grass in the most extreme areas. |

Eleven physical and chemical characteristics of soil were identified: alkaline hydrolyzable nitrogen (AN), available phosphorus (AP), available potassium (AK), exchangeable calcium ($Ca^{2+}$), exchangeable magnesium ($Mg^{2+}$), available copper (A-Cu), available zinc (A-Zn), available iron (A-Fe), pH, soil organic matter (SOM), and soil moisture content (W%).

### 2.3. Soil Sample Experimental Methods

The determination of soil physicochemical properties was undertaken using the methods mentioned by Lu (1999) as a reference [32]. Briefly, soil organic matter (SOM) was determined using the external heating method with potassium dichromate-concentrated sulfuric acid [33]. Soil pH was measured using the glass electrode method with a water-to-soil ratio of 2.5:1 [34]. Soil moisture content (W%) was determined by drying the sample in an oven at 105 °C [35]. The content of alkaline hydrolyzable nitrogen (AN) was assessed using the alkaline hydrolysis diffusion method. Available phosphorus (AP) was measured through the hydrochloric acid-ammonium fluoride extraction method followed by molybdenum-antimony colorimetry [36]. Available potassium (AK), exchangeable calcium ($Ca^{2+}$), and exchangeable magnesium ($Mg^{2+}$) contents were determined using ammonium acetate extraction and measured by Flame Atomic Absorption Spectrophotometrics [37]. Available copper (A-Cu), available zinc (A-Zn), and available iron (A-Fe) were measured using Diethylenetriamine Penta-acetic Acid (DTPA) extraction and Flame Atomic Absorption Spectrophotometrics to evaluate the available trace elements in the soil [38].

### 2.4. Soil Quality Evaluation Methods

Three phases are usually involved in the soil quality index method: (1) using data reduction to identify relevant soil indicators; (2) scoring the indicators that are chosen; and (3) creating a soil quality index [12,39].

Initially, principal component analysis (PCA) is employed to extract the principal components (PCs) with eigenvalues >1 in order to decrease the dimensionality of the chosen indicators. Groupings of indicators are created when loading factors on a primary component are above 0.5. If an indicator has loadings less than 0.5 on at least two principal components, it will be assigned to the group with the highest loading value. Alternatively, if an indicator has loadings ≥0.5 on multiple principal components, it will be grouped with the component that has a lower correlation with other indicators [40]. Since PCA only considers the loading of an indicator on a single principal component, the Norm value is calculated to prevent the loss of information of the indicator on other principal components with eigenvalues ≥1. The length of the indicator's vector constant mode in the multidimensional space made up of the principal components is represented by the Norm value. Greater combined loading of the indicator on all principal components is shown by a larger Norm value, which denotes a stronger capacity to comprehend the combined data [41].

$$N_{ik} = \sqrt{\sum_{i=1}^{k}\left(u_{ik}^{2}\cdot\lambda_{k}\right)} \tag{1}$$

where $N_{ik}$ is the composite loading of each variable in $i$ on the first $k$th principal components with eigenvalues ≥1; $u_{ik}$ is the loading of each indicator in $i$ on the $k$th principal component; and $\lambda_k$ is the eigenvalue of the kth principal component.

The indicators in each group whose Norm values fell within 10% of the maximum Norm value in the group were chosen to be included in the minimum data set (MDS) after the Norm values of each indicator within the group were calculated individually. When deciding which indicators to keep in a group, the Pearson correlation coefficient is used. If the correlation coefficient between the indicators is less than 0.5, all of the indicators are kept; if it equals or exceeds 0.5, the indicator with the larger Norm value is selected to be included in the MDS.

The second step involves establishing an affiliation function between the indicators and soil productivity, taking into account the positive and negative effects of the evaluation indicators on soil quality. An S-type affiliation function indicates a positive correlation between the evaluation indicators and soil function within a certain range. An inverse S-type affiliation function indicates a negative correlation. A parabolic function suggests an optimal range of suitability (Table 3). SOM, AN, AP, AK, $Ca^{2+}$, $Mg^{2+}$, A-Cu, A-Zn, and A-Fe were positively correlated with the quality of the tillage layer, defined as an S-type affiliation function. W% and pH had an optimal range of appropriateness with the quality of the tillage layer, defined as a parabolic affiliation function.

**Table 3.** Evaluation indicator type and membership function.

| Function Type | Membership Function | Evaluation Index | Membership Function Parameter | | | |
|---|---|---|---|---|---|---|
| | | | a | m | n | b |
| Type of parabolic linear | $u_{(x)} = \begin{cases} 1, m \geq x \geq n \\ \frac{x-a}{m-a}, a < x < m \\ \frac{x-n}{b-n}, b > x > n \\ 0, x \leq a; x \geq b \end{cases}$ | pH | 4.52 | 7 | 8 | 9.36 |
| | | W% | 0 | 8.5% | 15.5% | 55.74% |
| Type of S | $U_{(x)} = \begin{cases} 1, x \geq b \\ \frac{x-a}{b-a}, a < x < b \\ 0, x \leq a \end{cases}$ | SOM | 5.99 | | | 38.50 |
| | | AN | 10.08 | | | 346.08 |
| | | AP | 0.03 | | | 129.54 |
| | | AK | 24.04 | | | 735.51 |
| | | $Ca^{2+}$ | 50.52 | | | 8051.00 |
| | | $Mg^{2+}$ | 8.12 | | | 274.50 |
| | | A-Cu | 0.02 | | | 3.94 |
| | | A-Zn | 0.05 | | | 4.78 |
| | | A-Fe | 0.37 | | | 389.39 |
| Type of reverse S | $u_{(x)} = \begin{cases} 1, x \leq a \\ \frac{x-b}{a-b}, a < x < b \\ 0, x \geq b \end{cases}$ | / | | | | |

Note: The membership function is denoted by $u_{(x)}$, where x represents the actual value of the indicators; the lower and upper bounds of the indicators' critical values, respectively, are represented by a and b, which stand for the minimum and maximum values measured in the field; the optimal value for the indicator is represented by n, and its lower bound by m [42].

The common factor variance obtained in the process of principal component analysis can reflect the degree of contribution of the corresponding indicator to the overall variance. The larger its value, the greater the contribution to the overall variance. The weight of each indicator is equal to the proportion of its common factor variance to the sum of the common factor variances of all indicators [43].

Ultimately, the index of soil quality was computed. A higher number denotes greater soil quality. The soil quality index (SQI) incorporates the indices used to evaluate soil quality. The formula is as follows:

$$SQI = \sum_{i=1}^{n} w_i \cdot N_i \quad (2)$$

where $w_i$ is the weight of the ith evaluation indicator; $N_i$ is the degree of affiliation of the ith evaluation indicator; and n is the number of evaluation indicators.

Using the equidistant division approach, the soil quality of the red beds in the research region was divided into three groups: Class I ($0.66 \leq$ MDS-SQI $< 1$), most suitable for vegetation growth; Class II ($0.33 \leq$ MDS-SQI $< 0.66$), suitable for crops, albeit with some limitations; Class III (MDS-SQI $< 0.33$), characterized by serious limitations on vegetation growth.

## 2.5. Data Analysis

Data statistics and analysis were performed using Microsoft Excel 2016. IBM Statistics SPSS 26 was used for Pearson correlation and principal component analyses. Linear regression analysis and graph plotting were conducted using Origin, while ArcMap 10.8 was employed for mapping the study area and the location of sampling points. Elevation data come from Computer Network Information Center (2024) [44]; China map data come from China (2024) [45].

## 3. Results

### 3.1. Characteristics of Red Bed Soil in Different Land Use Types

The statistics for 11 soil properties across 6 different land use types in the study area are presented in Table 4. The average soil organic matter content did not vary significantly among different land use types, with agricultural land exhibiting the lowest average and red bed erosion land showing relatively higher levels. Soil pH ranged from 4.52 to 9.36 in the typical red bed area, with woodland having the lowest average pH (6.86, weakly acidic) and red bed erosion land having the highest (8.60, weakly alkaline). Soil water content varied notably among land use types, with agricultural land having the highest mean value (17.26%) and red bed erosion land showing the lowest (2.71%). Alkali hydrolyzable nitrogen (AN), available phosphorus (AP), and available potassium (AK) had their highest mean values in agricultural soils (90.92 mg/kg, 31.14 mg/kg, and 196.58 mg/kg, respectively), while red bed erosion land soil had the lowest mean values for AN and AP (16.86 mg/kg and 0.84 mg/kg, respectively). Exchangeable calcium was highest in shrubland (6774.34 mg/kg) and lowest in woodland (3838.73 mg/kg), whereas exchangeable magnesium was highest in agricultural land (93.53 mg/kg) and lowest in bare rock land (40.38 mg/kg).

**Table 4.** Descriptive statistics of soil characteristics of different land use types in the study area.

| Land Use Type | SOM (g/kg) | pH | W (%) | AN (mg/kg) | AP (mg/kg) | AK (mg/kg) | $Ca^{2+}$ (mg/kg) | $Mg^{2}$ (mg/kg) | A-Cu (mg/kg) | A-Zn (mg/kg) | A-Fe (mg/kg) |
|---|---|---|---|---|---|---|---|---|---|---|---|
| AL | 13.58 ± 6.20 | 7.57 ± 1.03 | 17.26 ± 11.25 | 90.92 ± 66.19 | 31.14 ± 27.32 | 196.58 ± 138.66 | 4760.17 ± 2064.34 | 93.53 ± 48.31 | 1.13± 0.88 | 1.22 ± 0.94 | 41.47 ± 71.63 |
| WL | 15.68 ± 6.23 | 6.86 ± 1.49 | 9.53 ± 4.13 | 65.53 ± 25.47 | 1.66 ± 1.48 | 100.14 ± 50.95 | 3838.73 ± 3169.05 | 66.51 ± 38.69 | 0.31 ± 0.21 | 0.82 ± 0.36 | 18.07 ± 32.40 |
| SL | 16.40 ± 3.72 | 8.31 ± 0.47 | 8.49 ± 3.22 | 33.28 ± 12.70 | 1.42 ± 0.80 | 105.28 ± 38.57 | 6774.34 ± 496.02 | 76.64 ± 33.15 | 0.19 ± 0.09 | 0.45 ± 0.25 | 1.88 ± 1.16 |
| GL | 14.82 ± 4.18 | 8.17 ± 0.54 | 8.59 ± 3.97 | 51.35 ± 11.45 | 15.50 ± 20.18 | 147.21 ± 76.21 | 6002.18 ± 1490.33 | 78.05 ± 31.13 | 0.67 ± 0.37 | 1.03 ± 0.33 | 7.38 ± 7.24 |
| BRL | 15.08 ± 7.68 | 7.37 ± 1.10 | 7.44 ± 3.16 | 42.05 ± 16.29 | 1.04 ± 0.35 | 58.33 ± 16.22 | 4750.24 ± 2525.74 | 40.38 ± 13.77 | 0.20 ± 0.13 | 0.43 ± 0.22 | 5.20 ± 4.78 |
| REL | 16.90 ± 2.64 | 8.60 ± 0.46 | 2.71 ± 1.95 | 16.86 ± 11.72 | 0.84 ± 0.51 | 104.48 ± 28.06 | 6589.87 ± 382.31 | 91.79 ± 61.73 | 0.07 ± 0.04 | 0.11 ± 0.06 | 0.79 ± 0.25 |

Note: Mean ± SD; AL: agricultural land, WL: woodland, SL: shrubland, GL: grassland, BRL: bare rock land, REL: red bed erosion land.

The mean values of trace elements available copper (A-Cu), available zinc (A-Zn), and available iron (A-Fe) were highest in agricultural land (1.13 mg/kg, 1.22 mg/kg, and 41.47 mg/kg, respectively) and lowest in red bed erosion land (0.07 mg/kg, 0.11 mg/kg, and 0.79 mg/kg, respectively).

### 3.2. Correlation Analysis of Soil Quality Indices

The correlation analysis among soil indicators is depicted in Figure 2. There were highly significant correlations ($p < 0.01$) observed among most of the soil evaluation indices. Specifically, pH exhibited highly significant correlations with $Ca^{2+}$, $Mg^{2+}$, W (%), AN, A-Zn, and A-Fe. Additionally, W (%) displayed highly significant positive correlations with AN, AP, AK, A-Cu, A-Zn, and A-Fe, suggesting a close relationship between the fundamental physicochemical properties of soil and nutrient content. Significant positive correlations were also found between AN, AP, AK, $Ca^{2+}$, $Mg^{2+}$, and trace elements in soil nutrient indices. Furthermore, highly significant positive correlations were observed between AN, AP, and AK, as well as between SOM, $Ca^{2+}$, and $Mg^{2+}$, indicating mutual influence among soil nutrient indices and trace elements. Notably, highly significant positive correlations were observed between the soil trace elements A-Cu, A-Zn, and A-Fe, suggesting close interaction among these trace elements.

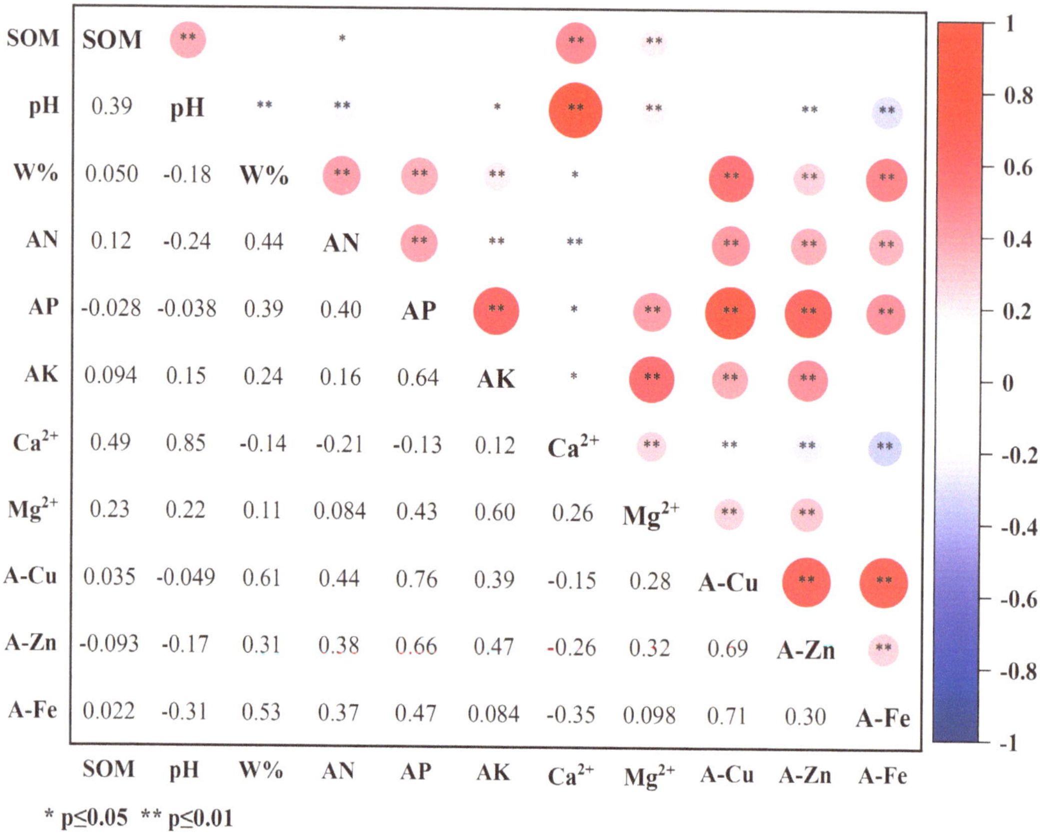

Figure 2. Pearson correlation coefficient matrix of indicators for soil quality evaluation in red bed soil.

### 3.3. Establishment of MDS Based on Principal Component Analysis

The soil indicators from various land use types underwent principal component analysis, and the findings are presented in Table 5. The eigenvalues of the three components of the soil quality evaluation indices exceeded 1, with PC1 explaining 36.887% of the variance, PC2 explaining 22.896%, and PC3 explaining 11.532%. Together, these three components contributed cumulatively to 71.315% of the variance, meeting the criteria for information extraction.

In PC1, soil water content, alkaline dissolved nitrogen, available phosphorus, available potassium, available copper, available zinc, and available iron exhibited loading factors ≥0.50, with available copper demonstrating the highest Norm value (1.819), and available phosphorus being within 10 percent of its Norm value. PC2 encompassed pH, exchangeable calcium, and exchangeable magnesium, with exchangeable calcium displaying the highest Norm value (1.515), and its Norm value being within 10% of pH. Although the loading factor of organic matter in both PC2 and PC3 was ≥0.50, it exhibited a significant positive correlation with pH, exchangeable calcium, and exchangeable magnesium. Therefore, in adherence to the principle of "if the loading of an index on different principal components is ≥0.50, then it will be classified into a group with lower correlation with other indexes", organic matter was separately categorized into PC3.

Therefore, the final selection for inclusion in the MDS comprised available phosphorus and available copper in PC1, pH and exchangeable calcium in PC2, and organic matter exclusively in PC3, totaling five indicators. The weights of these indicators in the MDS were calculated as follows: organic matter, 0.125; pH, 0.207; available phosphorus, 0.221; exchangeable calcium, 0.224; and available copper, 0.223, in that order (Table 6).

**Table 5.** Load matrix and Norm value of each indicator for soils in red bed desertification area.

| Evaluation Index | Grouping | PCA (Principal Component) | | | |
|---|---|---|---|---|---|
| | | PC1 | PC2 | PC3 | Norm |
| W (%) | 1 | 0.660 | −0.073 | 0.424 | 1.417 |
| AN (mg/kg) | 1 | 0.598 | −0.119 | 0.346 | 1.280 |
| AP (mg/kg) | 1 | 0.854 | 0.194 | −0.206 | 1.763 |
| AK (mg/kg) | 1 | 0.572 | 0.496 | −0.438 | 1.480 |
| A-Cu (mg/kg) | 1 | 0.897 | 0.072 | 0.156 | 1.819 |
| A-Zn (mg/kg) | 1 | 0.769 | 0.032 | −0.310 | 1.589 |
| A-Fe (mg/kg) | 1 | 0.691 | −0.251 | 0.393 | 1.514 |
| pH | 2 | −0.257 | 0.835 | 0.124 | 1.430 |
| $Ca^{2+}$ (mg/kg) | 2 | −0.314 | 0.855 | 0.208 | 1.515 |
| $Mg^{2+}$ (mg/kg) | 2 | 0.411 | 0.602 | −0.307 | 1.311 |
| SOM (g/kg) | 3 | −0.017 | 0.597 | 0.553 | 1.134 |
| Eigenvalue | | 4.058 | 2.519 | 1.269 | / |
| Percentage of explained variance/% | | 36.887 | 22.896 | 11.532 | / |
| Cumulative explanation percentage/% | | 36.887 | 59.783 | 71.315 | / |

**Table 6.** Load matrix and Norm value of each indicator for soils in red bed desertification area.

| Indicators | TDS | | MDS | |
|---|---|---|---|---|
| | Communality | Weight | Communality | Weight |
| SOM (g/kg) | 0.662 | 0.084 | 0.493 | 0.125 |
| pH | 0.779 | 0.099 | 0.819 | 0.207 |
| W (%) | 0.621 | 0.079 | | |
| AN (mg/kg) | 0.491 | 0.063 | | |
| AP (mg/kg) | 0.809 | 0.103 | 0.873 | 0.221 |
| AK (mg/kg) | 0.765 | 0.098 | | |
| $Ca^{2+}$ (mg/kg) | 0.873 | 0.111 | 0.886 | 0.224 |
| $Mg^{2+}$ (mg/kg) | 0.626 | 0.080 | | |
| A-Cu (mg/kg) | 0.833 | 0.106 | 0.883 | 0.223 |
| A-Zn (mg/kg) | 0.689 | 0.088 | 0.493 | 0.125 |
| A-Fe (mg/kg) | 0.695 | 0.089 | | |

### 3.4. Soil Quality Index Based on TDS and MDS

The soil quality index (SQI) was computed using the affiliation function in Table 1 based on the total data set (TDS) and the minimum data set (MDS), by the relationship between soil characteristics and soil functions (Figure 3). The results show that soil quality varies significantly under different land use patterns. Additionally, the trends of soil quality changes in TDS-SQI and MDS-SQI across different land use patterns are consistent. The TDS-SQI ranged from 0.1057 to 0.6322, with a mean value of 0.3083 $\pm$ 0.0903 and a coefficient of variation of 29.3%, indicating a moderate variation. Meanwhile, the MDS-SQI, based on the minimum data set, varied from 0.0253 to 0.6998, with a mean value of 0.3696 $\pm$ 0.1349 and a coefficient of variation of 36.49%, also indicating a moderate variation. Under different land use modes, the mean values of SQI were as follows: agricultural land (0.380) > grassland (0.342) > shrubland (0.310) > woodland (0.274) > bare rock land (0.263) > red bed erosion land (0.242).

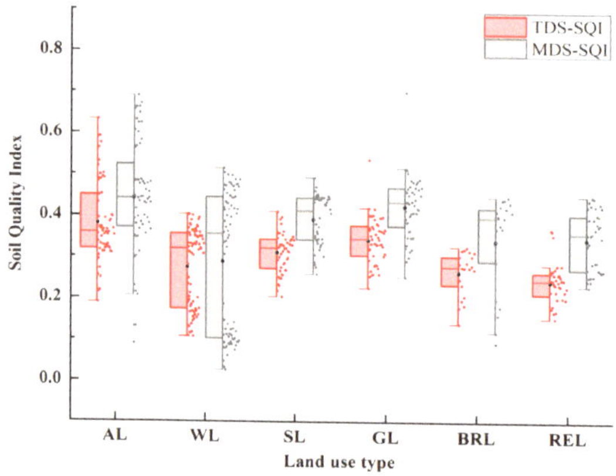


**Figure 3.** SQI of different land use types in red bed areas based on TDS and MDS (the whiskers at both ends represent the maximum and minimum values of a set of data, the solid dot in the middle of the line segment represents the average, and the upper and lower boundaries of the box represent the upper and lower quartiles of a set of data, respectively).

In the study area, 1.72%, 71.13%, and 27.15% of the sampling sites were classified into soil quality index Classes I, II, and III, respectively (Figure 4). Among the soils classified as Class III with severe limitations on vegetation growth, forest soils accounted for the highest percentage, followed by red bed erosion land. Concerning different land use types, 15%, 18.75%, and 12.5% of the sampling sites in agricultural, shrub, and grassland plots, respectively, exhibited Class III soil quality. Meanwhile, 28.75% and 38.95% of the sampling sites in bare rock land and red bed erosion land were classified as Class III, and 44.87% of the sampling sites in woodland had low-quality soil. These findings suggest that most soil quality in the study area is moderate, with soil quality being deficient in red bed erosion land, bare rock land, and woodland, often hindering vegetation growth. Agricultural soil quality is relatively higher in comparison.

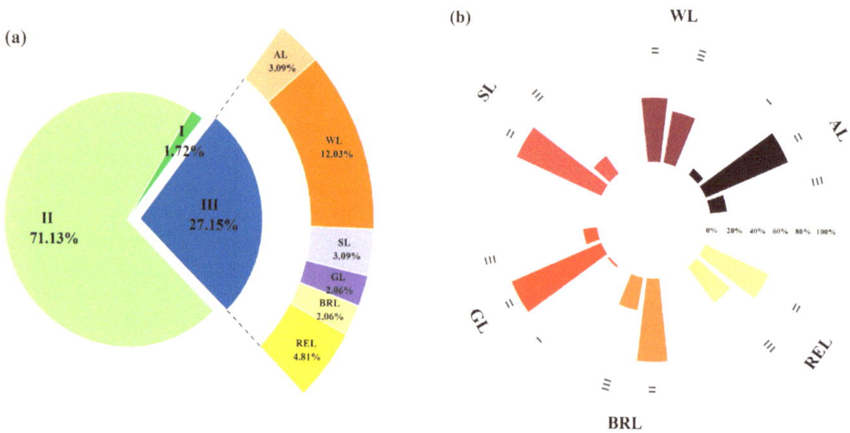


**Figure 4.** Soil quality status of different land use types in red bed areas based on MDS ((**a**) represents the proportion of all sample sites in grade I, II, and III soil quality; (**b**) represents the proportion of grade I, II, and III soil quality in each of the six different land use types).

Figure 5 shows the proportionate contributions of each indicator to the various land use types of soil. The total contribution of pH and $Ca^{2+}$ to soil quality under different land use patterns was the highest, 39.58% and 41.05%, respectively. In addition, the total contribution of SOM was 10.02%, which fluctuated above and below 10% in different land use types. The percentage contribution of A-Cu under different land use patterns was agricultural land (14.32%) > grassland (8.80%) > woodland (5.82%) > bare rock land (3.03%) > shrubland (2.54%) > red bed erosion land (0.83%). There were significant differences in the contribution of AP under different land use patterns, with 12.01% and 6.23% in agricultural land and grassland, respectively, and less than 1% in the rest of the land use patterns, which were, in descending order, woodland (0.96%) > shrubland (0.61%) > bare rock land (0.51%) > red bed erosion land (0.40%).

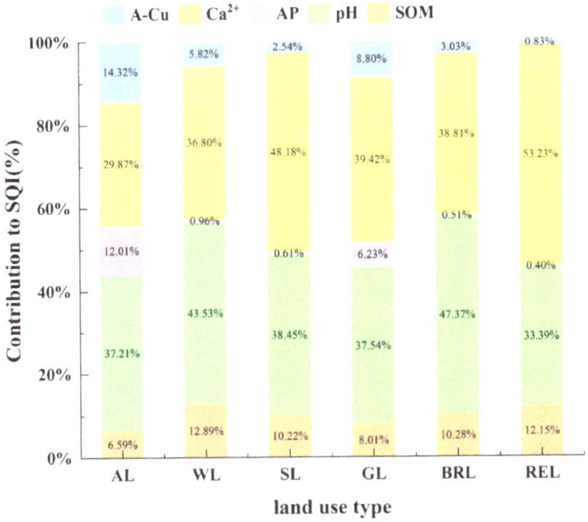


**Figure 5.** Percentage of the contribution of MDS index to soil SQI of different land use types.

## 3.5. Validate the Applicability of MDS

The computed total data set soil quality index (TDS-SQI) and minimal data set soil quality index (MDS-SQI) were examined using linear regression and a scatter plot in order to verify the applicability of MDS. As depicted in the fitting effect (Figure 6), a significant positive correlation was observed between the TDS-SQI and the MDS-SQI, with an $R^2$ of 0.797, indicating a relatively strong fitting effect. These results indicate that MDS can effectively replace TDS in calculating the soil quality index to assess the soil quality in the study area.

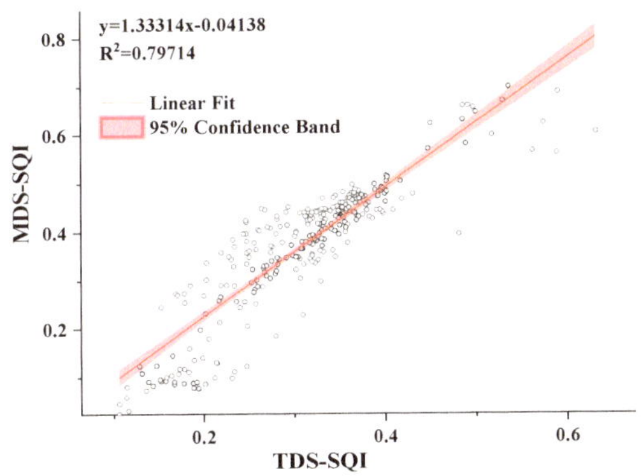


**Figure 6.** Relationship between MDS-SQI and TDS-SQI in the red bed desertification region.

## 4. Discussion

### 4.1. The MDS for Soil Quality Assessment in Humid Red Bed Basins

Numerous researchers have examined the effects of indicators including bulk density, soil structure, percentages of clay and sand, and conductivity on soil quality in order to evaluate the effects of soil physical and chemical features on nutrient cycling and uptake. However, only a few of these indicators are usually chosen for the minimal data set (MDS) for soil quality evaluation because of the relationships between these parameters [46,47]. In this study, the soil quality of different land use types in the red bed desertification area of the humid region was assessed using PCA and MDS methods. The results showed that a minimum data set consisting of SOM, pH, AP, $Ca^{2+}$, and A-Cu was the most suitable for evaluating soil quality in the region. SOM, pH, and AP were highly consistent with the MDS identified in many other soil quality assessment studies [48,49].

Soil pH is one of the most important factors in determining the quality of soil since it directly affects the chemical reactions and nutrient availability of the soil [50]. Huang et al. (2010) [11] concluded that pH is highly sensitive to land use changes, and it has been widely used as an essential indicator for assessing soil quality, particularly in forest ecosystems [48,51]. From the specific contribution of pH to the SQI of different land types in the study area (Figure 5), it was observed that the contribution of pH varied significantly in different land use types of soils, with the highest contribution of 43.53% in arborvitae woodlands. pH is one of the important indicators in the minimal data set created based on principal component analysis.

Moharana et al. (2019) [52] considered $Ca^{2+}$ and $Mg^{2+}$ as the main cations present in semi-arid soils. Pessoa et al. (2022)'s [53] study concluded that soil exchangeable and soluble ions were at low levels in native vegetation areas, while they were high in cultivated areas and areas of desertification. This suggests that anthropogenic factors may be the main influence on land salinity degradation. The conclusions of this paper are in agreement with these findings. From the characteristics of different land use types, the red bed erosion land with the most severe desertification has very low soil water content, but the highest exchangeable calcium and magnesium contents among all land use types. The exchangeable salty ion content in agricultural land that has been artificially fertilized is the second highest, while the exchangeable salty ion content of arboreal forest land soil, which is less affected by anthropogenic influences, is relatively the lowest.

In addition, the contents of Ca, Mg, Na, and K reflect the dissolution of easily weathered primary minerals [54]. The weak weathering resistance and strong erosion by flowing water of the red bed soft rock in the study area may be the reason for the higher content of exchangeable salty ions in the red bed eroded land compared to other land types. This indicates the significant role of mineral composition in regional soil fertility and productivity [55,56]. Therefore, it is essential to consider relevant indicators such as cation exchange capacity and exchangeable salinity ions when establishing a minimum data set for soil quality assessment in areas of desertification.

Available trace elements present in soil can be absorbed and utilized by plants, playing an irreplaceable role similar to macronutrients in plant growth and development [57]. However, many previous studies on soil quality have not considered the impact of these available micronutrients [58,59]. Liu et al. (1982) [60] emphasized that micronutrient content is crucial in soil quality assessment, especially when deficiencies limit crop growth. Tian et al. (2020) [61] also highlighted this point in their research. Therefore, in this study, available micronutrients were included in the soil quality assessment. Available copper, in particular, was incorporated into the minimum data set as a representative of these micronutrients, addressing the gap left by previous soil quality assessment studies.

Soil organic matter (SOM) and available phosphorus (AP) have been the most frequently included indicators in the minimum data set (MDS) for assessing soil quality in numerous prior studies. SOM is a primary source of plant mineral nutrients and influences the functional activity of soil biota by affecting the rate of soil nutrient cycling. Available phosphorus (AP) is the best indicator of phosphorus availability in soils and is frequently used in soil diagnosis and fertility assessment. The contribution of AP to the soil quality index (SQI) of soils under different land use types varies widely. As previously mentioned, it is relatively higher in land classes with higher SQI, such as cropland and grassland, and contributes less than 1% to other land classes. This variation may further indicate that anthropogenic factors, such as fertilizer application, significantly influence soil quality.

Overall, $Ca^{2+}$ had the greatest contribution to the SQI (41.05%), followed by pH (39.58%), SOM (10.02%), A-Cu (5.89%), and AP (3.45%). Our results confirm previous studies indicating that the MDS for the scientific assessment of soil quality should include indicators that (i) characterize nutrient retention (organic matter, exchangeable calcium); (ii) characterize available nutrients (available phosphorus, available copper); and (iii) correlate with alkali saturation (pH) [62].

### 4.2. Soil Quality Characteristics and Influencing Factors of Different Land Use Types in the Red Bed Region

Soil water vapor movement and physicochemical qualities, including pH and soil nutrients, are greatly impacted by varying land use patterns. These factors ultimately determine the quality of the soil [63,64]. These effects can be attributed to a variety of factors, including vegetation type, vegetation cover, and human activities [65]. Both the total data set (TDS) and the minimum data set (MDS) for the soil quality index (SQI) of the various land use types in this study showed the same trend, suggesting that MDS can effectively replace TDS for evaluating soil quality under various land use practices in typical red bed desertification areas.

The soil of the red bed erosion area is weakly alkaline, with the lowest average values of W%, AN, AP, A-Cu, A-Zn, and A-Fe among all land types. This aligns with its characteristics of extreme water shortage, low soil nutrient content, and severe desertification [66,67]. However, its SOM is unexpectedly high, likely due to government-led ecological restoration efforts such as regular irrigation and the application of organic fertilizers. Despite the increase in SOM, the red bed soil remains arid, and the levels of other nutrients have not significantly increased. This suggests that while human activities have improved some soil properties, it is challenging to achieve qualitative changes in a short period [68]. The poor water retention capacity of the red bed soil makes it difficult to quickly improve soil quality. The overall quality of the soil is influenced by the synergistic effects of its physical, chemical, and biological properties. Therefore, preventing and controlling desertification in the red beds is a long-term process that requires consideration from multiple perspectives.

The soil properties of the bare rock land are comparable to, albeit slightly better than, those of the red bed erosion land. This is attributed to the presence of a shallow, thin soil layer on the surface of the bare rock land, along with sporadic vegetation growth, which exerts a modest soil-fixing influence. However, if desertification persists unchecked, the gradual decline in vegetation cover may lead the bare rock land to degrade into red bed erosion land within a few years, resulting in further soil quality deterioration.

In contrast to the red bed erosion land, agricultural soils exhibit the highest levels of water content, alkaline dissolved nitrogen, available phosphorus, quick-acting potassium, exchangeable magnesium, available copper, available zinc, and available iron among the six land types. To ensure optimal crop yields, agricultural land soils undergo regular irrigation and application of fertilizers conducive to crop growth, underscoring the significant impact of anthropogenic land management practices on agricultural land quality. However, the average organic matter content of agricultural land was slightly lower compared to other sampling sites. This discrepancy can be attributed to the crop planting and growth processes, which promote the decomposition, transformation, leaching, and migration of organic matter, leading to its loss from the topsoil layer [69]. Additionally, frequent plowing processes may lead to the degradation of the soil structure, exacerbate the depletion of nutrients, and reduce the quality of the agricultural soil, which in turn affects crop yields.

In addition to agricultural soils, the soil quality of grassland and shrubland is better than that of other land use types. This is because some of the grasslands we sampled were formed from the abandonment of agricultural land for many years and retain rich nutrient elements even though they have not been artificially fertilized or irrigated. Additionally, compared to other land use types, the vegetation growing in grassland and shrubland usually has high adaptability and a fast growth rate. As a result, the soil in these areas tends to have a well-developed root system and a thicker layer of apoptotic material. This leads to a higher biomass input of organic matter, making the soil less susceptible to rapid erosion in a short period of time, thus preventing the formation of red bed deserts [70,71].

In addition, an interesting phenomenon observed in this study is that, contrary to many previous studies where forest land typically exhibits superior soil quality compared to other land use types due to factors such as less anthropogenic disturbance and abundant apomictic material, the soil quality of forest land in this study was among the lowest, only better than that of bare rock land and red bed erosion land. The percentage of soils with an SQI of III was also the highest for arboreal woodland. This could be attributed to the fact that arboreal woodlands in the region are predominantly planted with horsetail pine and artificial vegetation such as Taiwan acacia and new silver acacia, which are generally characterized by drought and barrenness tolerance. Augusto et al. (2002) [72] demonstrated that the decomposition of pine apomictic material produces high concentrations of $Al^{3+}$ and $H^{+}$ in the soil solution, which not only lowers the soil pH but also inhibits the uptake of $Ca^{2+}$ and $Mg^{2+}$ [73], reduces root growth, alters photosynthetic activity, and leads to nutrient imbalances in woodland species [51,74]. Additionally, the well-developed local agriculture and significant anthropogenic disturbances, such as fuel wood collection and turpentine extraction, contribute to the poor soil quality in forested areas of the region [27].

Regarding the contribution of indicators to soil quality within the minimum data set (Figure 5), $Ca^{2+}$ and pH jointly accounted for 80.63% of soil quality, highlighting the pivotal role of parent rock mineral composition in determining soil quality within the red bed desertification area. Additionally, pH exerts further control over soil quality by influencing soil chemical reactions and nutrient effectiveness. Furthermore, anthropogenic land management and utilization practices, exemplified by fertilizer application patterns and vegetation types represented by AP, A-Cu, and SOM, also significantly impact soil quality.

As illustrated in Figure 6, apart from grassland, there is a consistent decrease in soil water content across each land category, namely agricultural land, woodland, shrubland, bare rock land, and red bed erosion land. This decline corresponds to a gradual decrease in soil fertility and a subsequent decline in the soil quality index, suggesting a progressive reversal of vegetation succession within the red bed desertification area under study. Consequently, if timely preventive and control measures are not taken, soil erosion and land desertification in the area may be further aggravated.

### 4.3. *Soil Quality Improvement Recommendations for Red Bed Desertification Areas*

China has made remarkable efforts and achieved considerable success in combating desertification in arid regions [75]. However, the mechanisms and distribution patterns of desertification in arid areas differ from those in the hot and humid southern red bed areas [1]. Therefore, in the future, local red bed desertification control should focus on improving the soil quality of forested land in the basin. Simultaneously, regulating the soil quality of arable land through rational farming practices and appropriate fertilizer application is essential to prevent the degradation process, which could lead to a decline or even loss of arable land productivity due to over-construction and excessive management aimed at increasing crop yields [76]. For bare rock and red bed eroded land with a high degree of desertification, efforts should be made to increase vegetation cover. This should be done in conjunction with national policies on land transformation and utilization to prevent the further aggravation of red bed desertification, which in turn affects the security of the regional ecological environment.

Although this study has made some progress in assessing soil quality in red bed desertification areas, there are still some limitations. For example, in the analysis process, we primarily focus on chemical indicators and neglect important physical and biological indicators, which limits a comprehensive understanding of the multidimensional characteristics of soil quality in red bed desertification areas. Therefore, future research should consider a wider range of ecosystem indicators in the red bed region, including soil physical indicators (such as porosity and particle distribution), biological indicators (such as soil microbial diversity and activity), climate, topographic factors, and anthropogenic influences. By integrating these diverse indicators, we can conduct a more comprehensive soil quality analysis and deeply explore the dynamic mechanisms of the ecosystem in the red bed desertification area. Ultimately, we hope that through our current and future work, more scholars will focus on the ecological and environmental issues in the red bed desertification area, promoting the development of a more scientific and reliable soil quality assessment system. This, in turn, will provide strong support for soil management and land use decision-making in the red bed desertification area.

## 5. Conclusions

To evaluate the comprehensive quality of soils under six land use modes in the humid red bed desertification region of South China, this study selected 11 soil physicochemical indicators, verified the correlation between these indicators, and established the minimum data set (MDS) to calculate the soil quality index (SQI). The results showed that most of the soils in the study area were of medium quality. The soil quality of red bed erosion land and bare rock land, which had a high degree of desertification, was generally low, followed by arboreal forest land, while the quality of agricultural soil was relatively the highest. The significant difference in the contribution of the indicators in the MDS to soil quality indicates that the mineral composition of the host rock is the most important factor affecting soil quality in the red bed desertification area. Additionally, fertilizer application patterns, vegetation types, and other anthropogenic land management practices are also important factors affecting soil quality. The characteristics of the soil quality and the factors that influence various land use patterns in the humid red bed desertification region are revealed in this study, which is very important for the scientific assessment, application, and enhancement of soil quality in desertification regions.

**Author Contributions:** F.S.: fieldwork, sampling, indoor test analysis, data processing, and paper drafting. B.C.: project implementation, paper conception and revision, fieldwork, sampling. B.W.: field research and sampling, indoor test analysis, data processing, and paper conception. W.L.: field research and sampling, indoor test analysis, and data processing. C.Z.: field research, sampling, indoor testing, and analysis. J.F.: project implementation and fieldwork. B.Y.: project implementation, field research arrangement. G.X.: field sampling, in-door sample testing, and analysis. All authors have read and agreed to the published version of the manuscript.

**Funding:** This research was funded by National Key R&D Program of China: Theory and Method of Big Data and Knowledge Fusion of Cross-scale Multi-source Heterogeneous Geological Resources (2022YFF0800101) and Local Government Commissioned Project: Monitoring Analysis and Evaluation Project of Comprehensive Remediation for Nanxiong red beds Desertification (Phase II) (ZPSG2020ZB010).

**Data Availability Statement:** The data that have been used are confidential.

**Conflicts of Interest:** The authors declare no conflicts of interest.

## References

1. Yan, L.-B.; Kasanin-Grubin, M. Land degradation and management of red beds in China: Two case studies. *J. Mt. Sci.* **2019**, *16*, 2591–2604. [CrossRef]
2. Guo, Y.; Xie, Q.; Wen, J. Red beds distribution and engineering geological problem in China. *Hydrogeol. Eng. Geol.* **2007**, *6*, 67–71. [CrossRef]
3. Zhao, J.; He, X.; Shao, T. Material composition and microstructure of purple soil and purple mudstone in Chongqing area. *Acta Pedol. Sin.* **2012**, *49*, 212–219.
4. Jin, Y.; Peng, H.; Yan, L.; Xiang, Y.; Wang, X. Discussion on desertification of humid region in southern China. *Prog. Geogr.* **2015**, *34*, 772–780. [CrossRef]

5. Peng, H. Perspectives on the red beds landforms in humid area of southern China and some related problems. *Geogr. Res.* **2011**, *30*, 1739–1752. [CrossRef]
6. Doran, J.W.; Parkin, T.B. Defining and assessing soil quality. *Defin. Soil Qual. A Sustain. Environ.* **1994**, *35*, 1–21.
7. Ditzler, C.A.; Tugel, A.J. Soil quality field tools: Experiences of USDA-NRCS soil quality institute. *Agron. J.* **2002**, *94*, 33–38. [CrossRef]
8. Karlen, D.L.; Ditzler, C.A.; Andrews, S.S. Soil quality: Why and how? *Geoderma* **2003**, *114*, 145–156. [CrossRef]
9. Pierce, F.; Larson, W.; Dowdy, R.; Graham, W. Productivity of soils: Assessing long-term changes due to erosion. *J. Soil Water Conserv.* **1983**, *38*, 39–44.
10. Bünemann, E.K.; Bongiorno, G.; Bai, Z.; Creamer, R.E.; De Deyn, G.; De Goede, R.; Fleskens, L.; Geissen, V.; Kuyper, T.W.; Mäder, P. Soil quality—A critical review. *Soil Biol. Biochem.* **2018**, *120*, 105–125. [CrossRef]
11. Huang, T.; Yue, X.; Ge, X.; Wang, X. Evaluation of soil quality on gully region of loess plateau based on principal component analysis. *Agric. Res. Arid. Areas* **2010**, *28*, 141–147.
12. Shao, G.; Ai, J.; Sun, Q.; Hou, L.; Dong, Y. Soil quality assessment under different forest types in the Mount Tai, central Eastern China. *Ecol. Indic.* **2020**, *115*, 106439. [CrossRef]
13. Yu, P.; Han, D.; Liu, S.; Wen, X.; Huang, Y.; Jia, H. Soil quality assessment under different land uses in an alpine grassland. *Catena* **2018**, *171*, 280–287. [CrossRef]
14. Zhang, X.; Kong, F.; Jiang, Z. Construction and empirical analysis of a comprehensive evaluation method of coastal wetland soil quality based on ecological functions. *Environ. Sci.* **2022**, *43*, 2709–2718. [CrossRef]
15. Maurya, S.; Abraham, J.S.; Somasundaram, S.; Toteja, R.; Gupta, R.; Makhija, S. Indicators for assessment of soil quality: A mini-review. *Environ. Monit. Assess.* **2020**, *192*, 604. [CrossRef] [PubMed]
16. Abraham, J.S.; Sripoorna, S.; Dagar, J.; Jangra, S.; Kumar, A.; Yadav, K.; Singh, S.; Goyal, A.; Maurya, S.; Gambhir, G. Soil ciliates of the Indian Delhi Region: Their community characteristics with emphasis on their ecological implications as sensitive bio-indicators for soil quality. *Saudi J. Biol. Sci.* **2019**, *26*, 1305–1313. [CrossRef] [PubMed]
17. Moral, F.J.; Rebollo, F.J. Characterization of soil fertility using the Rasch model. *J. Soil Sci. Plant Nutr.* **2017**, *17*, 486–498. [CrossRef]
18. Liu, Z.-P.; Shao, M.-A.; Wang, Y.-Q. Spatial patterns of soil total nitrogen and soil total phosphorus across the entire Loess Plateau region of China. *Geoderma* **2013**, *197*, 67–78. [CrossRef]
19. Das, P.P.; Singh, K.R.; Nagpure, G.; Mansoori, A.; Singh, R.P.; Ghazi, I.A.; Kumar, A.; Singh, J. Plant-soil-microbes: A tripartite interaction for nutrient acquisition and better plant growth for sustainable agricultural practices. *Environ. Res.* **2022**, *214*, 113821. [CrossRef]
20. Luo, X.; Zhang, L.; Lin, Y.; Wen, D.; Hou, E. Nitrogen availability mediates soil organic carbon cycling in response to phosphorus supply: A global meta-analysis. *Soil Biol. Biochem.* **2023**, *185*, 109158. [CrossRef]
21. Soumare, A.; Djibril, S.; Diédhiou, A.G. Potassium sources, microorganisms and plant nutrition: Challenges and future research directions. *Pedosphere* **2023**, *33*, 105–115. [CrossRef]
22. Ciais, P.; Gervois, S.; Vuichard, N.; Piao, S.; Viovy, N. Effects of land use change and management on the European cropland carbon balance. *Glob. Chang. Biol.* **2011**, *17*, 320–338. [CrossRef]
23. Cotrufo, M.F.; Lavallee, J.M. Soil organic matter formation, persistence, and functioning: A synthesis of current understanding to inform its conservation and regeneration. *Adv. Agron.* **2022**, *172*, 1–66.
24. O'Sullivan, L.; Bampa, F.; Knights, K.; Creamer, R. Soil protection for a sustainable future: Options for a soil monitoring network for Ireland. *Soil Use Manag.* **2017**, *33*, 346–363. [CrossRef]
25. Bertini, S.C.B.; Azevedo, L.C.B.; de Carvalho Mendes, I.; Cardoso, E.J.B.N. Hierarchical partitioning for selection of microbial and chemical indicators of soil quality. *Pedobiologia* **2014**, *57*, 293–301. [CrossRef]
26. Jin, H.; Shi, D.; Chen, Z.; Liu, Y.; Lou, Y.; Yang, X. Evaluation indicators of cultivated layer soil quality for red soil slope farmland based on cluster and PCA analysis. *Trans. Chin. Soc. Agric. Eng.* **2018**, *34*, 155–164. [CrossRef]
27. Chen, J.; Fu, J.; Xu, G.; Yu, B.; Lai, X.; Hu, S.; Zhang, K.; Liu, J. Soil Quality Characteristics and Influencing Factors of Different Land Use in the Red Bed Landform Region of South China: Taking Nanxiong Basin as An Example. *Ecol. Environ.* **2022**, *31*, 918. [CrossRef]
28. Yan, P.; Lin, K.; Wang, Y.; Tu, X.; Bai, C.; Yan, L. Assessment of influencing factors on the spatial variability of SOM in the red beds of the Nanxiong basin of China, using GIS and geo-statistical methods. *ISPRS Int. J. Geo-Inf.* **2021**, *10*, 366. [CrossRef]
29. Yan, P.; Lin, K.; Wang, Y.; Zheng, Y.; Gao, X.; Tu, X.; Bai, C. Spatial interpolation of red bed soil moisture in Nanxiong basin, South China. *J. Contam. Hydrol.* **2021**, *242*, 103860. [CrossRef]
30. Wu, Z.; Peng, H. A study on the formation and development rules of the red beds in Guangdong Province. *Trop. Geogr.* **2006**, *26*, 207.
31. Shen, W.-S.; Cao, X.-Z.; Shen, F.-Y. Classification and grading of land degradation in China. *J. Ecol. Rural. Environ.* **2006**, *22*, 88–93.
32. Lu, R. *Soil Agrochemical Analytical Methods*; Chinese Agriculture and Technology Press Publisher: Beijing, China, 1999.
33. Walkley, A.; Black, I.A. An examination of the Degtjareff method for determining soil organic matter, and a proposed modification of the chromic acid titration method. *Soil Sci.* **1934**, *37*, 29–38. [CrossRef]
34. Anderson, J.M.; Ingram, J.S. Tropical soil biology and fertility: A handbook of methods. *Soil Sci.* **1994**, *157*, 265. [CrossRef]
35. Schmugge, T.; Jackson, T.; McKim, H. Survey of methods for soil moisture determination. *Water Resour. Res.* **1980**, *16*, 961–979. [CrossRef]

36. Bray, R.H.; Kurtz, L.T. Determination of total, organic, and available forms of phosphorus in soils. *Soil Sci.* **1945**, *59*, 39–46. [CrossRef]
37. Schollenberger, C.; Simon, R. Determination of exchange capacity and exchangeable bases in soil—Ammonium acetate method. *Soil Sci.* **1945**, *59*, 13–24. [CrossRef]
38. Norvell, W.; Lindsay, W. Reactions of DTPA chelates of iron, zinc, copper, and manganese with soils. *Soil Sci. Soc. Am. J.* **1972**, *36*, 778–783. [CrossRef]
39. Rinot, O.; Levy, G.J.; Steinberger, Y.; Svoray, T.; Eshel, G. Soil health assessment: A critical review of current methodologies and a proposed new approach. *Sci. Total Environ.* **2019**, *648*, 1484–1491. [CrossRef] [PubMed]
40. Wu, C.; Liu, G.; Huang, C.; Liu, Q. Soil quality assessment in Yellow River Delta: Establishing a minimum data set and fuzzy logic model. *Geoderma* **2019**, *334*, 82–89. [CrossRef]
41. Zhang, F.; Gao, Z.; Ma, Q.; Song, Z.; Li, G.; Li, X.; Su, Y. Construction of minimum data set for soil quality assessment in the Dunhuang Oasis. *Chin. J. Soil Sci.* **2017**, *48*, 1047–1054.
42. Jin, H.; Zhong, Y.; Shi, D.; Li, J.; Lou, Y.; Li, Y.; Li, J. Quantifying the impact of tillage measures on the cultivated-layer soil quality in the red soil hilly region: Establishing the thresholds of the minimum data set. *Ecol. Indic.* **2021**, *130*, 108013. [CrossRef]
43. Chen, Y.; Wang, H.; Zhou, J.-M.; Xing, L.; Zhu, B.; Zhao, Y.; Chen, X. Minimum data set for assessing soil quality in farmland of northeast China. *Pedosphere* **2013**, *23*, 564–576. [CrossRef]
44. Computer Network Information Center, C.A.o.S. ASTER GDEM 30M Resolution Digital Elevation Data. Available online: https://www.gscloud.cn/sources/index?pid=1&rootid=1 (accessed on 15 June 2024).
45. Ministry of Natural Resources of the People's Republic of China. China Standard Map (Revision No. GS (2024) 0650). Available online: https://cloudcenter.tianditu.gov.cn/administrativeDivision (accessed on 15 June 2024).
46. Yu, P.; Liu, S.; Zhang, L.; Li, Q.; Zhou, D. Selecting the minimum data set and quantitative soil quality indexing of alkaline soils under different land uses in northeastern China. *Sci. Total Environ.* **2018**, *616*, 564–571. [CrossRef] [PubMed]
47. Zhu, Z.; Chen, J.; Hu, H.; Zhou, M.; Zhu, Y.; Wu, C.; Zhu, L.; Jiang, X.; Wang, J. Soil quality evaluation of different land use modes in small watersheds in the hilly region of southern Jiangsu. *Ecol. Indic.* **2024**, *160*, 111895. [CrossRef]
48. Cheng, J.; Ding, C.; Li, X.; Zhang, T.; Wang, X. Soil quality evaluation for navel orange production systems in central subtropical China. *Soil Tillage Res.* **2016**, *155*, 225–232. [CrossRef]
49. Salazar, M.P.; Lozano, L.A.; Villarreal, R.; Irizar, A.B.; Barraco, M.; Polich, N.G.; Soracco, C.G. Capacity and Intensity Indicators to evaluate the effect of different crop sequences and cover crops on soil physical quality of two different textured soils from Pampas Region. *Soil Tillage Res.* **2022**, *217*, 105268. [CrossRef]
50. Schoenholtz, S.H.; Van Miegroet, H.; Burger, J. A review of chemical and physical properties as indicators of forest soil quality: Challenges and opportunities. *For. Ecol. Manag.* **2000**, *138*, 335–356. [CrossRef]
51. Bautista-Cruz, A.; Del Castillo, R.F.; Etchevers-Barra, J.D.; del Carmen Gutiérrez-Castorena, M.; Baez, A. Selection and interpretation of soil quality indicators for forest recovery after clearing of a tropical montane cloud forest in Mexico. *For. Ecol. Manag.* **2012**, *277*, 74–80. [CrossRef]
52. Moharana, P.; Singh, R.; Singh, S.; Tailor, B.; Jena, R.; Meena, M. Development of secondary salinity and salt migration in the irrigated landscape of hot arid India. *Environ. Earth Sci.* **2019**, *78*, 454. [CrossRef]
53. Pessoa, L.G.; Freire, M.B.d.S.; Green, C.H.; Miranda, M.F.; de A Filho, J.C.; Pessoa, W.R. Assessment of soil salinity status under different land-use conditions in the semiarid region of Northeastern Brazil. *Ecol. Indic.* **2022**, *141*, 109139. [CrossRef]
54. Dos Santos, J.C.B.; Le Pera, E.; de Souza Junior, V.S.; Corrêa, M.M.; de Azevedo, A.C. Gneiss saprolite weathering and soil genesis along an east-west regolith sequence (NE Brazil). *Catena* **2017**, *150*, 279–290. [CrossRef]
55. Câmara, E.R.G.; dos Santos, J.C.B.; de Araújo Filho, J.C.; Schulze, S.M.B.B.; Corrêa, M.M.; Ferreira, T.O.; de Sousa, J.E.S.; de Souza Júnior, V.S. Parent rock–pedogenesis relationship: How the weathering of metamorphic rocks influences the genesis of Planosols and Luvisols under a semiarid climate in NE Brazil. *Geoderma* **2021**, *385*, 114878. [CrossRef]
56. Oliveira Filho, J.d.S.; Barrozo, M.V.d.S.; Bastos, F.d.H.; Pereira, M.G. Effects of climate and land-use on physicochemical properties of gneiss-derived soils in tropical drylands: Evidence from northeastern Brazil. *Geoderma Reg.* **2023**, *33*, e00639. [CrossRef]
57. Dong, G.; Zhang, A.; Luo, G.; Xu, W.; Dai, L. Study on contents of available trace elements in oasis soil of Sangong river watershed. *Soils* **2009**, *41*, 726–732.
58. Ge, Y.; Murray, P.; Hendershot, W. Trace metal speciation and bioavailability in urban soils. *Environ. Pollut.* **2000**, *107*, 137–144. [CrossRef]
59. Thornton, I. Geochemistry and the mineral nutrition of agricultural livestock and wildlife. *Appl. Geochem.* **2002**, *17*, 1017–1028. [CrossRef]
60. Liu, Z.; Zhu, F.Q.; Tang, L.; Xu, J.; Yen, C. Geographical distribution of trace element-deficient soils in China. *Acta Pedol. Sin.* **1982**, *19*, 209–223.
61. Tian, K.; Zhang, B.; Zhang, H.; Huang, B.; Darilek, J.L.; Zhao, Y.; Yang, J. Evaluation of soil quality in major grain-producing region of the North China Plain: Integrating minimum data set and established critical limits. *Ecol. Indic.* **2020**, *117*, 106613. [CrossRef]
62. Macedo, R.S.; Lima, R.P.; de Almeida Alves Carneiro, K.; Moro, L.; Refati, D.C.; Campos, M.C.C.; Beirigo, R.M.; da Cruz, G.K.G.; de Sousa, A.A.P.; de Brito Neto, J.F. Assessment of Soil Quality of Smallholder Agroecosystems in the Semiarid Region of Northeastern Brazil. *Land* **2024**, *13*, 304. [CrossRef]

63. Cotrufo, M.F.; Haddix, M.L.; Kroeger, M.E.; Stewart, C.E. The role of plant input physical-chemical properties, and microbial and soil chemical diversity on the formation of particulate and mineral-associated organic matter. *Soil Biol. Biochem.* **2022**, *168*, 108648. [CrossRef]
64. Xiao, K.; Li, D.; Wen, L.; Yang, L.; Luo, P.; Chen, H.; Wang, K. Dynamics of soil nitrogen availability during post-agricultural succession in a karst region, southwest China. *Geoderma* **2018**, *314*, 184–189. [CrossRef]
65. Wang, X.; Liu, Z.; Xiong, K.; He, Q.; Li, Y.; Li, K. Characteristics and controlling factors of soil dissolved organic matter in the rainy season after vegetation restoration in a karst drainage area, South China. *Catena* **2022**, *217*, 106483. [CrossRef]
66. Huang, S.; Peng, X.; Huang, Q.; Zhang, W. Soil aggregation and organic carbon fractions affected by long-term fertilization in a red soil of subtropical China. *Geoderma* **2010**, *154*, 364–369. [CrossRef]
67. Zhang, H.; Wang, B.; Xu, M.; Fan, T. Crop yield and soil responses to long-term fertilization on a red soil in southern China. *Pedosphere* **2009**, *19*, 199–207. [CrossRef]
68. Li, X.; Wang, D.; Ren, Y.; Wang, Z.; Zhou, Y. Soil quality assessment of croplands in the black soil zone of Jilin Province, China: Establishing a minimum data set model. *Ecol. Indic.* **2019**, *107*, 105251. [CrossRef]
69. Jin, H.; Shi, D.; Lou, Y.B.; Zhang, J.; Ye, Q.; Jiang, N. Evaluation of the quality of cultivated-layer soil based on different degrees of erosion in sloping farmland with purple soil in China. *Catena* **2021**, *198*, 105048. [CrossRef]
70. Gan, F.; Shi, H.; Yan, Y.; Pu, J.; Dai, Q.; Gou, J.; Fan, Y. Soil quality assessment of karst trough valley under different bedrock strata dip and land-use types, based on a minimum data set. *Catena* **2024**, *241*, 108048. [CrossRef]
71. Huang, W.; Zong, M.; Fan, Z.; Feng, Y.; Li, S.; Duan, C.; Li, H. Determining the impacts of deforestation and corn cultivation on soil quality in tropical acidic red soils using a soil quality index. *Ecol. Indic.* **2021**, *125*, 107580. [CrossRef]
72. Augusto, L.; Ranger, J.; Binkley, D.; Rothe, A. Impact of several common tree species of European temperate forests on soil fertility. *Ann. For. Sci.* **2002**, *59*, 233–253. [CrossRef]
73. De Wit, H.A.; Eldhuset, T.D.; Mulder, J. Dissolved Al reduces Mg uptake in Norway spruce forest: Results from a long-term field manipulation experiment in Norway. *For. Ecol. Manag.* **2010**, *259*, 2072–2082. [CrossRef]
74. Wright, R.; Baligar, V.; Ahlrichs, J. The influence of extractable and soil solution aluminum on root growth of wheat seedlings. *Soil Sci.* **1989**, *148*, 293–302. [CrossRef]
75. Miao, L.; Moore, J.C.; Zeng, F.; Lei, J.; Ding, J.; He, B.; Cui, X. Footprint of research in desertification management in China. *Land Degrad. Dev.* **2015**, *26*, 450–457. [CrossRef]
76. Sui, H.; Song, G.; Liu, W.; Zhang, Y.; Su, R.; Wang, Q.; Ren, G.; Mi, Y. Spatiotemporal variation of cultivated land ecosystem stability in typical regions of Lower Liaohe Plain China based on stress-buffer-response. *Sci. Total Environ.* **2023**, *858*, 160213. [CrossRef] [PubMed]

*land*

MDPI

*Article*

# Sub-Regional Biophysical and Monetary Evaluation of Ecosystem Services: An Experimental Spatial Planning Implementation

**Anna Richiedei [1,*], Marialaura Giuliani [1] and Michele Pezzagno [2]**

1 Department of Civil Engineering, Architecture, Land, Environment and of Mathematics—DICATAM, University of Brescia, 25123 Brescia, Italy; marialaura.giuliani@unibs.it
2 Department of Civil Engineering, Architecture, Land, Environment and of Mathematics—DICATAM, University Research and Documentation Center on the UN Agenda 2030 (CRA2030), University of Brescia, 25123 Brescia, Italy; michele.pezzagno@unibs.it
* Correspondence: anna.richiedei@unibs.it

**Abstract:** Preserving soil is crucial for addressing the key challenges of the new millennium, like climate change and biodiversity loss. Spatial planning plays a pivotal role in stopping soil consumption and degradation, thereby safeguarding soils that provide valuable ecosystem services. With the advent of the System of Environmental-Economic Accounting by the UN, countries are developing a shared protocol for the biophysical and monetary quantification of ecosystem services. However, downscaling efforts are necessary and must be conditioned by the national context, policies, economic dynamics, and data availability. Therefore, this research proposes a soil quality assessment methodology based on its ecosystem value at the sub-regional level in northern Italy, building upon national guidelines. This study includes modeling and mapping outputs involving six ecosystem services through eight biophysical indicators and the monetary quantification of these services. Both assessments have been conducted over two time periods to highlight the impacts of land cover transformation.

**Keywords:** land planning; land cover changes; ecosystem quality; ecosystem accounting

## 1. Introduction

The Introduction provides an overview of the general background necessary to comprehend this study's findings. It outlines the following points:

- The concept of ecosystem services (Section 1.1);
- The loss of natural capital (Section 1.2);
- Ecosystems and the monetization of ecosystem services for natural resources conservation and management (Section 1.3);
- The payment of ecosystem services (Section 1.4);
- Downscaling global principles to local contexts (Section 1.5);
- Mapping ecosystem services: a literature review (Section 1.6).

### *1.1. Ecosystem Services: A Rapidly Growing Concept*

The definition of "Ecosystem Services" is 160 years old and is first attributed to Marsh. In his "Report on the Artificial Propagation of Fish", Marsh criticized members of a community for their "mistaken prejudices" against "birds, quadrupeds, and reptiles" because of the supposed damage these animals inflict upon crops. Instead, Marsh argued that they "much more than compensate the little injury they inflict upon the crops" by consuming "vast numbers of noxious insects" [1,2]. However, the topic of ecosystem services (ESs) has remained dormant for a long time.

Global interest in ecosystem health and the services ecosystems provide to humans has grown in both the public and private sectors, influencing research and policy [3] only after the publication of the Millennium Ecosystem Assessment (MEA) by the United Nations in 2005. Since then, the most commonly accepted definition of ecosystem services refers to the MEA and regards the benefits that people derive from ecosystems, namely "the support of sustainable human well-being that ecosystems provide" [4,5]. The MEA was followed by another initiative, The Economics of Ecosystems and Biodiversity (TEEB), which expanded awareness of ecosystem services, particularly highlighting the importance of biodiversity in decision-making at all levels [6,7].

Recognition of the priority function of ESs necessarily stems from an awareness of the importance of protecting and restoring ecosystems themselves as assets that can provide these services.

This priority is currently clearly embodied in many international and European strategies and policies. The European Biodiversity Strategy for 2030, based on the Aichi Biodiversity Targets, recognizes the valuation of biological diversity, as well as its protection and restoration, as a priority. The European Green Deal clearly outlines the European efforts to enhance and restore ecosystems since they are a prior carbon stockholder.

The Common Agricultural Policy plays a crucial role in promoting sustainable agricultural practices in Europe. Similarly, the UN Paris Agreement contributes to climate change mitigation since oceans and forests are the main basins for carbon stock.

The UN Agenda 2030 is pivotal for sustainable development, with formal reference in this context to Goals 14 (Life of Water) and 15 (Life on Earth), Goal 2 (Zero Hunger), Goal 6 (Clean Water and Sanitation), Goal 7 (Affordable and Clean Energy), Goal 11 (Sustainable Cities and Communities), Goal 12 (Responsible Consumption and Production), and Goal 13 (Climate Action) in terms of sustainable ecosystem management. Additionally, it holistically encompasses social and economic implications, which include Goal 1 (No Poverty), Goal 3 (Good Health and Well-Being), Goal 9 (Industry, Innovation, and Infrastructures), Goal 10 (Reduced Inequalities), and Goal 16 (Peace, Justice, and Strong Institutions).

More recently and incisively, the newly approved Nature Restoration Regulation seeks to enhance biodiversity and ecosystem resilience across the EU, aiming to restore at least 20% of land and sea areas by 2030 and all degraded ecosystems by 2050, including binding targets and an implementation framework for national restoration plans.

These documents are just a few examples showing the widespread acknowledgment of the importance of ESs in maintaining livelihoods and the dependency of natural land and agricultural systems on the soil. It is common, both here and in the literature, to refer to many ecosystem services as soil ecosystem services [8].

Various frameworks have been developed to support ecosystem conservation and enhancement. Authors emphasize that a shared understanding of how ecosystems underpin economic activity and human well-being is essential for effectively designing, implementing, and monitoring ecosystem restoration policies, as well as for informed planning, policymaking, and financial decisions [9]. Over the past decade, there has been growing interest in creating an inventory and spatially mapping the current state of ecosystems, as well as their potential to provide ESs. Recently, the System of Environmental-Economic Accounting—Ecosystem Accounting (SEEA EA) was finalized by the United Nations (UN) in 2021 as a framework designed to evaluate and monetize ecosystems and their services [10]. Developed through collaborative, international efforts, it builds on earlier versions from 2012, along with practical recommendations issued in 2017, which have guided ecosystem service measurement and valuation efforts [10].

### 1.2. Natural Capital: How to Face the Loss

Alongside ESs, natural capital is defined as the world's stock of natural assets, which supplies a wide range of goods and services, including natural income over time, which directly or indirectly creates value for people [11,12].

Costanza et al. [7] highlight the relationship between natural capital, human beings, and ecosystem services, noting that the latter can reach humans only after the interaction between natural capital and social capital.

Ecosystem services are mechanisms through which natural capital benefits humans, connecting these benefits to the socio-economic system. Biodiversity, particularly at the community level, plays a crucial role in supporting the productivity and stability of ecosystems, ensuring the quality and quantity of natural capital [13–15]. Maintaining the stock and diversity of natural capital is essential for sustaining the flow of ecosystem services, which are vital for current and future human prosperity [6]. The debate over substituting natural capital for human capital distinguishes weak and strong sustainability. Weak sustainability assumes that the two are interchangeable, focusing on maintaining total capital stock. In contrast, strong sustainability emphasizes natural capital's limited substitutability, advocating for its preservation to avoid net loss [16,17]. Biodiversity offsetting markets, with a "no net loss" strategy, aim to balance environmental damage by restoring habitats. However, this approach risks viewing diverse natural capital as interchangeable, potentially overlooking unique, irreplaceable ecological functions and neglecting critical, non-substitutable aspects of ecosystems [16].

Conversely, due to the delicate and complex equilibrium at the basis of natural capital preservation and its relationships with human systems through ecosystem services, assessing natural capital and its benefits should be highly site-specific. The development of markets or compensation systems must begin with an awareness of the area's challenges and potentials [18], rather than relying on a complete computation of interchangeable components. In this context, the spatial assessment and mapping of ecosystems and their services—including their monetary value—are essential. This mapping should involve homogeneous and interconnected territorial, environmental, and ecosystem realities, facilitating an integrated, informed, and context-specific management of natural resources. This same spirit and scope have driven the present research (yet from the choice of the case study at the wide-area scale) and should guide the use and interpretation of its results.

### 1.3. Ecosystems and ES Monetization for Enhancing Natural Resources Conservation and Management

The economic dimensions of ecosystems were first explored in the early 1990s [19], as concepts such as "natural capital" and "ecosystem services", and along with their valuation, gained prominence in response to global biodiversity loss and its associated social and economic impacts [5,11,20,21]. This period also marked a growing need to establish economic mechanisms that would promote conservation efforts, such as implementing Payment for Ecosystem Services (PESs) schemes [22,23]. In 1997, global ecosystem services were valued at USD 33 trillion per year, significantly exceeding THE global GDP [5]. By 2014, Costanza et al. adjusted this figure based on dollar revaluation to USD 145 trillion in 2007 but revised it to USD 124 trillion after accounting for land-use changes, indicating a net loss of USD 20.2 trillion since 1997. Beyond land use, pressures such as pollution and urban sprawl also degrade ecosystems, with urban expansion negatively impacting water quality, carbon sequestration, and ecosystem health [21–25]. Soil ecosystems are especially affected, with 60–70% of European soils degraded due to unsustainable practices, costing over USD 6 trillion annually in lost ecosystem services [26,27]

Ecosystem conservation is crucial for both environmental and economic sustainability. In 2018, Credit Suisse reported USD 52 billion in global conservation spending, far short of the USD 300–400 billion needed annually to safeguard natural capital like clean air, water, and biodiversity [28,29]. Despite high costs, the benefits of conservation outweigh the expenses, with a reported 100:1 benefit–cost ratio for global wildlife conservation [30]. Restoration, while valuable, often results in suboptimal outcomes compared to original ecosystems. The sustainable management and protection of natural capital remain the most cost-effective solutions to ensure long-term human well-being [7,31–33].

### *1.4. PES and Other Market Mechanisms*

Many authors have discussed the reasons and utility for valuing ES, which can be summarized as follows: improving an understanding and awareness of ecosystem services' role in our society; completing national and global accounts, such as national income and well-being accounts, or full-cost accounts; creating innovative institutional and market instruments allowing for sustainable ecosystem management, e.g., through the Payment of Ecosystem Services (PESs) or common asset trusts; and informing policy and decision-making, as well as territorial and urban planning in land-use transformations [7,34,35].

Scholars highlight a gap between scientific research on ecosystem services (ESs) and its practical impact on policy and urban planning. This gap may result from the traditional focus of ES research, which must integrate both natural and social sciences in a site-specific manner [36,37]. Salzman argues that passing environmental laws or policies alone is insufficient to improve ES provision; policies must be transformed into actionable laws and applied through individual decisions [38]. The key point is the need to integrate ESs into decision-making and ecosystem management, such as introducing market-based instruments like PESs [39–41]. These instruments aim to obtain economic resources from ES beneficiaries to compensate those responsible for managing the services (sellers or providers), ensuring the maintenance of the site and the environmental quality of the ecosystem [42,43]. The basic concept of the PESs scheme is as follows: if some ESs are defended by land owners as providers of market-based benefits and goods and, thus, have direct private revenue, other services have a good public characteristic but are not traded in the market [44–46]. Therefore, an effort is needed to find instruments to compensate land owners and managers for the production and conservation of these services [47–50]. PESs mechanisms are structured to offer monetary incentives to communities or individuals in order to adopt interventions and behaviors that improve the provision of Ess. These mechanisms are characterized by voluntary transactions between service users and providers [18].

Another option proposed by some authors is conservation easements (CEs), defined as legal contracts that result in restrictions on the use of private lands to preserve nature in exchange for potential tax benefits [51,52]. The primary goal is to prevent the future conversion of natural areas and low-intensity agricultural lands into more intense land uses, such as industrial or urban zones. Additionally, CEs aim to maintain and improve the provision of ESs, thereby benefiting the local and regional communities [53,54]. A notable example of a program reflecting elements of conservation easements is the Australian Biodiversity Conservation Strategy. Over the past few decades, it has aimed to double the value of complementary markets for ecosystem services by 2015. It sought to achieve this goal by increasing incentives for private sector participation through financial rewards for actions that protect or enhance biodiversity [55]. Similarly, in another part of the world, Canadian provinces allow a 100% property tax reduction for maintaining and protecting lands with features deemed essential for provincial natural heritage conservation [56].

Such solutions inevitably introduce complexities in balancing economic efficiency, social acceptance, and environmental benefits [57]. Significant challenges remain, particularly regarding landowners' willingness to participate, the management and monitoring of private conservation efforts, and the financial burdens faced by both governments and participants [58,59]. These financial and fiscal instruments, widely discussed in the literature and implemented in various countries, should be tailored to each nation's specific context, taking into account its internal markets and fiscal structures. For instance, in Italy, property tax exemptions for agricultural landowners already benefit direct farmers (https://www.finanze.gov.it/it/fiscalita/fiscalita-regionale-e-locale/Imposta-municipale-propria-IMU/disciplina-del-tributo/esenzioni/, last consultation: 31 October 2024). Thus, new incentives for adopting sustainable or conservation-oriented farming practices should be implemented through alternative fiscal or regulatory tools, such as economic incentives, rather than relying solely on property tax relief. This approach aligns with Italy's CAP Strategic Plan for 2023–2027, which has introduced 34 voluntary schemes designed to compensate farmers for the additional costs and income losses incurred by adopting more environmentally and climate-friendly practices. These practices include reducing fertilizer and pesticide use, implementing biodiversity-preserving techniques, and adopting soil conservation practices [60].

National targets for ecological conservation or restoration should ideally be translated into actionable goals at lower levels of government or the local scale through negotiations with ecosystem service (ES) "sellers", such as landowners, to enhance effectiveness, as suggested by Ding et al. [61]. For instance, in British Columbia, tax incentive programs for land conservation are being explored at the local government level, where authorities have been empowered to implement tax-based conservation initiatives [56]. To achieve this goal and to appropriately tailor the initiative based on the territorial criticalities and potentialities, it is evident that local authorities need to be provided with local assessment and mapping outputs regarding the condition of ecosystem services across the territory. This can help identify and value strategic or vulnerable areas, as well as the effect of potential policies, transformative scenarios, or management regimes [62,63].

### *1.5. Downscaling Global Principles to Local Context*

In response to the call for the rapid implementation of SEEA EA, the European Commission incorporated ecosystem accounts into its extension of Regulation 691/2011 on European Environmental Accounts. This extension, expected to include data on national ecosystem conditions and some biophysical ecosystem services, was approved by the European Council and Parliament in 2023 [17]. After the approval, the European statistical community proactively worked to address methodological gaps. In 2021, Eurostat formed a Task Force to support the coordination of legal, methodological, and practical aspects critical to the rollout of SEEA EA. Italy, for instance, is adapting SEEA EA in national projects through the National Institute of Statistics (ISTAT) and the National Institute for Environmental Protection and Research (ISPRA), focusing on quantifying and valuing ecosystem services for economic analysis [17]. This initiative reflects the broader international drive needed to integrate SEEA EA into policy frameworks and recognize the importance of monetary accounting of nature. Although SEEA EA was fully published in 2021, its preliminary version and decade-long development have fostered significant progress in ecosystem service valuation and early monetization efforts [10]. These efforts, often highlighted in the scientific and technical literature, provide valuable practices and insights that are essential to adapting global principles in local contexts during the ongoing transition phase. Downscaling efforts of SEEA EA are necessarily conditioned by the national context and policies that regulate land use and conservation strategies. These policies are influenced by specific urban geographies and regional economic dynamics that strongly impact nature degradation and must be assessed with tools that quantify the loss of ecosystem services in the national context. An interesting example is the set of guidelines for the biophysical and monetary assessment of ecosystem services [64], which were employed by the Italian National System for Environmental Protection (SNPA) in 2018. These guidelines serve as the primary methodological reference for this research, which focuses on a case study within the Italian context.

Considering trends in land-use dynamics, particularly soil sealing, reversing the current practice of consuming agricultural or natural land—rather than addressing degraded areas—also requires a monetary valuation of soil ecosystem services.

Although the cost of such transformations is substantial, it remains unclear without a coherent monetary valuation of ecosystem services (ESs), which must be grounded in biophysical quantification.

Monetizing natural resources is a strategy to address nature's "economic invisibility", which is a key factor that can lead to more sustainable use and the conservation of nature through efficient resource allocation. Many ecosystem goods and services are not reflected in market prices due to their free accessibility, leading to a lack of recognition of both their benefits and the costs of their degradation [17]. A robust valuation framework at the national and regional levels is essential to integrate environmental considerations into economic decisions, promoting sustainable use and the conservation of ecosystems [17].

Furthermore, the effective management of environmental resources in planning and decision-making requires understanding spatial dynamics to address modern challenges across various sectors, from urban planning to natural resource protection. As Morya and Punia [65] emphasize, "the physical assessment of ecosystem services is not sufficient to mitigate the increasing pressure of urbanization on resources".

Urban dynamics are complex and site-specific, just as the pressures on land are driven by both economic and social factors that characterize the territory. The research focuses on the emblematic case of Italy, where land consumption and ES protection remain largely unresolved, partly due to the lack of a national regulatory framework on these issues.

### 1.6. Mapping ES Exercises: A Literature Review for the Italian Context

To better understand the current state of the art in research and the ongoing debate among scholars, the topic of ES biophysical and monetary mapping has been examined through a literature review focused specifically on Italy. This review targeted local contexts within Italy, utilizing the Scopus and Web of Science databases. The results were systematically aggregated and are presented in the following section. The research began by focusing on exercises related to the biophysical mapping of ecosystem services in local case studies from Italy. In this context and referring to the European Union's Nomenclature of Territorial Units for Statistics (NUTS) framework, which categorizes regions from the national (NUTS0) to local levels (Regulation (EC) No. 1059/2003), the Provincial and Municipal levels (NUTS3 and LAU, respectively) correspond to the sub-regional, local scale within Italy's spatial planning framework.

During the literature review, scientific papers addressing soil ESs were examined, initially focusing on those that performed spatial mapping of either biophysical or monetary aspects. Subsequently, studies assessing the simultaneous mapping of both biophysical and monetary aspects were considered. Finally, papers evaluating multiple ESs concurrently, including scenario development, were selected as relevant to the research objectives.

Marine and water ES mapping exercises were excluded as the present research focuses on quantifying soil ESs. Similarly, papers at excessively broad or narrow scales (e.g., national assessments or neighboring ones) were excluded, as they did not fit with the identified spatial categories (NUTS3 and LAU). An overview of the results is presented in Table 1, which clearly shows that only five papers were found with the same features of the current research: La Notte, 2012; Häyhä et al., 2015; Manes et al., 2016; Marino et al., 2021; Sebastiani et al., 2021 [66–70].

**Table 1.** The literature review results. Source: own elaboration by authors.

| **Search 1. Biophysical assessment and mapping exercises of ecosystem services in the Italian context**<br>Search string: (biophysical AND ecosystem AND services) AND (mapping OR quantification OR evaluation OR valuation OR assessment OR accounting) AND TITLE-ABS-KEY (mapping OR spatial) AND TITLE-ABS-KEY (Italy OR Italian). | | |
|---|---|---|
| **Reference** | **Assessment features** | **Exclusion motivation** |
| **La Notte, 2012** [66] | 1 scenario<br>1 ES | |
| **Schirpke et al., 2012** [71] | | No ES mapping |
| **Petrosillo et al., 2013** [72] | | No ES mapping |
| **Ferrari and Geneletti, 2014** [73] | | Biophysical OR monetary assessment |
| **Rova et al., 2015** [74] | | Biophysical OR monetary assessment |
| **Häyhä et al., 2015** [67] | 1 scenario<br>>1 ES | |
| **Arcidiacono et al., 2016** [75] | | Only biophysical assessment |
| **Franzese et al., 2017** [76] | | Water ecosystems |
| **Picone et al., 2017** [77] | | Water ecosystems |
| **Mancini et al., 2018** [78] | | National scale |
| **Salata et al., 2019** [79] | | No ES mapping |
| **Sacchelli and Bernetti, 2019** [80] | | No ES mapping |
| **Salizzoni et al., 2020** [81] | | No ES mapping |
| **Capriolo et al., 2020** [82] | | National scale |
| **Buonocore et al., 2020** [83] | | Water ecosystems |
| **Buonocore et al., 2020** [84] | | Water ecosystems |
| **Pacetti et al., 2020** [85] | | Water ecosystems |
| **Masiero et al., 2022** [86] | | Too-small scale (neighbourhood scale) |
| **Marino et al., 2021** [69] | 1 scenario<br>>1 ES | |
| **Di Pirro et al., 2021** [87] | | No ES mapping |
| **Zulian et al., 2022** [88] | | National scale |
| **Boschetto et al., 2023** [89] | | National scale |
| **Marino et al., 2023** [90] | | Too-wide scale (national scale) |
| **Catucci et al., 2023** [91] | | Water ecosystems |
| **Pignatti et al., 2024** [92] | | No ES mapping |
| **Search 2. Monetary assessment and mapping exercises of ecosystem services in the Italian context**<br>**Search string: (monetary AND ecosystem AND services) AND (mapping OR quantification OR evaluation OR valuation OR assessment OR accounting) AND TITLE-ABS-KEY (mapping OR spatial) AND TITLE-ABS-KEY (Italy OR Italian)** | | |
| **Reference** | **Assessment features** | **Exclusion motivation** |
| **La Notte, 2012** [66] | 1 scenario<br>1 ES | |
| **Häyhä et al., 2015** [67] | 1 scenario<br>>1 ES | |
| **Pelerosso et al., 2016** [93] | | No ES mapping |
| **Manes et al., 2016** [68] | 1 scenario<br>1 ES | |
| **Schirpke et al., 2017** [94] | | No ES mapping |
| **Franzese et al., 2017** [76] | | Water ecosystems |
| **Picone et al., 2017** [77] | | Water ecosystems |
| **Appolloni et al., 2018** [95] | | Water ecosystems |
| **Buonocore et al., 2020** [83] | | Water ecosystems |

**Table 1.** *Cont.*

| Reference | Assessment features | Exclusion motivation |
|---|---|---|
| **Capriolo et al., 2020** [82] | | Too-wide scale (national scale) |
| **Sebastiani et al., 2021** [70] | 1 scenario<br>>1 ES | |
| **Lai et al., 2021** [96] | | Only biophysical assessment |
| **Marino et al., 2021** [69] | 1 scenario<br>>1 ES | |
| **Boschetto et al., 2023** [89] | | Too-wide scale (national scale) |
| **Search 3. Biophysical and monetary assessment and mapping exercises of ecosystem services in the Italian context**<br>Search string: (monetary AND ecosystem AND services) AND (mapping OR quantification OR evaluation OR valuation OR assessment OR accounting) AND TITLE-ABS-KEY (mapping OR spatial) AND TITLR-ABS-KEY (biophysical) AND TITLE-ABS-KEY (Italy OR Italian) or (biophysical AND ecosystem AND services) AND (mapping OR quantification OR evaluation OR valuation OR assessment OR accounting) AND TITLE-ABS-KEY (mapping OR spatial) AND TITLR-ABS-KEY (monetary) AND TITLE-ABS-KEY (Italy OR Italian) | | |
| **Reference** | **Assessment features** | **Exclusion motivation** |
| **La Notte, 2012** [66] | 1 scenario<br>1 ES | |
| **Häyhä et al., 2015** [67] | 1 scenario<br>>1 ES | |
| **Franzese et al., 2017** [76] | | Water ecosystems |
| **Picone et al., 2017** [77] | | Water ecosystems |
| **Buonocore et al., 2020** [83] | | Water ecosystems |
| **Capriolo et al., 2020** [82] | | Too-wide scale (national scale) |
| **Marino et al., 2021** [69] | 1 scenario<br>>1 ES | |
| **Boschetto et al., 2023** [89] | | Too-wide scale (national scale) |

Interestingly, none of these papers involved a multi-scenario and multi-service mapping exercise focused on a sub-regional territory. The term "sub-regional", as mentioned, refers to a geographical area smaller than a region (NUTS2). It strikes a balance between being large enough to address environmental and ecological issues comprehensively yet focused enough to avoid a perspective fragmented by local administrative borders. This scale is crucial for supporting integrated and comprehensive planning and offering a wide-area vision while enabling effective planning tools. These tools can involve strategic synergies between municipalities sharing common criticalities or territorial potentialities to enhance and preserve ecosystems. Additionally, a multi-scenario approach is vital to understanding the ecosystemic effects of territorial evolution, particularly the changes in land use. Finally, soil's role in providing diverse ecosystem services highlights the trade-offs in land-use decisions, making a multi-service and holistic perspective essential. Therefore, the aim of this paper is to fill the gap in the literature by proposing a multi-scenario, multi-service mapping approach that integrates both biophysical and economic dimensions, with the goal of enhancing sustainable soil management policies at the sub-regional level.

Section 2 provides an overview of the methodology employed for the assessment and mapping of a selection of ecosystem services in the case study of the Province of Brescia. The details of the case study, along with the mapping outputs (Figures) and Tables, are presented in Section 3. In Section 4, a critical evaluation of the results is presented. This is accompanied by their contextualization, drawing on dynamic information regarding the variations in ecosystem services over the specified time frame. Finally, Section 5 outlines the limitations and strengths of the analysis, as well as potential directions for further developments and applications of the research.

## 2. Materials and Methods

### 2.1. Ecosystem Services

The ecosystem services chosen for analysis in the Province of Brescia are listed in Table 2, with reference to the MEA classes and denominations [4].

**Table 2.** ES denomination and classification. Source: own denomination and MEA's categories (2005).

| ES Common Denomination | MEA ES Class and Denomination |
|---|---|
| Carbon storage and sequestration (CSS) | Regulating. Climate regulation |
| Habitat quality (HQ) | Supporting. Provisioning habitats |
| Crop pollination (CP) | Regulating. Pollination |
| Wood provision (WP) | Provisioning. Food and fibre |
| Particulate removal (PR) | Regulating. Air quality maintenance |
| Hydrological regime regulation (HRR) | Regulating. Water regulation |

These services can be included in the already cited category of soil ES (hereinafter simply referred to as ES), which should be assessed on the basis of soil properties like carbon content, nutrient cycling, and moisture retention. However, as is common in the literature, this analysis uses Land Use and Land Cover (LULC) as a proxy [8].

It is necessary to specify, in advance, that certain ecosystem services are more relevant to specific contexts: some are better suited to urban environments, such as microclimate regulation, while others are more applicable to rural or non-urban settings, like timber production. In particular, urban development relies heavily on surrounding rural areas for essential resources such as food, water, and raw materials [97]. However, given their higher population and activity density, the demand for ecosystem services is stronger in urban contexts for regulation services, whose supply is spatially constrained to the location where the demand exists (consider, for example, the service of hydrogeological regulation). Effective policies should focus on maintaining these linkages and protecting rural ecosystems to ensure equitable development for both areas [97]. As an initial analytical experiment, this research primarily focuses on the non-urban context. Notably, due to this focus, some ecosystem services have been excluded from this analysis. Microclimate regulation, which relates to the mitigation of the "urban heat island" effect caused by the accumulation of heat on artificial surfaces, was not examined as it primarily pertains to urban contexts, whereas this thesis focuses on rural areas. Similarly, water purification processes, particularly relevant near urban settlements where water quality is critical, were excluded as follows: this service, which involves the filtration of water as it infiltrates the soil, is more appropriately studied at broader geographical scales, such as watershed areas, rather than within the administrative boundaries of a province.

In any case, the justification for selecting these ecosystem services for the present analysis is closely tied to the territorial characteristics of the case study, as will be made explicit in Section 3: there, specific ecosystem services are presented in relation to their relevance within the context of the Province of Brescia.

For all ecosystem services, the analysis was conducted both in biophysical and monetary terms and in two distinct temporal scenarios—except for the pollination service, which was evaluated solely in monetary terms under a single scenario for reasons that will be discussed below. Biophysical quantification provides a tangible measure of ecosystem services in physical units, while economic quantification converts these benefits into monetary values, providing an estimate of the economic contribution of these services to society. As outlined in the SEEA EA guidelines [10], monetary estimates of ecosystem services are typically derived by assigning prices to individual ecosystem services and multiplying these prices by the physical quantities recorded in the ecosystem services flow account. In this analysis, however, habitat quality and pollination services follow a different methodology. The decision to use a dual scenario enhances the static and spatial analysis of ESs and trade-offs at specific moments, as well as their changes and evolution in response to land transformations. As will be clarified further below, due to modeling, parameter, and data restrictions, it was not possible to analyze ESs under identical temporal scenarios. Some scenarios refer to 2006–2012 (CSS, WP), while others refer to 2012–2018 (HQ, CP, PR, HRR).

### 2.2. *Biophysical Assessment of Ecosystem Services*

The methodology applied for the biophysical assessment of the chosen ES is hybrid, focusing on manageable models and tools that are not overly sector-specific. The application of such models and tools requires a multidisciplinary approach involving experts from various fields. The analysis was partially conducted in alignment with the Italian SNPA guidelines for the ecosystem services assessment, titled "Mapping and Impact Assessment of Land Consumption on Ecosystem Services: Methodological Proposals for the Land Consumption Report" [64]. This was performed using Q-GIS 3.34.5 software and the InVEST model. For the biophysical assessment of the remaining three services, the methodology from the LIFE+ "Making Good Natura" EU project was used [98] (Figure 1).

#### 2.2.1. The InVEST Model

InVEST is a suite of free, open-source software models developed by Stanford University as part of the Natural Capital Project (https://naturalcapitalproject.stanford.edu/software/InVEST, last consultation: 26 October 2024), specifically designed for territorial and town planning evaluation and the comparison of the impacts of land-use alternatives. It is also used by various regional authorities in Italy, such as the Lombardy Region [99,100]. It is particularly useful for assessing soil ecosystem services from a spatial planning perspective. It analyzes ESs through various sub-models, providing biophysical results (e.g., carbon sequestration in tons) and monetary values (e.g., net present value of carbon). In this research, InVEST was used for biophysical aspects, while monetary quantification was evaluated separately. The model requires specific calibration and high-quality data to function effectively [100], and the availability of detailed, open-source data is a major challenge at the planning level. Due to lower-resolution data, InVEST is better suited for regional planning, with sub-regional applications (such as the ones conducted here) needing additional calibration effort.

In this case, the input data and parameters required depend on the ecosystem service being analyzed and, therefore, on the InVEST sub-model used. Typically, a Land Use and Land Cover (LULC) map and calibration parameter tables are necessary. For the ESs analyzed in this paper, LULC maps were selected from the Corine Land Cover (CLC) database as referred to for the years 2006, 2012, and 2018, and from which only land cover (LC) information was used for the services analyzed. Parameters were set according to exercises in the literature, as close as possible to the spatial and morphological context of the case study. The ecosystem services selected for analysis in the Province of Brescia using InVEST software include (Figure 1) the following:

- Carbon sequestration and storage (CSS);
- Habitat quality (HQ);
- Crop pollination (CP).

#### 2.2.2. LIFE+ "Making Good Natura" EU Project—LIFE11 ENV/IT/000168 ("Project MGN")

For the biophysical quantification of ecosystem services that were not analyzed using InVEST, the ES-specific methodology from the "Making Public Goods Provision the Core

Business of Natura 2000" (LIFE+ MGN) project was utilized [98]. Launched in 2013 and supported by the European Commission under the LIFE+ program, the LIFE+ MGN primarily aims to enhance the governance of agro-forestry sites within this network by identifying key ecosystem services (ESs), conducting the biophysical quantification and economic valuation of various ES, developing and implementing PES frameworks, and evaluating site management effectiveness [101].

The project proposes a simplified (parametric) methodology for analyzing specific ecosystem services—in this research, used only for biophysical assessment—across 21 Natura 2000 sites. Although the findings pertain solely to the analyzed sites, the methodology (detailed in [98]) can be replicated in other contexts.

The ecosystem services examined using the LIFE+ MGN methodology include (Figure 1) the following:

- Wood provision (WP);
- Particulate removal (PR);
- Hydrological regime regulation (HRR).

### 2.3. Monetary Assessment of Ecosystem Services

The cited SNPA guidelines [64] also include an approach for the monetary assessment of ESs. The methodology proposed differs from service to service and relies on different parameters in the literature, sometimes with a very high range of values. The monetary assessment of ecosystem services in this analysis is aligned with it, as reported synthetically in Figure 1.

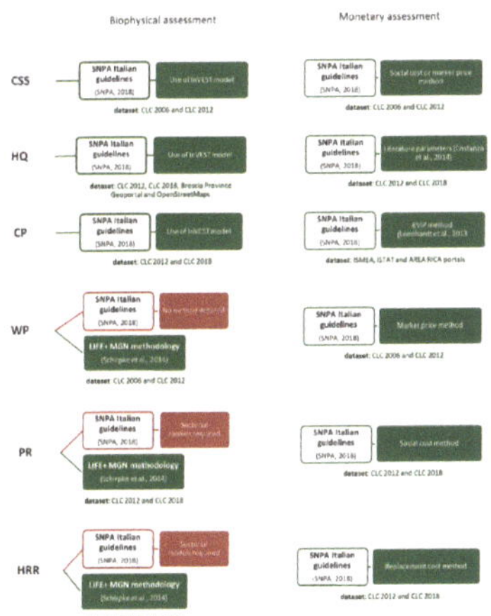


**Figure 1.** Sources and methodology followed for the biophysical and monetary assessment of six ecosystem services. The assessment path chosen is highlighted in green, while discarded approaches are shown in red. Source: elaborations by the authors following [64,98,102].

As explained in the following sections, this research, chosen chose to refine the method, narrowing the range on the basis of statistical evaluations and re-evaluating the coefficients by applying the monetary revaluation coefficient for the year 2024 using ISTAT (https://rivaluta.istat.it/, last consultation: 28 October 2024). Additionally, where deemed necessary, the proposed parameter references were also updated.

Please note that all results reported in the tables, both in Section 3 and in Appendix A, have been rounded to the nearest whole unit.

# 3. Results

## 3.1. Study Area

The case study of this biophysical and monetary assessment exercise is the Province of Brescia (NUTS3), located in Lombardy, Northern Italy (Figure 2). The Province of Brescia provides an ideal case study as it encompasses diverse territorial systems, from plains to lakes and mountains. The northern part features three main valleys—Valle Camonica, Valle Trompia, and Valle Sabbia—which have a distinct morphology compared to the southern plains. The province is also characterized by three main lakes—Lake Garda, Lake Iseo, and Lake Idro—further adding to its geographical complexity. Addressing this complexity requires an integrated planning and management approach, particularly when defining the relationships and priorities between different systems. These features support a wide range of ESs, making this territory an ideal case study for a comprehensive and integrated ES assessment. Agriculture is a key economic activity in the area, with notable products such as olives from Lake Garda and wines from renowned regions such as Franciacorta and Lugana. The area holds a variety of tourism-related services, ranging from lakes, mountains, and valleys to agricultural productivity, all while retaining cultural and historical significance in Brescia's landscape. Finally, northern forests play a vital role in carbon sequestration and timber production while serving as essential biodiversity hotspots. The areas near major rivers and wetlands hold significant ecological importance but are increasingly vulnerable to hydrogeological and natural risks, like those in mountainous zones. Environmental and socioeconomic pressures in Brescia mirror challenges in other regions of Italy. For example, the Province of Brescia is marked by intense industrial activity and heavy traffic, experiencing high levels of PM10, ozone, and GHG emissions, exacerbated by the Po Valley's stagnant air conditions and the Alpine range acting as a barrier [103,104]. Urbanization has significantly increased land consumption, particularly in the plains, emphasizing the urgent need for strategic analysis to safeguard critical ecosystems and natural resources. These aspects also specifically justify the selection of the ecosystem services analyzed in this study. Given these characteristics, using Brescia as a case study is both significant and comprehensive, as it provides valuable insights for developing planning frameworks applicable to other regions facing similar territorial complexities and diverse environmental, economic, and social dynamics. It can be stated that the varied features of Brescia Province reflect the heterogeneity of the Italian landscape, making it a microcosm for broader national challenges and enhancing the application of the method to other contexts.

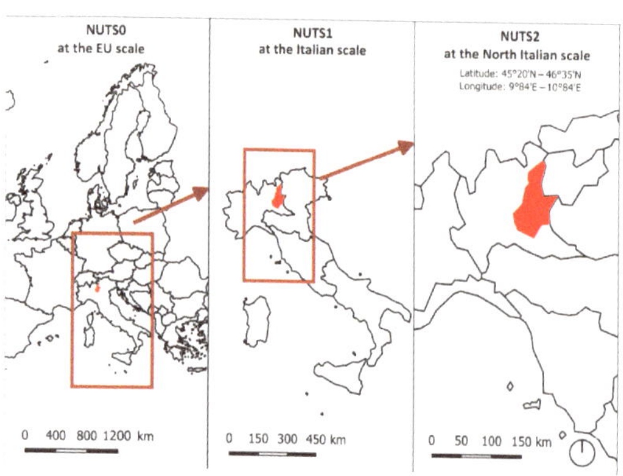


**Figure 2.** Administrative framework and spatial collocation of the case study (Brescia Province, NUT3). Source: own elaboration by authors already in [105].

## 3.2. ES Assessment

### 3.2.1. Carbon Storage and Sequestration

The forestry sector holds one of the highest potentials for cost-effective $CO_2$ reduction compared to other mitigation activities by absorbing carbon dioxide from the atmosphere and storing it in woody biomass (both aboveground and belowground), litter, and soil [106,107]. The factors influencing the amount of carbon stored and the future uptake potential across different compartments vary, with climate and soil consumption being predominant, particularly in areas with forest cover or high natural integrity [108,109]

In the case of the Province of Brescia, approximately one-third of the area is covered by forests, mainly concentrated in the northern valleys of Valle Camonica, Valle Trompia, and Valle Sabbia, making this ecosystem service especially relevant for the region.

#### Biophysical Assessment

Carbon storage and sequestration biophysical analysis were performed using the InVEST software, as proposed in [64], specifically the "Carbon storage and sequestration" sub-model. InVEST estimates the carbon content of various land-use categories based on storage parameters (tC/ha) for the four primary carbon pools recognized by the Intergovernmental Panel on Climate Change [110]: aboveground biomass, belowground biomass, soil, and dead organic matter. For the carbon storage assessment, the software requires input to an LC map for each scenario analyzed and carbon storage parameters for each LC category. Carbon sequestration is then calculated from the differences in LC among reference years, namely from differences in carbon storage. Carbon stored in permanent crops and tree plantations was not considered.

For this analysis, the Corine Land Cover (CLC) 2012 dataset was used, rasterized at a 10-meter resolution using QGIS software; the 2018 mapping was avoided due to the absence of a V-level classification for forestry land uses, which was considered necessary in this case for a precise ES assessment. Carbon storage parameters (tC/ha) for forest types have been obtained from the National Inventory of Forests and Forest Carbon Pools (INFC), for which the correspondence between CLC categories was established based on Marchetti et al.'s work [111] (see Table A1 in Appendix A). More precisely, the values for aboveground biomass were derived from the INFC 2015 database, while belowground biomass was not calculated due to a lack of data in the INFC databases. Soil carbon values were calculated by subtracting the litter carbon content in the INFC 2005 database (declared as a quite static parameter and, thus, not uploaded in the 2015 version) from the total soil carbon in the INFC 2015 (litter carbon + soil carbon). Similarly, dead organic matter values were calculated by summing litter and fine necro-mass (not uploaded in the 2015 version) with coarse deadwood in the INFC 2015 database. See Table A2 in Appendix A for the detailed parameters used for each carbon pool and forest type.

The same analysis was performed for 2006, using the CLC 2006 dataset and the same parameters as before, with the aim of isolating the effects of land-use change on carbon storage and sequestration services, which is the purpose of this analysis.

Carbon storage and sequestration biophysical mapping outputs are reported in Figure 3 for both the years considered, together with the variation in time, which in this case represents the carbon sequestration service.

#### Monetary Assessment

The monetary valuation of carbon sequestration and storage can be performed through two main approaches, as demonstrated in SNPA guidelines [64]: the social cost and the market value of emission permits.

The social cost approach estimates the global damage avoided by carbon sequestration. In this paper, it was assumed that the unit monetary value of EUR 101.85/tC proposed in the Interagency Working Group on Social Cost of Greenhouse Gases, United States Government [112], would be accurate, adjusted for inflation to EUR 145.74/ton of carbon for 2024 (through Rivaluta ISTAT, https://rivaluta.istat.it/, last consultation: 28 October 2024).

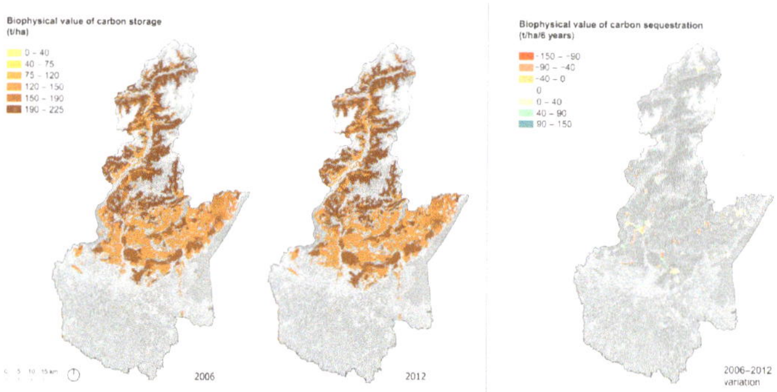


**Figure 3.** Maps of the biophysical assessment of carbon storage in 2012 and 2006 and carbon sequestration in 2006–2012. Source: own elaboration by authors.

For the market value approach, the price of carbon emission permits in the Italian voluntary market was used, with a current unit monetary value of EUR 28.00/tC [113].

The total economic value for carbon storage and sequestration (EUR) is calculated by multiplying the total stored or sequestered carbon obtained by the biophysical assessment by these unit monetary values.

In Figure 4, a map of the unit monetary values (EUR/ha) of carbon storage in 2012 with the social cost method is proposed, together with the results for the monetary assessment using the LC typology with both methods. See Table A3 in Appendix A for more detailed information about the parameters and results.

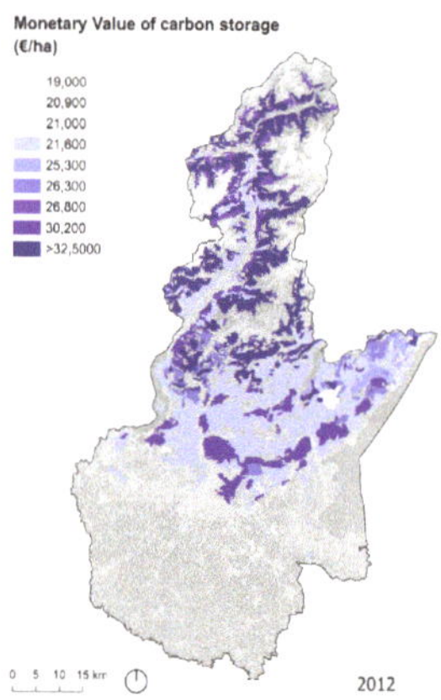


| CLC−Code | Monetary value (EUR) | | | |
|---|---|---|---|---|
| | Social cost | | Market price | |
| | 2006 | 2012 | 2006 | 2012 |
| 3111<br>31311 | 0 | 0 | 0 | 0 |
| 3112<br>31312 | 28,178,392 | 28,690,726 | 5,413,716 | 5,512,147 |
| 3116<br>31316 | 2,599,637 | 4,138,623 | 499,450 | 795,124 |
| 3117<br>31317 | 2,429,369 | 2,219,941 | 466,738 | 426,502 |
| 3113<br>31313 | 1,350,367,087 | 1,316,529,931 | 259,436,520 | 252,935,626 |
| 3122<br>31322 | 121,954,868 | 110,717,221 | 23,430,330 | 21,271,320 |
| 3115<br>31315 | 222,101,202 | 246,927,355 | 42,670,740 | 47,440,414 |
| 3124<br>31324 | 155,095,925 | 151,788,574 | 29,797,488 | 29,162,070 |
| 3114<br>31314 | 700,830,111 | 653,458,796 | 134,645,554 | 125,544,437 |
| 3123<br>31323 | 1,496,008,421 | 1,514,695,932 | 287,417,564 | 291,007,864 |
| 3121<br>31321 | 0 | 0 | 0 | 0 |
| Total | 4,079,565,011 | 4,029,167,098 | 783,778,100 | 774,095,504 |

**Figure 4.** Mapping and results of the monetary assessment of carbon storage and sequestration in 2006 and 2012. The monetary assessment of carbon sequestration can be derived by subtracting the values between the two years. The map on the left shows the unitary monetary values related to LC in 2012, obtained using the social cost method. Source: own elaborations.

### 3.2.2. Habitat Quality

Above other pressure forms, soil sealing, especially through road construction, poses significant threats not only by reducing habitat areas but also by creating barriers to species migration and movement [114].

In the case of Brescia Province, a region with a rich variety of ecosystems, including agricultural lands, forests, and wetlands, all of which are vital for local biodiversity, the increasing pressure from urbanization and infrastructure construction has threatened to fragment habitats and diminish their ecological quality, jeopardizing the integrity of the Ecological Network [115].

#### Biophysical Assessment

The assessment of the habitat quality ecosystem service was conducted using the "Habitat quality" sub-model within the InVEST software suite, as suggested in [64]. The model operates under the premise that regions with higher habitat quality can support a greater diversity of native species; conversely, reductions in both habitat extent and quality lead to diminished species persistence [116]. InVEST calculates habitat quality by assessing the capacity of land cover types to sustain life and by evaluating the impact of different threats (such as urbanization or road networks), and habitat sensitivity to those threats. The model incorporates different spatial data layers to map these threats and quantify their influence on habitat integrity. The primary input required is a raster map, with the Corine Land Cover (CLC) 2018 dataset used in this analysis, rasterized at a 10-meter resolution using QGIS software. Two additional tables—the "sensitivity table" and the "threat table"—are essential for running the model. The sensitivity table assigns habitat values on a scale from 0 (non-habitable) to 1 (optimal habitat) while also indicating each habitat type's vulnerability to different threats. These parameters were sourced from the work of Sallustio et al. [117], which provides a comprehensive analysis of habitat values across Italy. The habitat types were previously associated with the LC map starting from the description of each field, seeking a logical link between the various categories (see Table A4 in Appendix A; for the sensitivity table, see Table A5). Simultaneously, the threat table (see Table A6 in Appendix A) details the characteristics of each threat, including its spatial extent (the radius of influence in kilometers), its weight or impact on habitat quality, and its decay pattern (whether effects decrease exponentially or linearly with distance). In particular, the radius of influence and the impact weights have on the habitat were obtained for each threat and habitat type identified by Sallustio et al. [117], while the decay pattern was assumed to be linear for safety. Each threat is required as raster input data; in this analysis, intensive agriculture, extensive agriculture, the soil consumed, and roads were considered, as outlined by Sallustio et al. [117]. Specifically, intensive and extensive agriculture, as well as the soil consumed, were obtained from the Brescia Province Geoportal, while road and railway network data were extracted from OpenStreetMap (OSM) and classified as proposed by Sallustio et al. [117] in order to maintain consistent reference parameters. A key parameter in the model is the semi-saturation constant (k), which modulates how habitat quality decreases as degradation intensifies. A low k value (0.05, the default in InVEST) means that even small increases in degradation will significantly reduce habitat quality. The same analysis was performed for 2012, using the CLC 2012 dataset and the same parameters as before, with the aim of isolating the effects of land-use change on habitat quality service.

The output consists of two raster maps, shown in Figures 5 and 6 for the two years considered, along with the period variation: one representing habitat quality and the other showing the level of degradation. These outputs provide relative values from 0 to 1, allowing for comparative analysis between the different locations but not yielding direct biophysical measurements. Further processing would be required to translate these results into more specific ecological metrics.

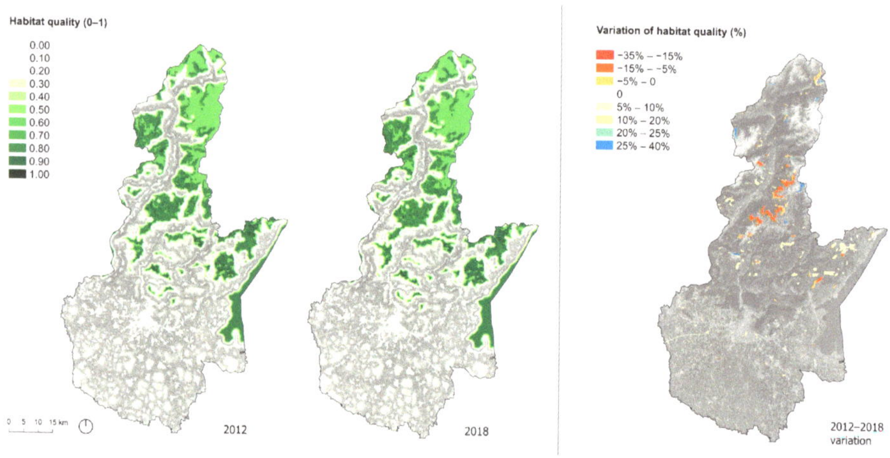


**Figure 5.** Maps of the biophysical assessment of habitat quality in 2012 and 2018. Source: own elaboration by authors.

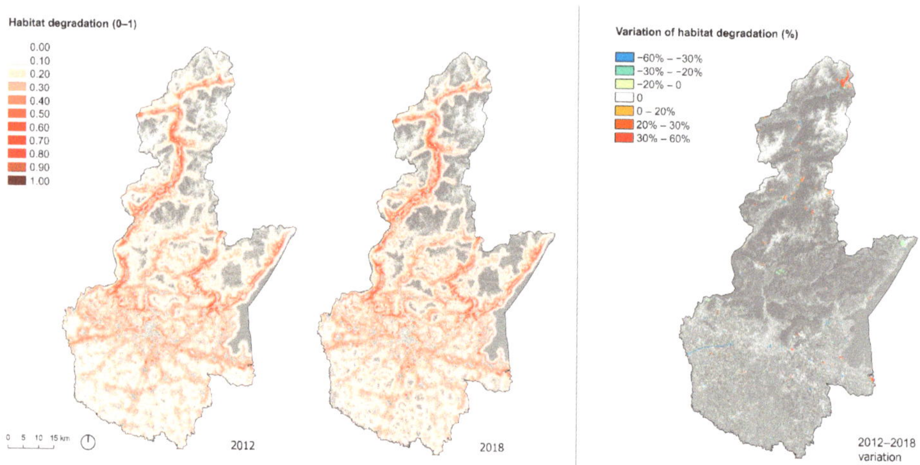


**Figure 6.** Maps of habitat degradation in 2012 and 2018. Source: own elaboration by authors.

Monetary Assessment

To assess the monetary value of habitat quality, the methodology proposed in the SNPA guidelines [64] follows the approach developed by Costanza et al. [5,7], which estimates economic value based on various parameters related to different habitat types for certain ecosystems. In the SNPA proposal, a range of coefficients was derived for the missing ecosystems, using the surfaces across the Italian territory as weights [64]. In the present analysis, the complete list of coefficients thus obtained (in EUR/ha) was adjusted for inflation using the ISTAT revaluation coefficient for 2018–2024. To calculate the total monetary value of the service, the unit monetary value (in EUR/hectare) is multiplied by the area of the specific land cover type, expressed in hectares.

In Figure 7, a map of the unit monetary values (EUR/ha) of habitat quality in 2018 is provided, together with the results for the monetary assessment by LC typology. See Table A7 in Appendix A for more detailed information about the parameters and results.

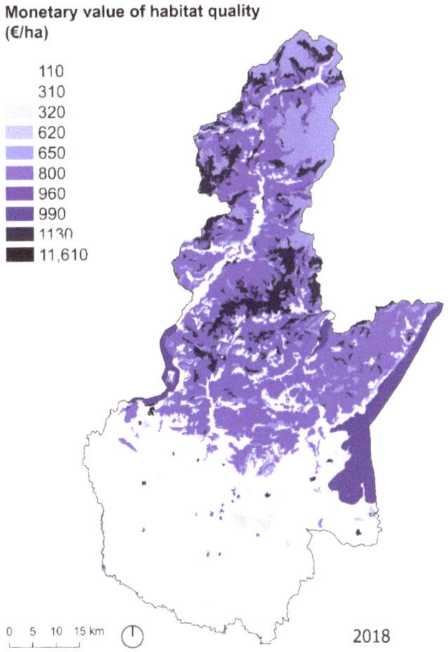


| CLC Code | Habitat category | Monetary value (EUR) 2012 | 2018 |
|---|---|---|---|
| 111 | Buildings and other artificial areas or impervious soils | 113,133 | 113,133 |
| 112 | Buildings and other artificial areas or impervious soils | 2,968,272 | 3,015,384 |
| 121 | Buildings and other artificial areas or impervious soils | 937,006 | 982,302 |
| 122 | Buildings and other artificial areas or impervious soils | 10,469 | 34,506 |
| 124 | Open urban areas | 295,812 | 297,735 |
| 131 | Open urban areas | 543,231 | 588,099 |
| 133 | Open urban areas | 31,729 | 29,165 |
| 141 | Grasslands | 91,676 | 61,117 |
| 142 | Grasslands | 490,069 | 837,532 |
| 211 | Intensive agricultural lands | 44,673,671 | 4,440,085 |
| 221 | Extensive agricultural lands | 340,716 | 430,834 |
| 222 | Extensive agricultural lands | 0 | 15,431 |
| 223 | Extensive agricultural lands | 321,582 | 315,410 |
| 231 | Grasslands | 5,120,263 | 5,033,115 |
| 241 | Extensive agricultural lands | 33,331 | 33,331 |
| 242 | Intensive agricultural lands | 3,101,014 | 2,979,417 |
| 243 | Intensive agricultural lands | 6,904,447 | 6,938,703 |
| 311 | Broadleaf forests | 61,230,302 | 69,942,130 |
| 312 | Conifer forests | 32,470,262 | 36,730,145 |
| 313 | (Broadleaf forests + Conifer forests)/2 | 30,441,975 | 16,636,931 |
| 321 | Grasslands | 30,616,322 | 31,271,634 |
| 322 | Shrublands | 3,588,206 | 4,816,965 |
| 323 | Shrublands | 42,305 | 84,609 |
| 324 | Inland unvegetated or sparsely vegetated areas | 11,032,512 | 10,555,932 |
| 332 | Inland unvegetated or sparsely vegetated areas | 9,576,004 | 8,682,252 |
| 333 | Inland unvegetated or sparsely vegetated areas | 7,499,941 | 8,012,428 |
| 334 | Inland unvegetated or sparsely vegetated areas | 0 | 34,601 |
| 335 | Inland unvegetated or sparsely vegetated areas | 1,282,850 | 1,261,959 |
| 411 | Wetlands | 2,995,148 | 2,704,920 |
| 511 | Water bodies | 66,994 | 66,994 |
| 512 | Water bodies | 23,543,563 | 23,573,120 |
| Total | | 280,362,804 | 280,480,684 |

**Figure 7.** Mapping and results of the monetary assessment of habitat quality in 2012 and 2018. The map on the left shows unit monetary values related to LC in 2018. Source: own elaborations by authors.

### 3.2.3. Crop Pollination

A European assessment indicates that approximately 9.2% of bee species are at risk of extinction due to land consumption, agricultural intensification, the widespread use of pesticides, herbicides, and fertilizers, as well as habitat fragmentation, which affects the pollination network [117,118]. The enhancement and protection of this service, which is so essential for environmental balance and human activities, has become even more important today.

This ecosystem service is particularly crucial for the Province of Brescia, where agricultural areas cover about half of its total surface, mainly in the southern region, which corresponds to the Po Valley. However, this area faces significant land consumption, which threatens the integrity of its pollination ecosystem service.

#### Biophysical Assessment

To assess the biophysical value of the crop pollination ecosystem service, an analysis was conducted using the InVEST software, specifically the "Crop pollination" sub-model, as described in [64]. The software requires raster input data, for which the CLC 2018 dataset was used, and was rasterized in QGIS with a grid size of 10 m. The model also required two CSV tables as the input: a "biophysical table" and a "pollinator table". The biophysical table includes a nesting availability index and a flower abundance index for each LC class. The required parameters were obtained from the "JRC ESTIMAP: Ecosystem services mapping at European scale" report [119], which provides general parameters for Europe. It is considered a suitable reference for this case study. The pollinator table maps the characteristics of the analyzed pollinator species, in this case only including the "solitary bee", as in [119]. Parameters include the following: a nesting suitability index (set to one for a generic substrate); a foraging activity index calculated from temperature and solar radiation; the medium distance traveled by the analyzed species (set to 200 m, as in [119]); and a relative abundance value (set to one since only one pollinator type was analyzed). Parameters were obtained with the methodology reported in [119], using values of T (temperature) = 12.59 °C and R (radiation) = 177.9 W/m$^2$ obtained from the ARPA database of the Lombardy Region using the average annual daily mean values from 1 January 2023 to 1 January 2024 (https://www.arpalombardia.it/temi-ambientali/meteo-e-clima/form-richiesta-dati/, last consultation: 8 January 2024).

The same analysis was performed for 2012, using the CLC 2012 dataset and the same parameters as before, with the aim of isolating the effects of land-use change on the crop pollination service.

The software generated two output maps, reported in Figures 8 and 9 for the two years considered, along with the period variation. One map represents the abundance of the generic bee, while the other indicates its potential supply based on relative abundance, habitat suitability, and the floral resources available, taking into account the bee's travel capacity. The values in these maps range from 0 to 1, enabling comparative analysis between pixels, but they do not provide a direct biophysical analysis without further processing.

#### Monetary Assessment

For the monetary assessment of pollination services, as outlined in the SNPA guidelines [64], this methodology adopted the approach proposed by Leonhardt et al. [102], specifically the calculation of the overall economic value of pollination (EVIP) based on the global valuation of the pollination service in relation to the agricultural production value. The EVIP was determined using the methodology described in [102], deriving the crop-dependency ratio on pollinators from Klein et al. [120].

For the valuation of the parameter $Q_{ict}$, ISTAT data for Brescia Province were used (https://esploradati.istat.it/databrowser/#/it/dw/categries/IT1,Z1000AGR,1.0/AGR_CRP/DCSP_COLTIVAZIONI/IT1,101_1015_DF_DCSP_COLTIVAZIONI_1,1.0, last consultation: 28 October 2024). Concerning $P_{ict}$, data were obtained from the ISMEA (Istituto di Servizi per il Mercato Agricolo Alimentare) and the AREA Rica (Analisi dei Risultati Economici Aziendali) portals.

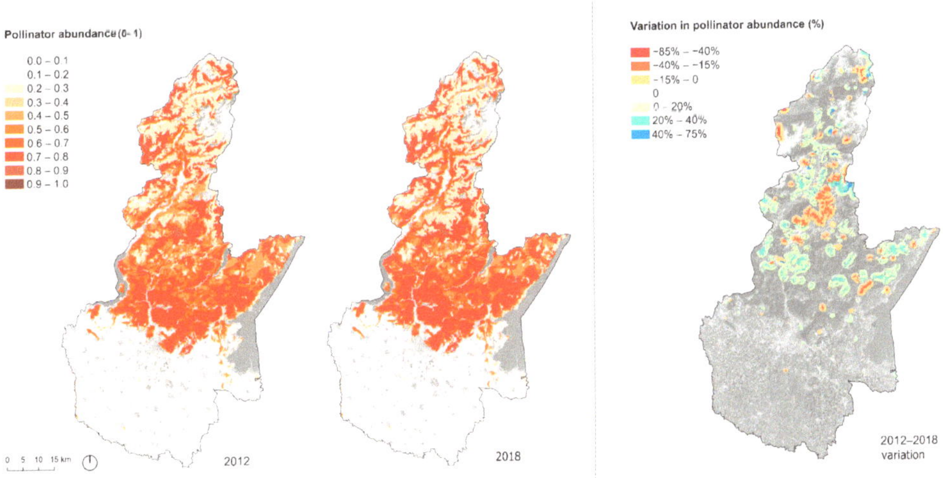


**Figure 8.** Maps of pollinator abundance in 2012 and 2018. Source: own elaboration by authors.

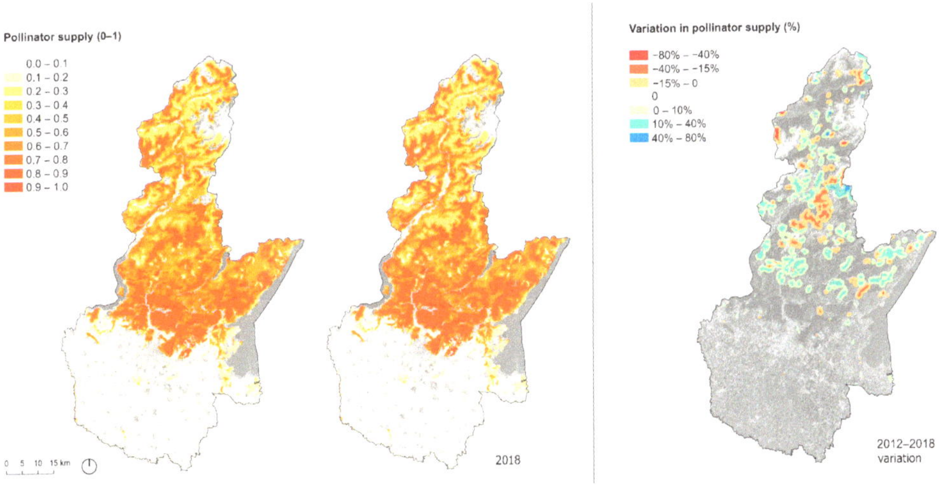


**Figure 9.** Maps of the biophysical assessment of crop pollination in 2012 and 2018. Source: own elaboration by authors.

In this case, it was not possible to generate a map for the monetary valuation of the service, as there is no agricultural land-use map available for Brescia Province. The CLC dataset only provides broad land-use categories without specifying individual crop types, which can vary frequently across the region. Consequently, the monetary valuation could not be geo-referenced. However, Table 3 presents the results of the monetary assessment performed by cultivation type. For more detailed information about the parameters and results, see Table A8 in Appendix A.

**Table 3.** Monetary value of crop pollination for crops dependent on the service based on the 2023 production in Brescia Province. Source: own elaboration by authors from ISMEA, AREA Rica, and ISTAT databases.

| Crop Type | Total Production (t/Year) | Unit Monetary Value for Crop Production (EUR/t) | Dependency Ratio | Monetary Value for Crop Pollination (EUR/Year) |
|---|---|---|---|---|
| Apple | 1811 | 729 | 0.65 | 858,142 |
| Chestnut | 1075.9 | 1920 | 0.25 | 516,480 |
| Cocumber | 205 | 291 | 0.65 | 38,776 |
| Green Bean | 2474.5 | 1097 | 0.05 | 135,754 |
| Lemon | 8 | 760 | 0.05 | 304 |
| Melon | 1845 | 597 | 0.95 | 1,046,392 |
| Peach | 788 | 512 | 0.65 | 262,246 |
| Pear | 185 | 1423 | 0.65 | 171,116 |
| Pepper | 234 | 493 | 0.05 | 5768 |
| Strawberry | 371 | 3023 | 0.25 | 280,383 |
| Plum | 72 | 308 | 0.65 | 14,414 |
| Tomato | 44,886 | 676 | 0.05 | 1,517,147 |
| Watermelon | 1280 | 164 | 0.95 | 199,424 |
| Zucchini | 8087.8 | 503 | 0.95 | 3,864,755 |
| **Total** | | | | 8,911,046 |

#### 3.2.4. Wood Provision

To evaluate this ecosystem service, the annual biomass of timber produced in a specific region must be estimated [98]. Land-use changes in forested areas can entirely eliminate this service, while other influencing factors include climate conditions, management practices, and the prevalence of diseases like the Bosforo bacterium, which has affected Northern Italy. The assessment of wood provision is particularly significant for Brescia Province, where forests cover approximately one-third of the territory, primarily in the northern region. This area encompasses three major valleys: Valle Camonica, Valle Trompia, and Valle Sabbia.

Biophysical Assessment

The "LIFE+ MGN" parametric methodology was applied to biophysically assess wood provision ecosystem services [98]. The annual biomass increment values were obtained from the INFC 2015 database, which provides specific growth rates for various forest types based on CLC classifications. As with carbon storage and sequestration service, the CLC 2012 dataset was used and rasterized at a 10-meter resolution using QGIS software. A correspondence between CLC classes and INFC classifications was established based on Marchetti et al.'s work [111]. Similar to the previous analysis, only tall tree forest areas were considered. The annual biomass increment values for each forest type were sourced from the INFC 2015 database. This approach estimates the wood provision service by calculating the annual production of forest biomass that can be sustainably harvested without distinguishing between different timber product types. The assessment focused on the volume of biomass produced annually within the limits of natural regeneration [98].

The same analysis was performed for 2006, using the CLC 2006 dataset and identical parameters, with the goal of isolating the effects of land-use change on the wood provision service.

Figure 10 reports the biophysical mapping results of wood provision in 2006 and 2012, along with the variation in the ES during the period.

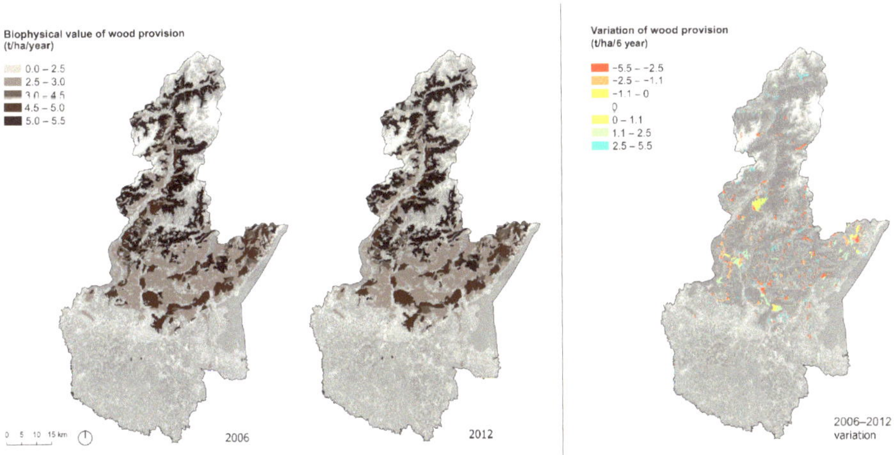


**Figure 10.** Maps of the biophysical assessment of wood provision in 2006 and 2012. Source: own elaborations by authors.

See Table A9 in Appendix A for further details about the parameters used.

Monetary Assessment

For the monetary assessment of wood provision, the SNPA guidelines [64] recommend using market prices for the assessed biomass volumes.

In the absence of specific data for the Province of Brescia, data from the ISPAT portal for the Trentino Alto Adige region were utilized. This portal provides valuable insights into regional timber utilization percentages and average prices based on the latest update from 2021. According to these findings, timber products are primarily used for construction (76.5%), while the remaining 23.5% is allocated for firewood. The respective average prices are EUR 63/$m^3$ for construction timber and EUR 50/$m^3$ for firewood. These prices are weighted averages derived from the production of softwood and hardwood in Trentino Alto Adige. By calculating a weighted average of these results, the overall price was estimated at EUR 60/$m^3$, which was adjusted for inflation to approximately EUR 70/$m^3$. The monetary value of wood production was determined by multiplying the wood increment coefficient (at the basis of the biophysical assessment) by the area of the respective land cover type, yielding wood provision values in tons or cubic meters. These results were then multiplied by the unit monetary value (EUR/t or EUR/$m^3$) to obtain the total economic value in euros. Figure 11 presents a map of the unit monetary values (EUR/ha/year) of the wood provision in 2012, alongside the results of the monetary assessment by LC typology. See Table A9 in Appendix A for more detailed information about the parameters and results.

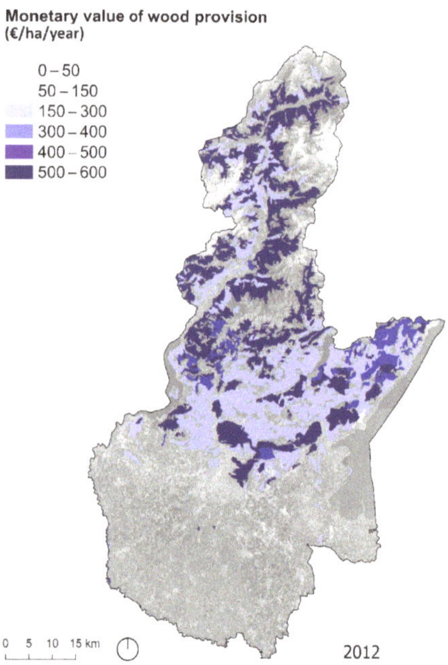


| CLC—Code | Monetary value (EUR/year) 2006 | 2012 |
|---|---|---|
| **3124** | 1,466,710 | 1,435,700 |
| **31323** | 7,905,310 | 7,285,670 |
| **3123** | 17,942,820 | 18,887,960 |
| **3122** | 280,700 | 549,500 |
| **31322** | 2,134,720 | 1,642,970 |
| **3121** | 0 | 0 |
| **31321** | 0 | 0 |
| **3125** | 0 | 0 |
| **31325** | 0 | 0 |
| **3115** | 2,995,510 | 3,124,520 |
| **31315** | 782,320 | 1,076,320 |
| **3112** | 864,920 | 881,230 |
| **31312** | 0 | 0 |
| **3114** | 11,152,610 | 10,132,290 |
| **31314** | 1,160,670 | 1,348,060 |
| **3113** | 14,769,930 | 13,706,560 |
| **31313** | 3,188,010 | 3,801,910 |
| **3116** | 51,660 | 81,760 |
| **31316** | 0 | 0 |
| **3117** | 52,850 | 48,720 |
| **31317** | 0 | 0 |
| **3111** | 0 | 0 |
| **31311** | 0 | 0 |
| Total | 64,748,740 | 64,003,170 |

**Figure 11.** Mapping and results of the monetary assessment of wood provision in 2006 and 2012. The map on the left shows unitary monetary values related to LC in 2012. Source: own elaborations.

### 3.2.5. Particulate Removal

Particulate and ozone removal by forests is a vital ecosystem service, particularly in the spatial context of this analysis, as exposure to air pollutants is the leading environmental risk factor in Europe [121]. Italy, in particular, has the highest estimated number of premature deaths attributed to air pollution [122].

This air quality service plays a crucial role in maintaining urban air quality and is equally significant for broader regions like Brescia Province. Forest ecosystems in this province are extensive, covering about one-third of its surface, predominantly in the northern areas, including Valle Camonica, Valle Trompia, and Valle Sabbia. These forests are particularly valuable in mitigating pollution levels, which can be notably high in the region.

#### Biophysical Assessment

The capacity of forest ecosystems to remove pollutants was assessed using the "LIFE+ MGN" parametric methodology, which estimates PM10 removal based on average capture coefficients specific to different vegetation types [98]. The CLC 2018 dataset was utilized for the LC map, rasterized at a 10-meter resolution using QGIS software. PM10 sequestration values were assigned to various forest types following the parameters provided by Schirpke et al. [98]. Specifically, broadleaf forests were assigned a value of 160 kg/ha/year, while the coniferous forest was assigned a value of 490 kg/ha/year. Mixed forests were assigned a value of 325 kg/ha/year, calculated as the average of the previous two.

The same analysis was conducted for 2012 using the CLC 2012 dataset and the same parameters to isolate the effects of land-use change on the capacity of forests to remove

PM10. Figure 12 presents the biophysical mapping results for particulate removal in 2012 and 2018, with variation in the ES during this period.

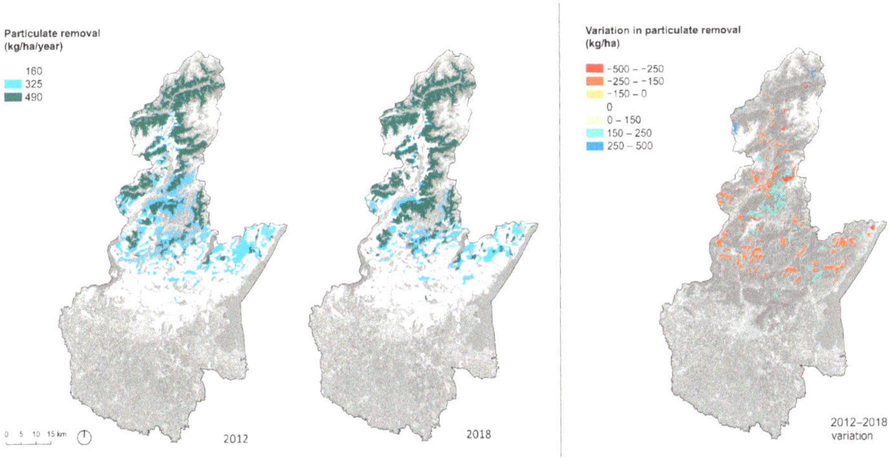


**Figure 12.** Maps of the biophysical assessment of particulate removal in 2012 and 2018. Source: own elaborations by authors.

For more detailed information about the parameters used, refer to Table A10 in Appendix A.

Monetary Assessment

The monetary assessment of particulate removal was conducted based on the societal costs of PM10 pollution, which include health and environmental impacts, as outlined in the SNPA guidelines [64]. The range of unit monetary values (EUR/ha) from the European Environment Agency (EEA) was used, as referenced in the SNPA report, adjusted to 2024 values to account for inflation [121]. Due to the broad range of available values, a filtering approach was employed to narrow the range. In the absence of site-specific criteria for further refinement, a statistical method was adopted: values below the 25th percentile and above the 75th percentile were excluded. This resulted in a refined range of unit monetary values set between EUR 500/ha (for broadleaf forests) and EUR 700/ha (for coniferous forests), with mixed forests assigned a mid-range value of EUR 600/ha. The total monetary value of particulate removal in 2018 was calculated by multiplying these unit monetary values by the respective forest areas in Brescia Province. The results, categorized by land cover typology, are presented in Figure 13, alongside a map illustrating the spatial distribution of unit monetary values across the province (EUR/ha/year). For further details about the parameters and results, refer to Table A10 in Appendix A.

### 3.2.6. Hydrological Regime Regulation

In the Province of Brescia, hydrological regulation plays a vital role due to the region's diverse geomorphology, which encompasses mountainous, hilly, and lowland areas. These landscapes are particularly vulnerable to hydrogeological risks, including floods and landslides, especially during extreme weather events. The region's balance between agricultural land, natural areas, and increasing urbanization underscores the critical need to monitor and preserve water infiltration processes. Effective spatial planning in this region must prioritize strategies that enhance water infiltration and minimize soil sealing, both of which are essential for sustainable water management and mitigating hydrological hazards [98].

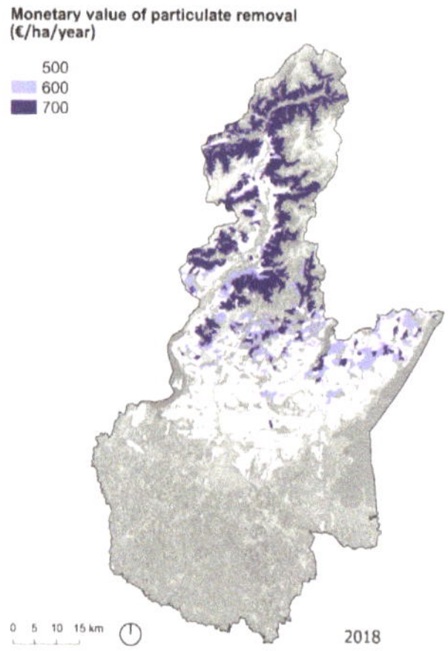


| CLC | | Monetary value (EUR/year) | |
|---|---|---|---|
| Code | Description | 2006 | 2012 |
| 311 | Broadleaf forests: | 38,097,500 | 43,520,072 |
| 312 | Coniferous forests | 28,284,200 | 14,491,946 |
| 313 | Mixed forests | 22,729,200 | 27,424,051 |
| Total | | 89,110,900 | 85,436,059 |

**Figure 13.** Mapping and results of the monetary assessment of particulate removal in 2012 and 2018. The map on the left shows unitary monetary values (absorption coefficients) related to LC in 2018. Source: own elaborations by authors.

Biophysical Assessment

The soil's capacity to regulate the hydrological regime was assessed using the "LIFE+ MGN" parametric methodology, which simplifies the evaluation by assigning coefficients (ranging from 0 to 5) for water retention based on land cover typologies [98]. The CLC 2018 dataset was utilized for the LC map and rasterized at a 10-meter resolution using QGIS software. Retention coefficients for the LC typologies were sourced from Nedkkov e Burkhard [123] and aligned with the CLC classification (refer to Table A11 in Appendix A). Nedkkov e Burkhard also provided literature-based insights into the water retention capacity of different land cover types, expressed as a percentage of average annual precipitation (%). In particular, by comparison with the work of Tate [124], the dimensionless LC coefficients were first converted into annual absorption percentages and subsequently converted into the annual absorbed volume by hectare ($m^3$/ha/year).

A similar analysis was conducted for 2012, using the CLC 2012 dataset and identical parameters to isolate the effects of land-use change on the hydrological regime regulation service. Figure 14 presents the biophysical mapping results for the hydrological regime regulation in 2012 and 2018, along with the observed variations in the ES during this period. Additional details on the parameters used can be found in Table A11 in Appendix A.

Monetary Assessment

For the monetary assessment of this service, the SNPA guidelines [64] were followed, assigning a unit monetary value of EUR 750 /$m^3$/100 years for flood prevention infrastructure. This value was updated for 2024 using the ISTAT monetary revaluation coefficient, resulting in EUR 10.27/$m^3$/year. The water volume absorbed by various land types, as determined in the biophysical assessment, was multiplied by the surface area of the corresponding land cover type, expressed in hectares (ha). This calculation yielded the total water absorption volume, expressed in cubic meters ($m^3$), which is critical for estimating the overall monetary value of the ecosystem service. The results of this assessment are presented in Figure 15 for 2012 and 2018. Additionally, Figure 15 includes a map of the unit monetary values (EUR/ha/year) for 2018. Further details on the parameters and results can be found in Table A11 in Appendix A.

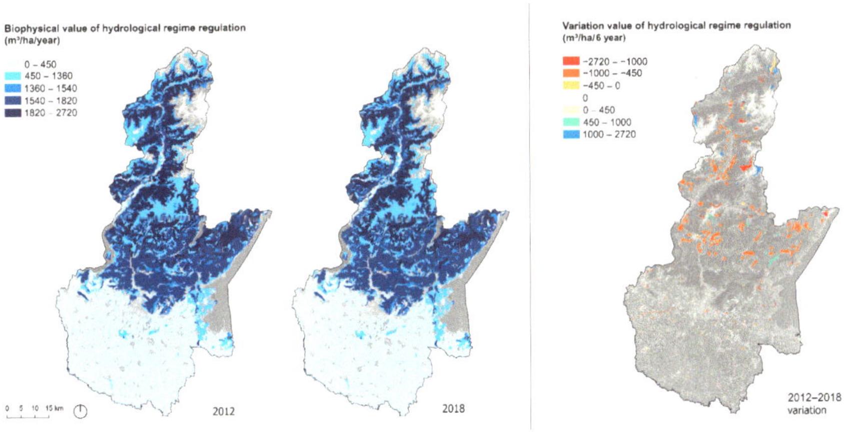


**Figure 14.** Maps of the biophysical assessment of hydrological regime regulation in 2012 and 2018. Source: own elaboration by authors.

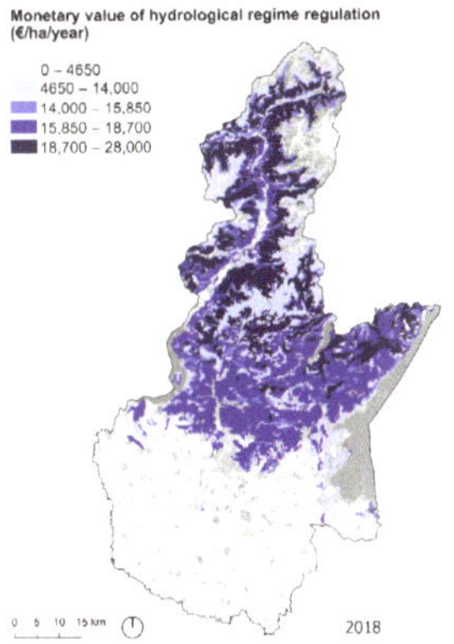


| CLC Code | Absorption coefficient | | Monetary value 2012 | Monetary value 2018 | CLC Code | Absorption coefficient | | Monetary value 2012 | Monetary value 2018 |
|---|---|---|---|---|---|---|---|---|---|
| 111 | 0% | 0 | 0 | 0 | 335 | 0% | 0 | 0 | 0 |
| 112 | 0% | 0 | 0 | 0 | 411 | 0% | 0 | 0 | 0 |
| 121 | 0% | 0 | 0 | 0 | 511 | 0% | 0 | 0 | 0 |
| 122 | 0% | 0 | 0 | 0 | 512 | 0% | 0 | 0 | 0 |
| 124 | 0% | 0 | 0 | 0 | 2111 | 5% | 1 | 668,560,055 | 664,465,406 |
| 131 | 0% | 0 | 0 | 0 | 2112 | 5% | 1 | 425,178 | 425,178 |
| 133 | 0% | 0 | 0 | 0 | 3111 | 20% | 4 | 5,196,209 | 4,691,541 |
| 141 | 0% | 0 | 0 | 0 | 3112 | 20% | 4 | 28,280,088 | 76,017,924 |
| 142 | 0% | 0 | 0 | 0 | 3113 | 20% | 4 | 896,009,642 | 925,224,300 |
| 221 | 17% | 3 | 8,730,322 | 11,039,428 | 3114 | 20% | 4 | 357,697,322 | 410,500,527 |
| 222 | 17% | 3 | 0 | 395,395 | 3115 | 20% | 4 | 131,288,394 | 203,624,112 |
| 223 | 17% | 3 | 8,240,032 | 8,081,874 | 3116 | 20% | 4 | 3,700,897 | 4,299,022 |
| 224 | 17% | 3 | 63,263 | 63,263 | 3117 | 20% | 4 | 1,999,980 | 2,467,265 |
| 231 | 15% | 2 | 63,145,711 | 62,112,138 | 3121 | 30% | 5 | 0 | 1,033,573 |
| 241 | 15% | 2 | 754,229 | 754,229 | 3122 | 30% | 5 | 30,839,578 | 4,6147,629 |
| 242 | 15% | 2 | 140,356,393 | 134,839,349 | 3123 | 30% | 5 | 940,048,429 | 915,745,501 |
| 243 | 15% | 2 | 312,502,133 | 314,024,558 | 3124 | 30% | 5 | 157,801,426 | 313,870,918 |
| 322 | 15% | 2 | 52,167,492 | 69,975,672 | 3131 | 30% | 5 | 511,171,586 | 306,663,843 |
| 323 | 15% | 2 | 614,557 | 614,557 | 3132 | 30% | 5 | 546,983,486 | 271,634,106 |
| 324 | 17% | 3 | 267,302,836 | 255,725,670 | 3211 | 15% | 2 | 277,863,477 | 294,959,330 |
| 332 | 0% | 0 | 0 | 0 | 3212 | 15% | 2 | 99,907,382 | 90,954,406 |
| 333 | 0% | 0 | 0 | 0 | 3231 | 15% | 2 | 0 | 614,557 |
| 334 | 0% | 0 | 0 | 0 | Total | | | 5,511,650,093 | 5,390,965,269 |

**Figure 15.** Mapping and results of the monetary assessment of the hydrological regime regulation in 2012 and 2018. The map on the left shows the unitary monetary values related to LC in 2018. Source: own elaborations by authors.

### 3.3. Overview and Evaluation of Results

The analyses revealed a total monetary loss solely attributed to land-use changes in the Province of Brescia, amounting to over EUR 51 million for the period 2006–2018 (with CSS assessed using the social cost method) and nearly EUR 125 million for the period 2012–2018. This variation is primarily due to differences in the ecosystem services analyzed rather than changes in land use across the periods examined. Figure 16 schematically illustrates the percentage change in the biophysical contribution of ecosystem services during the considered periods for the entire Province of Brescia. It shows that the most significant negative variation (representing a loss of ecosystem services) occurred for particulate removal, which decreased by −12.14% between 2012 and 2018. The only service that recorded a positive percentage variation was habitat quality, which will be discussed in more detail later. This service showed a slight increase of +0.04% in the period 2012–2018. This value, however, was derived from Figure 7 rather than the biophysical assessment in order to assign a value for each LC type. As shown in Figure 16, this positive change does not correlate to a proportional monetary loss (or gain). The service that exhibited the largest variation in monetary contribution (negative, indicating a loss) was hydrological regime regulation. Despite a −2.19% decrease in its biophysical contribution during 2012–2018, this service accounted for a total monetary loss of approximately EUR 121 million for the Province of Brescia. This loss is equivalent to roughly 0.2% of the province's GDP in 2018, with GDP calculated based on the per capita GDP for the Lombardy Region in 2018 (Territorial Economic Accounts 2019–2021 Lombardy (https://www.polis.lombardia.it/wps/wcm/connect/70f033b4-6ffb-468f-87ee-bd3676e8b233/WP-02+-+Conti+economici+territoriali+Lombardia_+2017-2019_Ancona_edgen2021.pdf?MOD=AJPERES&CACHEID=ROOTWORKSPACE-70f033b4-6ffb-468f-87ee-bd3676e8b233-nG.lDP6, last consultation: 2 November 2024).

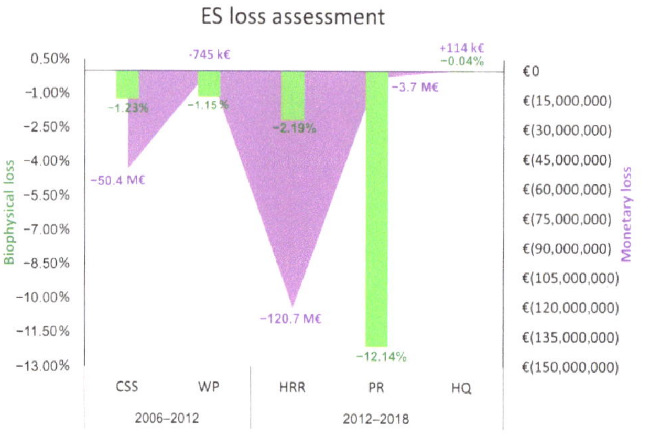


| ES | Biophysical loss in the period (%) | Monetary loss in the period (EUR) |
|---|---|---|
| CSS | −1.23% | −50,398,100 |
| WP | −1.15% | −745,500 |
| HRR | −2.19% | −120,684,800 |
| PR | −12.14% | −3,675,600 |
| HQ | 0.04% | 113,600 |

Note: CSS: carbon storage and sequestration; WP: wood provision; HRR: hydrological regime regulation; PR: particulate removal; HQ: habitat quality. Please note that the reported losses refer to different time periods, as detailed in Figure 16.

**Figure 16.** Total biophysical and monetary loss in Brescia Province for the analyzed periods and ecosystem services. Source: own elaboration by authors.

## 4. Discussion

### 4.1. Critical Analysis of the Results

The analysis examined both the biophysical and monetary quantification of six ecosystem services: carbon storage and sequestration (CSS), habitat quality (HQ), crop pollination (CP), wood provision (WP), particulate removal (PR), and hydrological regime regulation (HRR). Each service was evaluated using different methods, reflecting variations in the available data and the specific requirements of each assessment.

The assessments relied on LC maps as proxies for soil data, particularly for soil-related ecosystem services, in line with established methodologies [8]. This approach assumes that the LC classification accurately represents variations in soil-related ecosystem services.

The time periods used for each assessment were determined by data availability and methodological requirements. For CSS and WP, the years 2006 and 2012 were selected. For HQ, CP, PR, and HRR, the years 2012 and 2018 were chosen. The different timeframes were necessary because some datasets, especially the Corine Land Cover (CLC) database, did not provide a detailed land cover (LC) map for 2018, which is crucial for assessing certain services.

Biophysical assessments for CSS, HQ, and CP employed modeling approaches using the InVEST model [64], while other services (WP, PR, and HRR) were assessed using a simpler parametric approach based on LC classes, following guidelines from the existing literature [98]. The use of InVEST enabled more detailed assessments, particularly for services such as HQ and CP.

The biophysical assessment resulted in additional maps for specific services. For example, the InVEST model provided a habitat degradation map for HQ and a map of pollinator abundance for CP. These maps added significant value to the biophysical analysis, enabling a more detailed understanding of ecosystem service dynamics.

The monetary valuation followed the guidelines provided by the SNPA [64], which involved associating parametric values with LC classes. These values were adjusted for inflation or updated when necessary. Five monetary maps were produced, corresponding to the most recent years for each service: 2012 for CSS and WP and 2018 for HQ, PR, and HRR.

A significant limitation in the monetary assessment arose for CP. Due to the absence of a crop land-use map for the Province of Brescia, the results for CP could not be mapped in monetary terms. Additionally, agricultural production data for the analysis were only available for the 2023–24 period from ISTAT, which made estimates for earlier years impossible.

Given these strengths and weaknesses, the critical interpretation of the results focuses on several key aspects. First, it examines the use of the InVEST model, particularly in the construction of habitat quality. This model generates output based on the type of input information (data) provided, which is often influenced by both the local value system and the planning objectives (Section 4.1.1). Additionally, the importance of a cross-critical interpretation of the outputs generated by the sub-models is emphasized to accurately assess the phenomena and the impact of land transformation actions on habitats (Section 4.1.2). Finally, the text highlights the necessity of verifying the LC data structure at different historical thresholds to avoid misinterpretations of ongoing phenomena, including the need to validate the outputs through detailed checks on known cases (Section 4.1.3).

#### 4.1.1. InVEST Model: Habitat Quality

A critical analysis of the InVEST model's operation for the evaluated ecosystem services reveals that, despite its more complex modeling approach compared to simple parametric associations, the sub-models for carbon storage and sequestration (CSS) and crop pollination (CP) (perhaps as they are not fully utilized in this analysis) calculate the required services using only geo-referenced LC data. These sub-models associate site-specific and ad hoc-calculated parameters with this information. The only service that incorporates an additional layer of geo-referenced information beyond LC is habitat quality (HQ), which yields results that are more substantial, informative, and nuanced in interpretation. This enhancement can be closely linked to spatial planning objectives. For example, agriculture can be considered both a threat and a habitat by planners. Ultimately, the threat inputs required by the model are still based on LC data. Variables reflecting climatic variations across the territory, or those related to point or diffuse pollution, disturbances, or other local threats, were neither included in the model nor supported. Nevertheless, it is crucial to interpret the results for the HQ service in the context of the input data concerning threats to habitat quality in the area. As detailed by Sallustio et al. [117] and elaborated in Section 3, the threats considered in this analysis for the habitat quality assessment include the following: road and rail infrastructures, classified by functional type; intensive and extensive agriculture; and the soil consumed. Figure 17 illustrates these threats. In the Province of Brescia, a traditionally agricultural region, the most significant contribution to habitat threats, in distributive terms, arose from cultivated fields. However, the impact parameters may vary depending on the specific land use (habitat) considered.

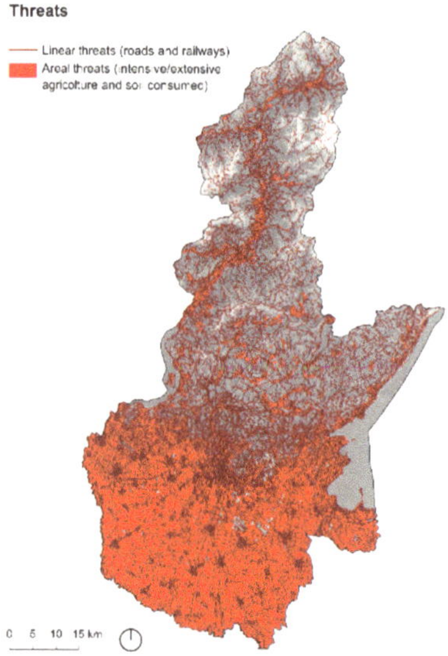


**Figure 17.** Threats considered in Brescia Province as inputs for the habit quality model. Source: own elaborations by authors.

This variation in impact explains the observed differences in HQ across the province. For instance, the habitat value in Po Valley (southern zone) is significantly lower compared to the hill and mountain regions further north. In contrast, the habitat degradation values follow an opposite pattern (see Figures 5 and 6 in Section 3).

#### 4.1.2. Cross-Critical Interpretation of the Model Outputs

Furthermore, a cross-critical interpretation of the ES maps is essential. For example, our analysis revealed a linear feature in the southwestern part of the province. A comparison of the HQ maps from 2012 and 2018 (in Figure 18) showed an increase in habitat quality in this area. This feature corresponds to the BreBeMi highway (A35), which was already completed by 2018 but did not exist in 2012, as confirmed by the LC maps.

This observed increase was likely due to the fact that, by 2018, the road was classified as "non-habitat" (Hj = 0) by the model. As a result, the degradation map value for that cell became zero, and no additional impacts were calculated. When a cell is classified as "non-habitat" by the model, it is not influenced by any threats within that cell since it is considered unsuitable for habitats. However, the suitability of adjacent habitat cells in 2018 may still contribute positively to the HQ value. In contrast, in 2012, the cell was still classified as a habitat and was, thus, impacted by mapped threats, resulting in a degradation value above zero and a lower habitat quality value.

These aspects highlight how the HQ model is more complex than other InVEST models. This increased complexity leads to a greater sensitivity to the quality of input data, requiring a critical a posteriori interpretation of the results. As demonstrated in the example above, such complexity can result in values that might initially seem illogical. In light of these considerations, a detailed analysis of the results suggests that the apparent improvement in habitat quality across the province from 2012 to 2018 is likely a consequence of this complexity.

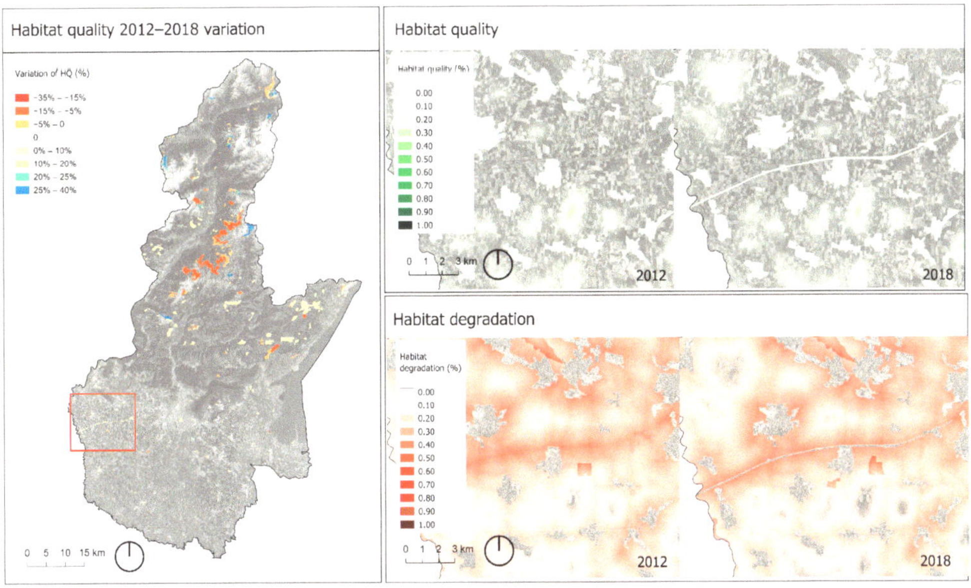


**Figure 18.** Details of habitat quality and habitat degradation maps in 2012 and 2018 in the area of highway construction (BreBeMi). Source: own elaborations by authors.

#### 4.1.3. Land Cover Updates or Changes

For the remaining ecosystem services analyzed, the results can be more easily interpreted by examining the variation in LC across the area. For example, the percentage change in areas associated with LC classes from the CLC database is shown for the three years analyzed across the entire Province of Brescia (Figure 19). The most critical classes for evaluation are those related to forest land use, such as carbon storage and sequestration (CSS), wood provision (WP), and particulate removal (PR) services, which were derived exclusively from the assessment due to the variation in these classes. In our case, the most significant differences observed related to mixed forests (code 313) compared to broad-leaved forests (311) and coniferous forests (312). The total area designated for these three covers appeared to be unchanged across the three years considered. Notably, the absence of the mixed forest class in 2006 coincides with an increase in both broad-leaved and coniferous forests. It remains unclear whether this discrepancy arose from errors in the CLC 2006 input layer (despite class 313 being included in the legend) or if it reflects an actual increase in this forest type in subsequent years. Regardless, this variation likely played a substantial role in altering the provision of ecosystem services in the study area.

As emphasized throughout, the goal of this analysis should extend beyond simply raising awareness among policymakers about the importance of ESs. Future assessments should prioritize the preservation of these services. Moreover, these evaluations must provide policymakers with the necessary tools to assess land-use decisions. More importantly, they should highlight the crucial role and potential of ecosystems in the study area, examining the impacts of transformations on the provision of essential ecosystem services not only for environmental health but also for human well-being, society, and the economy.

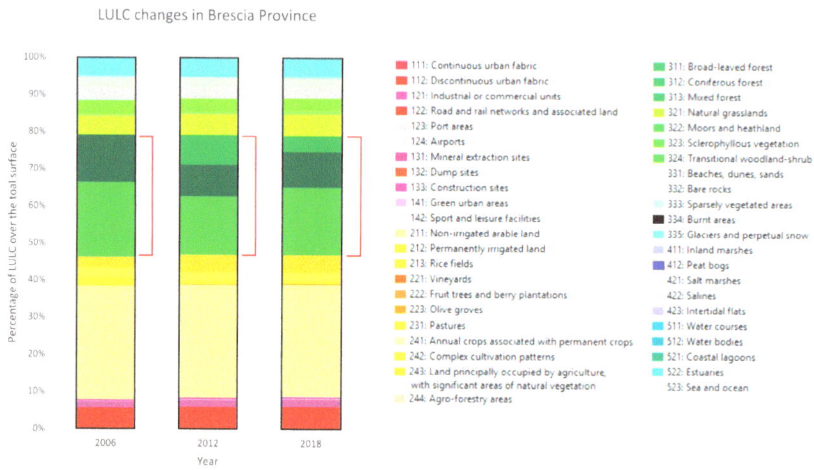


**Figure 19.** Variation in LC in Brescia Province during the years considered, with the most relevant variations highlighted by a red bracket. Source: own elaborations by authors.

In addition to the cumulative analysis, it is vital to interpret the spatial distribution of the results across the territory, from which contextualized conclusions can be drawn. These interpretations can guide planning, conservation, and land restoration strategies, supported by the economic and fiscal incentive tools discussed in the Introduction, which are well-documented in the literature.

## 5. Conclusions

The analysis presented in this paper introduces novel features, both in terms of the spatial context and the local scale (NUTS3) at which it was conducted. Despite limitations stemming from the scarcity and weaknesses of several input datasets—which constrained the analysis of ecosystem services to three different time scales and hindered the creation of a comprehensive depiction of their variations—the results offer valuable insights for local policymakers. Additionally, these findings open avenues for exploring potential tools to support territorial resource management and planning. The monetary valuation of ecosystem services, which complements the biophysical assessment, proved particularly beneficial in this context. It has the potential to engage stakeholders not only as direct beneficiaries of ESs managed in the form of private goods (e.g., food or timber supply for landowners) but also as indirect beneficiaries of the preservation of other ESs within the territory [57,125]. Moreover, governance models such as network governance provide promising frameworks for supporting PES programs. These models emphasize the "horizontal" structure of decision-making, fostering collective and inclusive processes that are sensitive to power dynamics among stakeholders. Such approaches are particularly valuable in the complex and uncertain contexts related to ecosystem services [18,126].

This inclusive governance model aligns with the evolving direction of ES trade-off analysis, emphasizing the importance of stakeholder engagement and the real flows of ecosystem services. In contrast, simplified overlay analyses often fail to capture the intricate dynamics of ecosystem processes [18]. According to several authors, trade-offs among ESs most frequently occur between agricultural supply services and other services, such as those related to the regulation of hydrological regimes [65], or among forest ecosystem services (FESs). These FESs include carbon sequestration, the provision of timber and other raw materials, PM10 sequestration, and hydrological regulation [57]. Achieving a balance in the management of FESs is particularly complex and delicate, leading to the coining of the term Balanced Forest Ecosystem Services Management (BFESM) to describe this challenge [57].

Future analyses must delve deeper into these trade-offs and differentiate between the theoretical and actual flows of ecosystem services. The latter involves the intersection between the provision of ecosystem services (as assessed in the current study) and the presence of beneficiaries who demand and utilize these services [127]. This type of assessment is conceptually crucial. As noted by Costanza et al. [7], ecosystems cannot deliver benefits to people without the interplay of human capital (people), social capital (communities), and built capital (infrastructure and buildings). Additionally, it is well-documented in the literature that limiting assessments to theoretical flows of ecosystem services without accounting for actual flows can result in overestimations [128]. However, it is acknowledged that recent studies attempting to bridge this gap—by overlaying service delivery with service utilization—often lack theoretical, terminological, and methodological consistency [127,128]. These evaluations are inherently complex. While the delivery of ESs was effectively tested in the current case study, incorporating utilization metrics could provide more critical insights for future spatial planning research.

Burkhard's matrix model, published in 2014 [129,130], has garnered significant recognition for assessing the supply and demand of ecosystem services across various European and Asian regions [123,131,132]. However, applying matrices calibrated for overly generic spatial contexts—such as Burkhard's matrices for the European continent—introduces uncertainties in the results [133]. Furthermore, as previously mentioned, simple overlay methods fail to accurately map and value ecosystem services due to their inability to capture the complex spatial and temporal dynamics of service flows, which can vary significantly depending on scale and environmental factors [127,134–138].

Future research could focus on integrating assessments of the supply and demand for ecosystem services to provide a clearer understanding of the land-use dynamics, needs, and trade-offs that emerge between different ecosystem services.

Thus, the qualitative insights derived from Burkhard's framework can guide future analytical developments, emphasizing regions with a high ecosystem services demand or those experiencing transitions toward increased demand due to ongoing land cover changes.

**Author Contributions:** Conceptualization, M.P., A.R. and M.G.; methodology, M.P. and A.R.; formal analysis, M.G.; investigation, M.P., A.R. and M.G.; data curation, M.G.; writing—original draft preparation, A.R. and M.G.; writing—review and editing, M.P., A.R. and M.G.; visualization, M.G.; supervision, M.P. and A.R.; project administration, M.P. All authors have read and agreed to the published version of the manuscript.

**Funding:** Part of this research was supported by the EU Horizon Europe RIA Framework Programme (called HORIZON-MISS-2022-SOIL-01), under the project "Literacy boost through an Operational Educational Ecosystem of Societal actors on Soil health" LOESS (Grant Number 101112707).

**Data Availability Statement:** The dataset is available upon request to the authors.

**Acknowledgments:** The authors thank Aurora Mantelli for the support and the technical-graphic processing activities which are, with further refinements, presented in this paper.

**Conflicts of Interest:** The authors declare no conflicts of interest.

## Appendix A.

### *Appendix A.1. Carbon Storage and Sequestration*

**Table A1.** Correspondence between CLC LC codes (2006/2012) and INFC forest inventory codes. Source: adaptation from [111].

| INFC | | CLC | |
|---|---|---|---|
| **Code** | **Description** | **Code** | **Description** |
| 1 | Larch and Swiss pine forests | 3124 | Forests predominantly of larch and/or Swiss pine |
| 2 | Spruce forests | 31323 | Mixed coniferous and broadleaf forests predominantly of fir |
| | | 3123 | Forests predominantly of fir |
| 3 | Silver fir forests | 31323 | Mixed coniferous and broadleaf forests predominantly of fir |
| | | 3123 | Forests predominantly of fir |
| 4 | Scots pine and mountain pine forests | 3122 | Forests predominantly of oro-Mediterranean and mountain pines |
| | | 31322 | Mixed coniferous and broadleaf forests predominantly of oro-Mediterranean and mountain pines |
| 5 | Black pine, Corsican pine, and Bosnian pine forests | 3122 | Forests predominantly of oro-Mediterranean and mountain pines |
| | | 31322 | Mixed coniferous and broadleaf forests predominantly of oro-Mediterranean and mountain pines |
| 6 | Mediterranean pine forests | 3121 | Forests predominantly of Mediterranean pines and cypresses |
| | | 31321 | Mixed coniferous and broadleaf forests predominantly of Mediterranean pines and cypresses |
| 7 | Other coniferous forests, pure or mixed | 3125 | Forests and former plantations predominantly of exotic conifers |
| | | 31325 | Mixed coniferous and broadleaf forests predominantly of exotic conifers |
| 8 | Beech forests | 3115 | Forests predominantly of beech |
| | | 31315 | Mixed coniferous and broadleaf forests predominantly of beech |
| 9 | Sessile oak, Downy oak, and English oak forests | 3112 | Forests predominantly of deciduous oaks |
| | | 31312 | Mixed coniferous and broadleaf forests predominantly of deciduous oaks |
| 10 | Turkey oak, Hungarian oak, Macedonian oak, Valonia oak forests | 3112 | Forests predominantly of deciduous oaks |
| | | 31312 | Mixed coniferous and broadleaf forests predominantly of deciduous oaks |
| 11 | Chestnut forest | 3114 | Forests predominantly of chestnut |
| | | 31314 | Mixed coniferous and broadleaf forests predominantly of chestnut |
| 12 | Hop-hornbeam and hornbeam forests | 3113 | Mixed forests predominantly of other native broadleaf specie |
| | | 31313 | Mixed coniferous and broadleaf forests predominantly of other native broadleaf species |
| 13 | Hygrophilous forests | 3116 | Forests predominantly of hygrophilous species |
| | | 31316 | Mixed coniferous and broadleaf forests predominantly of hygrophilous species |
| 14 | Other deciduous forests | 3117 | Forests and former plantations predominantly of exotic broadleaf species |
| | | 31317 | Mixed coniferous and broadleaf forests predominantly of exotic species |
| 15 | Holm oak forests | 3111 | Forests predominantly of oaks and other evergreen broadleaf species |
| | | 31311 | Mixed coniferous and broadleaf forests predominantly of oaks and other evergreen broadleaf species |
| 16 | Cork oak forests | 3111 | Forests predominantly of oaks and other evergreen broadleaf species |
| | | 31311 | Mixed coniferous and broadleaf forests predominantly of oaks and other evergreen broadleaf species |
| 17 | Other evergreen broadleaf forests | | / |
| 18 | Artificial poplar plantations | | / |
| 19 | Other broadleaf plantations | | / |
| 20 | Conifer plantations | | / |
| 21 | Subalpine shrublands | | / |
| 22 | Temperate shrublands | | / |
| 23 | Mediterranean scrub and shrublands | | / |

**Table A2.** C content (t/ha) in each of the four carbon pools considered for each CLC LC class in 2006 and 2012. Source: own elaborations from INFC 2005 and 2015 databases.

| CLC—Code | CLC Description | Aboveground Biomass (tC/ha) | Belowground Biomass (tC/ha) | Soil Organic Matter (tC/ha) | Dead Organic Matter (tC/ha) |
|---|---|---|---|---|---|
| 3111 | Forests predominantly of oaks and other evergreen broadleaf species | 0.0 | 0.0 | 0.0 | 0.0 |
| 3112 | Forests predominantly of deciduous oaks | 55.7 | 0.0 | 65.7 | 8.8 |
| 3113 | Mixed forests predominantly of other native broadleaf specie | 38.5 | 0.0 | 105.9 | 3.2 |
| 3114 | Forests predominantly of chestnut | 84.4 | 0.0 | 106.3 | 16.2 |
| 3115 | Forests predominantly of beech | 86.1 | 0.0 | 85.2 | 8.2 |
| 3116 | Forests predominantly of hygrophilous species | 43.3 | 0.0 | 86.8 | 12.6 |
| 3117 | Forests and former plantations predominantly of exotic broadleaf species | 54.7 | 0.0 | 82.1 | 6.9 |
| 3121 | Forests predominantly of Mediterranean pines and cypresses | 0.0 | 0.0 | 0.0 | 0.0 |
| 3122 | Forests predominantly of oro-Mediterranean and mountain pines | 59.5 | 0.0 | 102.0 | 11.0 |
| 3123 | Forests predominantly of fir | 102.6 | 0.0 | 95.9 | 24.5 |
| 3124 | Forests predominantly of larch and/or Swiss pine | 71.2 | 0.0 | 101.4 | 11.9 |
| 3125 | Forests and former plantations predominantly of exotic conifers | 67.5 | 0.0 | 74.4 | 9.0 |
| 31311 | Mixed coniferous and broadleaf forests predominantly of oaks and other evergreen broadleaf species | 0.0 | 0.0 | 0.0 | 0.0 |
| 31312 | Mixed coniferous and broadleaf forests predominantly of deciduous oaks | 55.7 | 0.0 | 65.7 | 8.8 |
| 31313 | Mixed coniferous and broadleaf forests predominantly of other native broadleaf species | 38.5 | 0.0 | 105.9 | 3.2 |
| 31314 | Mixed coniferous and broadleaf forests predominantly of sweet chestnut | 84.4 | 0.0 | 106.3 | 16.2 |
| 31315 | Mixed coniferous and broadleaf forests predominantly of beech | 86.1 | 0.0 | 85.2 | 8.2 |
| 31316 | Mixed coniferous and broadleaf forests predominantly of hygrophilous species | 43.3 | 0.0 | 86.8 | 12.6 |
| 31317 | Mixed coniferous and broadleaf forests predominantly of exotic species | 54.7 | 0.0 | 82.1 | 6.9 |
| 31321 | Mixed coniferous and broadleaf forests predominantly of Mediterranean pines and cypresses | 0.0 | 0.0 | 0.0 | 0.0 |
| 31322 | Mixed coniferous and broadleaf forests predominantly of oro-Mediterranean and mountain pines | 59.5 | 0.0 | 102.0 | 11.0 |
| 31323 | Mixed coniferous and broadleaf forests predominantly of fir | 102.6 | 0.0 | 95.9 | 24.5 |
| 31325 | Mixed coniferous and broadleaf forests predominantly of exotic conifers | 67.5 | 0.0 | 74.4 | 9.0 |

**Table A3.** Biophysical surfaces and monetary values for CSS for each CLC LC class in 2006 and 2012. Source: own elaborations by authors from INFC 2005 and 2015 databases.

| CLC—Code | Total C Stored (t/ha) | Surface (ha) | | Monetary Value | | | |
|---|---|---|---|---|---|---|---|
| | | | | Social Cost (EUR) | | Market Price (EUR) | |
| | | 2006 | 2012 | 2006 | 2012 | 2006 | 2012 |
| 3111<br>31311 | 0 | 308 | 315 | 0 | 0 | 0 | 0 |
| 3112<br>31312 | 130.2 | 1485 | 1512 | 28,178,392 | 28,690,726 | 5,413,716 | 5,512,147 |
| 3116<br>31316 | 142.7 | 125 | 199 | 2,599,637 | 4,138,623 | 499,450 | 795,124 |
| 3117<br>31317 | 143.7 | 116 | 106 | 2,429,369 | 2,219,941 | 466,738 | 426,502 |
| 3113<br>31313 | 147.6 | 62,775 | 61,202 | 1,350,367,087 | 1,316,529,931 | 259,436,520 | 252,935,626 |
| 3122<br>31322 | 172.5 | 4851 | 4404 | 121,954,868 | 110,717,221 | 23,430,330 | 21,271,320 |
| 3115<br>31315 | 179.5 | 8490 | 9439 | 222,101,202 | 246,927,355 | 42,670,740 | 47,440,414 |
| 3124<br>31324 | 184.5 | 5768 | 5645 | 155,095,925 | 151,788,574 | 29,797,488 | 29,162,070 |
| 3114<br>31314 | 206.9 | 23,242 | 21,671 | 700,830,111 | 653,458,796 | 134,645,554 | 125,544,437 |
| 3123<br>31323 | 223 | 46,031 | 46,606 | 1,496,008,421 | 1,514,695,932 | 287,417,564 | 291,007,864 |
| 3121<br>31321 | 0 | 0 | 0 | 0 | 0 | 0 | 0 |
| Total | | | | 4,079,565,011 | 4,029,167,098 | 783,778,100 | 774,095,504 |

*Appendix A.2. Habitat Quality*

**Table A4.** Correspondence between CLC LC codes (2018) and habitat categories introduced in [117]. Source: own elaboration by authors.

| CLC—Code | CLC Description | Habitat Category Assigned |
|---|---|---|
| 111 | Continuous urban fabric | Buildings and other artificial areas or impervious soils |
| 112 | Discontinuous urban fabric | Buildings and other artificial areas or impervious soils |
| 121 | Industrial or commercial units and public facilities | Buildings and other artificial areas or impervious soils |
| 122 | Road and rail networks and associated land | Buildings and other artificial areas or impervious soils |
| 123 | Port areas | Buildings and other artificial areas or impervious soils |
| 124 | Airports | Open urban areas |
| 131 | Mineral extraction sites | Open urban areas |
| 132 | Dump sites | Open urban areas |
| 133 | Construction sites | Open urban areas |
| 141 | Green urban areas | Grasslands |
| 142 | Sport and leisure facilities | Grasslands |
| 211 | Non-irrigated arable land | Intensive agricultural lands |
| 212 | Permanently irrigated arable land | Intensive agricultural lands |
| 213 | Rice fields | Extensive agricultural lands |

**Table A4.** *Cont.*

| CLC—Code | CLC Description | Habitat Category Assigned |
|---|---|---|
| 221 | Vineyards | Extensive agricultural lands |
| 222 | Fruit tree and berry plantations | Extensive agricultural lands |
| 223 | Olive groves | Extensive agricultural lands |
| 231 | Pastures, meadows and other permanent grasslands under agricultural use | Grasslands |
| 241 | Annual crops associated with permanent crops | Extensive agricultural lands |
| 242 | Complex cultivation patterns | Intensive agricultural lands |
| 243 | Land principally occupied by agriculture, with significant areas of natural vegetation | Intensive agricultural lands |
| 244 | Agro-forestry areas | Extensive agricultural lands |
| 311 | Broad-leaved forest | Broadleaves forests |
| 312 | Coniferous forest | Conifer forests |
| 313 | Mixed forest | (Broadleaves forests + Conifer forests)/2 |
| 321 | Natural grassland | Grasslands |
| 322 | Moors and heathland | Shrublands |
| 323 | Sclerophyllous vegetation | Shrublands |
| 324 | Transitional woodland/shrub | Inland unvegetated or sparsely vegetated areas |
| 331 | Beaches, dunes, and sand plains | Beaches, dune and, sands |
| 332 | Bare rock | Inland unvegetated or sparsely vegetated areas |
| 333 | Sparsely vegetated areas | Inland unvegetated or sparsely vegetated areas |
| 334 | Burnt areas | Inland unvegetated or sparsely vegetated areas |
| 335 | Glaciers and perpetual snow | Inland unvegetated or sparsely vegetated areas |
| 411 | Inland marshes | Wetlands |
| 412 | Peatbogs | Wetlands |
| 421 | Coastal salt marshes | Wetlands |
| 422 | Salines | Water bodies |
| 511 | Water courses | Water bodies |
| 512 | Water bodies | Water bodies |
| 521 | Coastal lagoons | Water bodies |
| 522 | Estuaries | Water bodies |

**Table A5.** Sensitivity table for each CLC LC class in 2018 with reference to the threats considered. Source: [117].

| CLC—Code | Habitat Suitability | Sensitivity to Threats | | | | | | | |
|---|---|---|---|---|---|---|---|---|---|
| | | Intensive Agriculture | Extensive Agriculture | Soil Consumed | Railways | Road 1 | Road 2 | Road 3 | Road 4 |
| 111 | 0.09 | 0.00 | 0.00 | 0.00 | 0.00 | 0.00 | 0.00 | 0.00 | 0.00 |
| 112 | 0.09 | 0.00 | 0.00 | 0.00 | 0.00 | 0.00 | 0.00 | 0.00 | 0.00 |
| 121 | 0.09 | 0.00 | 0.00 | 0.00 | 0.00 | 0.00 | 0.00 | 0.00 | 0.00 |
| 122 | 0.09 | 0.00 | 0.00 | 0.00 | 0.00 | 0.00 | 0.00 | 0.00 | 0.00 |
| 123 | 0.09 | 0.00 | 0.00 | 0.00 | 0.00 | 0.00 | 0.00 | 0.00 | 0.00 |
| 124 | 0.27 | 0.31 | 0.21 | 0.56 | 0.46 | 0.56 | 0.52 | 0.46 | 0.19 |
| 131 | 0.27 | 0.31 | 0.21 | 0.56 | 0.46 | 0.56 | 0.52 | 0.46 | 0.19 |
| 132 | 0.27 | 0.31 | 0.21 | 0.56 | 0.46 | 0.56 | 0.52 | 0.46 | 0.19 |
| 133 | 0.27 | 0.31 | 0.21 | 0.56 | 0.46 | 0.56 | 0.52 | 0.46 | 0.19 |
| 141 | 0.86 | 0.75 | 0.52 | 0.72 | 0.60 | 0.80 | 0.71 | 0.63 | 0.42 |

**Table A5.** *Cont.*

| CLC—Code | Habitat Suitability | Sensitivity to Threats | | | | | | | |
|---|---|---|---|---|---|---|---|---|---|
| | | Intensive Agriculture | Extensive Agriculture | Soil Consumed | Railways | Road 1 | Road 2 | Road 3 | Road 4 |
| 142 | 0.86 | 0.75 | 0.52 | 0.72 | 0.60 | 0.80 | 0.71 | 0.63 | 0.42 |
| 211 | 0.26 | 0.00 | 0.12 | 0.51 | 0.44 | 0.61 | 0.54 | 0.47 | 0.24 |
| 212 | 0.26 | 0.00 | 0.12 | 0.51 | 0.44 | 0.61 | 0.54 | 0.47 | 0.24 |
| 213 | 0.52 | 0.54 | 0.00 | 0.62 | 0.51 | 0.71 | 0.61 | 0.55 | 0.26 |
| 221 | 0.52 | 0.54 | 0.00 | 0.62 | 0.51 | 0.71 | 0.61 | 0.55 | 0.26 |
| 222 | 0.52 | 0.54 | 0.00 | 0.62 | 0.51 | 0.71 | 0.61 | 0.55 | 0.26 |
| 223 | 0.52 | 0.54 | 0.00 | 0.62 | 0.51 | 0.71 | 0.61 | 0.55 | 0.26 |
| 231 | 0.86 | 0.75 | 0.52 | 0.72 | 0.60 | 0.80 | 0.71 | 0.63 | 0.42 |
| 241 | 0.52 | 0.54 | 0.00 | 0.62 | 0.51 | 0.71 | 0.61 | 0.55 | 0.26 |
| 242 | 0.26 | 0.00 | 0.12 | 0.51 | 0.44 | 0.61 | 0.54 | 0.47 | 0.24 |
| 243 | 0.26 | 0.00 | 0.12 | 0.51 | 0.44 | 0.61 | 0.54 | 0.47 | 0.24 |
| 244 | 0.52 | 0.54 | 0.00 | 0.62 | 0.51 | 0.71 | 0.61 | 0.55 | 0.26 |
| 311 | 0.93 | 0.67 | 0.47 | 0.77 | 0.65 | 0.85 | 0.77 | 0.66 | 0.40 |
| 312 | 0.82 | 0.63 | 0.44 | 0.76 | 0.61 | 0.84 | 0.76 | 0.68 | 0.39 |
| 313 | 0.87 | 0.65 | 0.46 | 0.77 | 0.63 | 0.85 | 0.77 | 0.67 | 0.40 |
| 321 | 0.86 | 0.75 | 0.52 | 0.72 | 0.60 | 0.80 | 0.71 | 0.63 | 0.42 |
| 322 | 0.81 | 0.72 | 0.51 | 0.69 | 0.60 | 0.78 | 0.71 | 0.63 | 0.39 |
| 323 | 0.81 | 0.72 | 0.51 | 0.69 | 0.60 | 0.78 | 0.71 | 0.63 | 0.39 |
| 324 | 0.55 | 0.51 | 0.35 | 0.61 | 0.46 | 0.61 | 0.57 | 0.52 | 0.30 |
| 331 | 0.74 | 0.68 | 0.51 | 0.86 | 0.67 | 0.81 | 0.46 | 0.69 | 0.50 |
| 332 | 0.55 | 0.51 | 0.35 | 0.61 | 0.46 | 0.61 | 0.57 | 0.52 | 0.30 |
| 333 | 0.55 | 0.51 | 0.35 | 0.61 | 0.46 | 0.61 | 0.57 | 0.52 | 0.30 |
| 334 | 0.55 | 0.51 | 0.35 | 0.61 | 0.46 | 0.61 | 0.57 | 0.52 | 0.30 |
| 335 | 0.55 | 0.51 | 0.35 | 0.61 | 0.46 | 0.61 | 0.57 | 0.52 | 0.30 |
| 411 | 0.96 | 0.80 | 0.59 | 0.79 | 0.64 | 0.84 | 0.74 | 0.69 | 0.44 |
| 412 | 0.96 | 0.80 | 0.59 | 0.79 | 0.64 | 0.84 | 0.74 | 0.69 | 0.44 |
| 421 | 0.96 | 0.80 | 0.59 | 0.79 | 0.64 | 0.84 | 0.74 | 0.69 | 0.44 |
| 422 | 0.83 | 0.76 | 0.53 | 0.72 | 0.51 | 0.72 | 0.64 | 0.60 | 0.36 |
| 511 | 0.83 | 0.76 | 0.53 | 0.72 | 0.51 | 0.72 | 0.64 | 0.60 | 0.36 |
| 512 | 0.83 | 0.76 | 0.53 | 0.72 | 0.51 | 0.72 | 0.64 | 0.60 | 0.36 |
| 521 | 0.83 | 0.76 | 0.53 | 0.72 | 0.51 | 0.72 | 0.64 | 0.60 | 0.36 |
| 522 | 0.83 | 0.76 | 0.53 | 0.72 | 0.51 | 0.72 | 0.64 | 0.60 | 0.36 |

**Table A6.** Threat table for each threat considered. Source: [117].

| Threat | OSM Classification | Maximum Distance | Weight |
|---|---|---|---|
| Intensive agriculture | / | 1.60 | 0.69 |
| Extensive agriculture | / | 0.60 | 0.42 |
| Soil consumed | / | 1.70 | 0.79 |
| Railways | railways | 1.60 | 0.62 |
| Road 1 | highways, major roads, primary roads | 1.50 | 0.86 |
| Road 2 | secondary roads, tertiary roads | 1.00 | 0.69 |
| Road 3 | residential roads, service roads | 0.90 | 0.61 |
| Road 4 | dirt roads, bridleways | 0.30 | 0.28 |

**Table A7.** Surfaces and monetary values of HQ for each CLC LC class in 2012 and 2018. Source: own elaborations by authors.

| CLC—Code | Habitat Category | Surface (ha) | | Unit Monetary Value (EUR/ha) | Monetary Value (EUR) | |
|---|---|---|---|---|---|---|
| | | 2012 | 2018 | | 2012 | 2018 |
| 111 | Buildings and other artificial areas or impervious soils | 1059 | 1059 | 106.83 | 113,133 | 113,133 |
| 112 | Buildings and other artificial areas or impervious soils | 27,785 | 28,226 | 106.83 | 2,968,272 | 3,015,384 |
| 121 | Buildings and other artificial areas or impervious soils | 8771 | 9195 | 106.83 | 937,006 | 982,302 |
| 122 | Buildings and other artificial areas or impervious soils | 98 | 323 | 106.83 | 10,469 | 34,506 |
| 124 | Open urban areas | 923 | 929 | 320.49 | 295,812 | 297,735 |
| 131 | Open urban areas | 1695 | 1835 | 320.49 | 543,231 | 588,099 |
| 133 | Open urban areas | 99 | 91 | 320.49 | 31,729 | 29,165 |
| 141 | Grasslands | 81 | 54 | 1131.8 | 91,676 | 61,117 |
| 142 | Grasslands | 433 | 740 | 1131.8 | 490,069 | 837,532 |
| 211 | Intensive agricultural lands | 144,753 | 143,869 | 308.62 | 44,673,671 | 44,400,851 |
| 221 | Extensive agricultural lands | 552 | 698 | 617.24 | 340,716 | 430,834 |
| 222 | Extensive agricultural lands | | 25 | 617.24 | 0 | 15,431 |
| 223 | Extensive agricultural lands | 521 | 511 | 617.24 | 321,582 | 315,410 |
| 231 | Grasslands | 4524 | 4447 | 1131.8 | 5,120,263 | 5,033,115 |
| 241 | Extensive agricultural lands | 54 | 54 | 617.24 | 33,331 | 33,331 |
| 242 | Intensive agricultural lands | 10,048 | 9654 | 308.62 | 3,101,014 | 2,979,417 |
| 243 | Intensive agricultural lands | 22,372 | 22,483 | 308.62 | 6,904,447 | 6,938,703 |
| 311 | Broadleaves forests | 76,195 | 87,036 | 803.6 | 61,230,302 | 69,942,130 |
| 312 | Conifer forests | 40,406 | 45,707 | 803.6 | 32,470,262 | 36,730,145 |
| 313 | (Broadleaves forests + Conifer forests)/2 | 37,882 | 20,703 | 803.6 | 30,441,975 | 16,636,931 |
| 321 | Grasslands | 27,051 | 27,630 | 1131.8 | 30,616,322 | 31,271,634 |
| 322 | Shrublands | 3732 | 5010 | 961.47 | 3,588,206 | 4,816,965 |
| 323 | Shrublands | 44 | 88 | 961.47 | 42,305 | 84,609 |
| 324 | Inland unvegetated or sparsely vegetated areas | 16,899 | 16,169 | 652.85 | 11,032,512 | 10,555,932 |
| 332 | Inland unvegetated or sparsely vegetated areas | 14,668 | 13,299 | 652.85 | 9,576,004 | 8,682,252 |
| 333 | Inland unvegetated or sparsely vegetated areas | 11,488 | 12,273 | 652.85 | 7,499,941 | 8,012,428 |
| 334 | Inland unvegetated or sparsely vegetated areas | | 53 | 652.85 | 0 | 34,601 |
| 335 | Inland unvegetated or sparsely vegetated areas | 1965 | 1933 | 652.85 | 1,282,850 | 1,261,959 |
| 411 | Wetlands | 258 | 233 | 11,609.1 | 2,995,148 | 2,704,920 |
| 511 | Water bodies | 68 | 68 | 985.21 | 66,994 | 66,994 |
| 512 | Water bodies | 23,897 | 23,927 | 985.21 | 23,543,563 | 23,573,120 |
| **Total** | | | | | 280,362,804 | 280,480,684 |

*Appendix A.3. Crop Pollination*

**Table A8.** Monetary value of crop pollination for crops dependent on the service based on 2023 production in Brescia Province. Source: own elaborations by authors from ISMEA, AREA Rica, and ISTAT databases.

| Crop | Production (t) | | | Crop Production | | Crop Pollination | |
|---|---|---|---|---|---|---|---|
| | Outdoor | Greenhouse | Total | Unit Monetary Value (EUR/t) | Monetary Value (EUR) | Dependency Ration | Monetary Value (EUR) |
| Apple | 1811 | 0 | 1811 | 729 | 1,320,219 | 0.65 | 858,142 |
| Chestnut | 1075.9 | 0 | 1076 | 1920 | 2,065,920 | 0.25 | 516,480 |
| Cucumber | 25 | 180 | 205 | 291 | 59,655 | 0.65 | 38,776 |
| Green Bean | 2364.5 | 110 | 2475 | 1097 | 2,715,075 | 0.05 | 135,754 |

**Table A8.** *Cont.*

| Crop | Production (t) | | | Crop Production | | Crop Pollination | |
|---|---|---|---|---|---|---|---|
| | Outdoor | Greenhouse | Total | Unit Monetary Value (EUR/t) | Monetary Value (EUR) | Dependency Ration | Monetary Value (EUR) |
| Lemon | 8 | 0 | 8 | 760 | 6080 | 0.05 | 304 |
| Melon | 595 | 1250 | 1845 | 597 | 1,101,465 | 0.95 | 1,046,392 |
| Peach | 788 | 0 | 788 | 512 | 403,456 | 0.65 | 262,246 |
| Pear | 185 | 0 | 185 | 1423 | 263,255 | 0.65 | 171,116 |
| Pepper | 120 | 114 | 234 | 493 | 115,362 | 0.05 | 5768 |
| Strawberry | 71 | 300 | 371 | 3023 | 1,121,533 | 0.25 | 280,383 |
| Plum | 72 | 0 | 72 | 308 | 22,176 | 0.65 | 14,414 |
| Tomato | 44,436 | 450 | 44,886 | 676 | 30,342,936 | 0.05 | 1,517,147 |
| Watermelon | 880 | 400 | 1280 | 164 | 209,920 | 0.95 | 199,424 |
| Zucchini | 7637.8 | 450 | 8089 | 503 | 4,068,767 | 0.95 | 3,864,755 |
| **Total** | | | | | | | 8,911,046 |

## *Appendix A.4. Wood Provision*

**Table A9.** Biophysical, surfaces and monetary value of wood provision for each CLC LC in 2006 and 2012. Source: own elaborations from INFC 2005 and 2015 databases.

| INFC | | CLC—Code | Annual Tree Biomass Increment per Surface | | Surface (ha) | | Annual Tree Biomass Increment ($m^3$/Year) | | Monetary Value (EUR/Year) | |
|---|---|---|---|---|---|---|---|---|---|---|
| Code | Description | | (t/ha/Year) | ($m^3$/ha/Year) | 2006 | 2012 | 2006 | 2012 | 2006 | 2012 |
| 1 | Larch and Swiss pine forests | 3124 | 2.4 | 3.6 | 5771 | 5649 | 20,953 | 20,510 | 1,466,710 | 1,435,700 |
| 2 and 3 | Spruce forests and Silver fir forests | 31323 | 5.3 | 8.0 | 14,085 | 12,981 | 112,933 | 104,081 | 7,905,310 | 7,285,670 |
| | | 3123 | | | 31,969 | 33,653 | 256,326 | 269,828 | 17,942,820 | 18,887,960 |
| 4 and 5 | Scots pine and Mountain pine forests and Black pine, Corsican pine, and Bosnian pine forests | 3122 | 4.7 | 7.1 | 564 | 1104 | 4010 | 7850 | 280,700 | 549,500 |
| | | 31322 | | | 4289 | 3301 | 30,496 | 23,471 | 2,134,720 | 1,642,970 |
| 6 | Mediterranean pine forests | 3121 | 0 | 0.0 | 0 | 0 | 0 | 0 | 0 | 0 |
| | | 31321 | | | 0 | 0 | 0 | 0 | 0 | 0 |
| 7 | Other coniferous forests, pure or mixed | 3125 | 2.7 | 4.1 | 0 | 0 | 0 | 0 | 0 | 0 |
| | | 31325 | | | 0 | 0 | 0 | 0 | 0 | 0 |
| 8 | Beech forests | 3115 | 4.2 | 6.4 | 6735 | 7025 | 42,793 | 44,636 | 2,995,510 | 3,124,520 |
| | | 31315 | | | 1759 | 2420 | 11,176 | 15,376 | 782,320 | 1,076,320 |
| 9 and 10 | Sessile oak, Downy oak, and English oak forests and Turkey oak, Hungarian oak, Macedonian oak, Valonia oak forests | 3112 | 5.5 | 8.3 | 1485 | 1513 | 12,356 | 12,589 | 864,920 | 881,230 |
| | | 31312 | | | 0 | 0 | 0 | 0 | 0 | 0 |
| 11 | Chestnut forest | 3114 | 5 | 7.6 | 21,063 | 19,136 | 159,323 | 144,747 | 11,152,610 | 10,132,290 |
| | | 31314 | | | 2192 | 2546 | 16,581 | 19,258 | 1,160,670 | 1,348,060 |
| 12 | Hop-hornbeam and hornbeam forests | 3113 | 2.7 | 4.1 | 51,657 | 47,938 | 210,999 | 195,808 | 14,769,930 | 13,706,560 |
| | | 31313 | | | 11,150 | 13,297 | 45,543 | 54,313 | 3,188,010 | 3,801,910 |
| 13 | Hygrophilous forests | 3116 | 3.9 | 5.9 | 125 | 198 | 738 | 1168 | 51,660 | 81,760 |
| | | 31316 | | | 0 | 0 | 0 | 0 | 0 | 0 |
| 14 | Other deciduous forests | 3117 | 4.3 | 6.5 | 116 | 107 | 755 | 696 | 52,850 | 48,720 |
| | | 31317 | | | 0 | 0 | 0 | 0 | 0 | 0 |
| 15 and 16 | Holm oak forests and Cork oak forests | 3111 | 0 | 0.0 | 272 | 278 | 0 | 0 | 0 | 0 |
| | | 31311 | | | 37 | 37 | 0 | 0 | 0 | 0 |
| Total | | | | | | | | | 64,748,740 | 64,003,170 |

### Appendix A.5. Particulate Removal

**Table A10.** Biophysical surfaces and monetary value of particulate removal for each CLC LC class in 2012 and 2018. Source: own elaborations by authors.

| CLC | | Absorption Coefficient | Unit Monetary Value | Surface (ha) | | Monetary Value (EUR/Year) | |
|---|---|---|---|---|---|---|---|
| Code | Description | (kg/(ha Year)) | (EUR/ha/Year) | 2006 | 2012 | 2006 | 2012 |
| 311 | Broadleaf forests | 160 | 500 | 76,195 | 87,040 | 38,097,500 | 43,520,000 |
| 312 | Coniferous forests | 490 | 700 | 40,406 | 20,703 | 28,284,200 | 14,492,100 |
| 313 | Mixed forests | 325 | 600 | 37,882 | 45,707 | 22,729,200 | 27,424,200 |
| Total | | | | | | 89,110,900 | 85,436,300 |

### Appendix A.6. Hydrological Regime Regulation Table

**Table A11.** Biophysical surfaces and monetary value of hydrological regime regulation for each CLC LC class in 2012 and 2018. Source: own elaborations by authors from ARPA Lombardia data.

| CLC—Code | Absorption Coefficient | | Average Annual Absorption (m/Year) | Water Volume Absorbed ($m^3$/ha/Year) | Surface (ha) | | Monetary Value (EUR/Year) | |
|---|---|---|---|---|---|---|---|---|
| | | | | | 2012 | 2018 | 2012 | 2018 |
| 111 | 0% | 0 | 0.000 | 0 | 1059 | 1059 | 0 | 0 |
| 112 | 0% | 0 | 0.000 | 0 | 27,786 | 28,226 | 0 | 0 |
| 121 | 0% | 0 | 0.000 | 0 | 8771 | 9195 | 0 | 0 |
| 122 | 0% | 0 | 0.000 | 0 | 98 | 323 | 0 | 0 |
| 124 | 0% | 0 | 0.000 | 0 | 923 | 929 | 0 | 0 |
| 131 | 0% | 0 | 0.000 | 0 | 1692 | 1835 | 0 | 0 |
| 133 | 0% | 0 | 0.000 | 0 | 99 | 91 | 0 | 0 |
| 141 | 0% | 0 | 0.000 | 0 | 81 | 54 | 0 | 0 |
| 142 | 0% | 0 | 0.000 | 0 | 433 | 740 | 0 | 0 |
| 221 | 17% | 3 | 0.154 | 1540 | 552 | 698 | 8,730,322 | 11,039,428 |
| 222 | 17% | 3 | 0.154 | 1540 | 0 | 25 | 0 | 395,395 |
| 223 | 17% | 3 | 0.154 | 1540 | 521 | 511 | 8,240,032 | 8,081,874 |
| 224 | 17% | 3 | 0.154 | 1540 | 4 | 4 | 63,263 | 63,263 |
| 231 | 15% | 2 | 0.136 | 1360 | 4521 | 4447 | 63,145,711 | 62,112,138 |
| 241 | 15% | 2 | 0.136 | 1360 | 54 | 54 | 754,229 | 754,229 |
| 242 | 15% | 2 | 0.136 | 1360 | 10,049 | 9654 | 140,356,393 | 134,839,349 |
| 243 | 15% | 2 | 0.136 | 1360 | 22,374 | 22,483 | 312,502,133 | 314,024,558 |
| 322 | 15% | 2 | 0.136 | 1360 | 3735 | 5010 | 52,167,492 | 69,975,672 |
| 323 | 15% | 2 | 0.136 | 1360 | 44 | 44 | 614,557 | 614,557 |
| 324 | 17% | 3 | 0.154 | 1540 | 16,901 | 16,169 | 267,302,836 | 255,725,670 |
| 332 | 0% | 0 | 0.000 | 0 | 14,668 | 13,299 | 0 | 0 |
| 333 | 0% | 0 | 0.000 | 0 | 11,487 | 12,273 | 0 | 0 |
| 334 | 0% | 0 | 0.000 | 0 | 0 | 53 | 0 | 0 |
| 335 | 0% | 0 | 0.000 | 0 | 1965 | 1933 | 0 | 0 |
| 411 | 0% | 0 | 0.000 | 0 | 258 | 233 | 0 | 0 |
| 511 | 0% | 0 | 0.000 | 0 | 68 | 68 | 0 | 0 |
| 512 | 0% | 0 | 0.000 | 0 | 23,900 | 23,927 | 0 | 0 |
| 2111 | 5% | 1 | 0.045 | 450 | 144,663 | 143,777 | 668,560,055 | 664,465,406 |
| 2112 | 5% | 1 | 0.045 | 450 | 92 | 92 | 425,178 | 425,178 |
| 3111 | 20% | 4 | 0.182 | 1820 | 278 | 251 | 5,196,209 | 4,691,541 |
| 3112 | 20% | 4 | 0.182 | 1820 | 1513 | 4067 | 28,280,088 | 76,017,924 |

**Table A11.** *Cont.*

| CLC—Code | Absorption Coefficient | | Average Annual Absorption (m/Year) | Water Volume Absorbed ($m^3$/ha/Year) | Surface (ha) | | Monetary Value (EUR/Year) | |
|---|---|---|---|---|---|---|---|---|
| | | | | | 2012 | 2018 | 2012 | 2018 |
| 3113 | 20% | 4 | 0.182 | 1820 | 47,937 | 49,500 | 896,009,642 | 925,224,300 |
| 3114 | 20% | 4 | 0.182 | 1820 | 19,137 | 21,962 | 357,697,322 | 410,500,527 |
| 3115 | 20% | 4 | 0.182 | 1820 | 7024 | 10,894 | 131,288,394 | 203,624,112 |
| 3116 | 20% | 4 | 0.182 | 1820 | 198 | 230 | 370,0897 | 4,299,022 |
| 3117 | 20% | 4 | 0.182 | 1820 | 107 | 132 | 1,999,980 | 2,467,265 |
| 3121 | 30% | 5 | 0.272 | 2720 | 0 | 37 | 0 | 1,033,573 |
| 3122 | 30% | 5 | 0.272 | 2720 | 1104 | 1652 | 30,839,578 | 46,147,629 |
| 3123 | 30% | 5 | 0.272 | 2720 | 33,652 | 32,782 | 940,048,429 | 915,745,501 |
| 3124 | 30% | 5 | 0.272 | 2720 | 5649 | 11,236 | 157,801,426 | 313,870,918 |
| 3131 | 30% | 5 | 0.272 | 2720 | 18,300 | 10,978 | 511,171,586 | 306,663,843 |
| 3132 | 30% | 5 | 0.272 | 2720 | 19,581 | 9724 | 546,983,486 | 271,634,106 |
| 3211 | 15% | 2 | 0.136 | 1360 | 19,894 | 21,118 | 277,863,477 | 294,959,330 |
| 3212 | 15% | 2 | 0.136 | 1360 | 7153 | 6512 | 99,907,382 | 90,954,406 |
| 3231 | 15% | 2 | 0.136 | 1360 | 0 | 44 | 0 | 614,557 |
| Total | | | | | | | 5,511,650,093 | 5,390,965,269 |

## References


1. Marsh, G.P.; Vogt, K.; Haime, J.; Kellogg, E.C. *Report Made Under Authority of the Legislature of Vermont, on the Artificial Propagation of Fish*; Free Press Print: Burlington, VT, USA, 1857.
2. Craig, R.K. George Perkins Marsh: Anticipating the Anthropocene. *SSRN J.* **2019**, *346*, 6. [CrossRef]
3. Cox, L.M.; Almeter, A.L.; Saterson, K.A. Protecting Our Life Support Systems: An Inventory of U.S. Federal Research on Ecosystem Services. *Ecosyst. Serv.* **2013**, *5*, 163–169. [CrossRef]
4. *Ecosystems and Human Well-Being: Synthesis*; Millennium Ecosystem Assessment (Program), Ed.; Island Press: Washington, DC, USA, 2005; ISBN 978-1-59726-040-4.
5. Costanza, R.; d'Arge, R.; de Groot, R.; Farber, S.; Grasso, M.; Hannon, B.; Limburg, K.; Naeem, S.; O'Neill, R.V.; Paruelo, J.; et al. The Value of the World's Ecosystem Services and Natural Capital. *Nature* **1997**, *387*, 253–260. [CrossRef]
6. *The Economics of Ecosystems and Biodiversity: Ecological and Economic Foundations*, 1st ed.; Kumar, P., Ed.; Routledge: Oxfordshire, UK, 2012; ISBN 978-1-136-53880-3.
7. Costanza, R.; De Groot, R.; Sutton, P.; Van Der Ploeg, S.; Anderson, S.J.; Kubiszewski, I.; Farber, S.; Turner, R.K. Changes in the Global Value of Ecosystem Services. *Glob. Environ. Change* **2014**, *26*, 152–158. [CrossRef]
8. Adhikari, K.; Hartemink, A.E. Linking Soils to Ecosystem Services—A Global Review. *Geoderma* **2016**, *262*, 101–111. [CrossRef]
9. Farrell, C.A.; Aronson, J.; Daily, G.C.; Hein, L.; Obst, C.; Woodworth, P.; Stout, J.C. Natural Capital Approaches: Shifting the UN Decade on Ecosystem Restoration from Aspiration to Reality. *Restor. Ecol.* **2022**, *30*, e13613. [CrossRef]
10. United Nations. *System of Environmental-Economic Accounting—Ecosystem Accounting (SEEA EA)*; Pre-edited Text Subject to Official Editing; United Nations: New York, NY, USA, 2021.
11. Costanza, R.; Daly, H.E. Natural Capital and Sustainable Development. *Conserv. Biol.* **1992**, *6*, 37–46. [CrossRef]
12. Smith, A.C.; Harrison, P.A.; Pérez Soba, M.; Archaux, F.; Blicharska, M.; Egoh, B.N.; Erős, T.; Fabrega Domenech, N.; György, Á.I.; Haines-Young, R.; et al. How Natural Capital Delivers Ecosystem Services: A Typology Derived from a Systematic Review. *Ecosyst. Serv.* **2017**, *26*, 111–126. [CrossRef]
13. Nunes, P.A.L.D.; Van Den Bergh, J.C.J.M. Economic Valuation of Biodiversity: Sense or Nonsense? *Ecol. Econ.* **2001**, *39*, 203–222. [CrossRef]
14. Turner, R.K.; Van Den Bergh, J.C.J.M.; Söderqvist, T.; Barendregt, A.; Van Der Straaten, J.; Maltby, E.; Van Ierland, E.C. Ecological-Economic Analysis of Wetlands: Scientific Integration for Management and Policy. *Ecol. Econ.* **2000**, *35*, 7–23. [CrossRef]
15. Sustainable Development Solutions Network; The Institute for European Environmental Policy. *The 2020 Europe Sustainable Development Report: Meeting the Sustainable Development Goals in the Face of the COVID-19 Pandemic*; Sustainable Development Solutions Network (SDSN) EU: Paris, France; Bruxelles, Belgium, 2020.
16. O'Neill, J. What Is Lost through No Net Loss. *Econ. Philos.* **2020**, *36*, 287–306. [CrossRef]
17. Femia, A.; Fullone, F.; Lecardane, G. *Ambiente e Territorio: Strumenti e Metodi per un'Analisi del Consumo di Risorse e Degli Ecosistemi*; Istituto Nazionale di Statistica (ISTAT): Rome, Italy, 2023.

18. Reed, M.S.; Allen, K.; Attlee, A.; Dougill, A.J.; Evans, K.L.; Kenter, J.O.; Hoy, J.; McNab, D.; Stead, S.M.; Twyman, C.; et al. A Place-Based Approach to Payments for Ecosystem Services. *Glob. Environ. Change* **2017**, *43*, 92–106. [CrossRef]
19. De Castro Pardo, M.; Martínez, P.F.; Martínez, J.M.G.; Martín, J.M.M. Modelling Natural Capital: A Proposal for a Mixed Multi-Criteria Approach to Assign Management Priorities to Ecosystem Services. *CE* **2020**, *14*, 22–37. [CrossRef]
20. Öberg, C.; Alexander, A.T. The Openness of Open Innovation in Ecosystems—Integrating Innovation and Management Literature on Knowledge Linkages. *J. Innov. Knowl.* **2019**, *4*, 211–218. [CrossRef]
21. Grimm, N.B.; Faeth, S.H.; Golubiewski, N.E.; Redman, C.L.; Wu, J.; Bai, X.; Briggs, J.M. Global Change and the Ecology of Cities. *Science* **2008**, *319*, 756–760. [CrossRef] [PubMed]
22. IPBES. *Global Assessment Report on Biodiversity and Ecosystem Services of the Intergovernmental Science-Policy Platform on Biodiversity and Ecosystem Services*; Zenodo; IPBES: Bonn, Germany, 2019.
23. Shah, M.I.; Abbas, S.; Olohunlana, A.O.; Sinha, A. The Impacts of Land Use Change on Biodiversity and Ecosystem Services: An Empirical Investigation from Highly Fragile Countries. *Sustain. Dev.* **2023**, *31*, 1384–1400. [CrossRef]
24. Nowak, D.J.; Walton, J.T. Projected Urban Growth (2000–2050) and Its Estimated Impact on the US Forest Resource. *J. For.* **2005**, *103*, 383–389. [CrossRef]
25. Song, X.-P.; Hansen, M.C.; Stehman, S.V.; Potapov, P.V.; Tyukavina, A.; Vermote, E.F.; Townshend, J.R. Global Land Change from 1982 to 2016. *Nature* **2018**, *560*, 639–643. [CrossRef] [PubMed]
26. European Commission; Directorate General for Research and Innovation. *Caring for Soil Is Caring for Life: Ensure 75% of Soils Are Healthy by 2030 for Food, People, Nature and Climate: Report of the Mission Board for Soil Health and Food*; Publications Office: Luxembourg, 2020.
27. Sutton, P.C.; Anderson, S.J.; Costanza, R.; Kubiszewski, I. The Ecological Economics of Land Degradation: Impacts on Ecosystem Service Values. *Ecol. Econ.* **2016**, *129*, 182–192. [CrossRef]
28. Jack, B.K.; Kousky, C.; Sims, K.R.E. Designing Payments for Ecosystem Services: Lessons from Previous Experience with Incentive-Based Mechanisms. *Proc. Natl. Acad. Sci. USA* **2008**, *105*, 9465–9470. [CrossRef]
29. Huwyler, F.; Käppeli, J.; Tobin, J. *Conservation Finance. From Niche to Mainstream: The Building of an Institutional Asset Class*; Credit Suisse, IUCN: Gland, Switzerland; Rockefeller Foundation: New York, NY, USA; McKinsey: Chicago, IL, USA, 2016.
30. Balmford, A.; Bruner, A.; Cooper, P.; Costanza, R.; Farber, S.; Green, R.E.; Jenkins, M.; Jefferiss, P.; Jessamy, V.; Madden, J.; et al. Economic Reasons for Conserving Wild Nature. *Science* **2002**, *297*, 950–953. [CrossRef] [PubMed]
31. International Institute for Environment and Development. *Reaping the Rewards: Financing Land Degradation Neutrality*; United Nations Convention to Combat Desertification: Bonn, Germany, 2015.
32. Crouzeilles, R.; Curran, M.; Ferreira, M.S.; Lindenmayer, D.B.; Grelle, C.E.V.; Rey Benayas, J.M. A Global Meta-Analysis on the Ecological Drivers of Forest Restoration Success. *Nat. Commun.* **2016**, *7*, 11666. [CrossRef] [PubMed]
33. Ekins, P.; Simon, S.; Deutsch, L.; Folke, C.; De Groot, R. A Framework for the Practical Application of the Concepts of Critical Natural Capital and Strong Sustainability. *Ecol. Econ.* **2003**, *44*, 165–185. [CrossRef]
34. Chee, Y.E. An Ecological Perspective on the Valuation of Ecosystem Services. *Biol. Conserv.* **2004**, *120*, 549–565. [CrossRef]
35. Daily, G.; Postel, S.; Bawa, K.; Kaufman, L. *Nature's Services: Societal Dependence on Natural Ecosystems*; Bibliovault OAI Repository, the University of Chicago Press: Chicago, IL, USA, 1997.
36. Chan, K.M.A.; Balvanera, P.; Benessaiah, K.; Chapman, M.; Díaz, S.; Gómez-Baggethun, E.; Gould, R.; Hannahs, N.; Jax, K.; Klain, S.; et al. Why Protect Nature? Rethinking Values and the Environment. *Proc. Natl. Acad. Sci. USA* **2016**, *113*, 1462–1465. [CrossRef]
37. Ruhl, J.; Salzman, J.; Arnold, C.A.; Craig, R.; Hirokawa, K.; Olander, L.; Palmer, M.; Ricketts, T.H. Connecting Ecosystem Services Science and Policy in the Field. *Front. Ecol. Environ.* **2021**, *19*, 519–525. [CrossRef]
38. Salzman, J. Teaching Policy Instrument Choice in Environmental Law: The Five P's. *23 Duke Environ. Law Policy Forum* **2013**, *23*, 363–376.
39. Bateman, I.J.; Harwood, A.R.; Mace, G.M.; Watson, R.T.; Abson, D.J.; Andrews, B.; Binner, A.; Crowe, A.; Day, B.H.; Dugdale, S.; et al. Bringing Ecosystem Services into Economic Decision-Making: Land Use in the United Kingdom. *Science* **2013**, *341*, 45–50. [CrossRef]
40. Pirard, R.; Lapeyre, R. Classifying Market-Based Instruments for Ecosystem Services: A Guide to the Literature Jungle. *Ecosyst. Serv.* **2014**, *9*, 106–114. [CrossRef]
41. Wunder, S. When Payments for Environmental Services Will Work for Conservation. *Conserv. Lett.* **2013**, *6*, 230–237. [CrossRef]
42. Marion, J.L. A Review and Synthesis of Recreation Ecology Research Supporting Carrying Capacity and Visitor Use Management Decisionmaking. *J. For.* **2016**, *114*, 339–351. [CrossRef]

43. Schirpke, U.; Scolozzi, R.; Da Re, R.; Masiero, M.; Pellegrino, D.; Marino, D. Enhancing Outdoor Recreation and Biodiversity through Payments for Ecosystem Services: Emerging Potentials from Selected Natura 2000 Sites in Italy. *Environ. Dev. Sustain.* **2020**, *22*, 2045–2067. [CrossRef]
44. *Ecosystems and Human Well-Being: A Manual for Assessment Practitioners*; Ash, N., Ed.; Island Press: Washington, DC, USA, 2010; ISBN 978-1-59726-711-3.
45. Heal, G. *Nature and the Marketplace: Capturing the Value of Ecosystem Services*; Island Press: Washington, DC, USA, 2000; ISBN 978-1-55963-795-4.
46. *Mainstreaming the Economics of Nature: A Synthesis of the Approach, Conclusions and Recommendations of Teeb*; UNEP, Ed.; The Economics of Ecosystems & Biodiversity; UNEP: Geneva, Switzerland, 2010; ISBN 978-3-9813410-3-4.
47. Martin-Ortega, J.; Ojea, E.; Roux, C. Payments for Water Ecosystem Services in Latin America: A Literature Review and Conceptual Model. *Ecosyst. Serv.* **2013**, *6*, 122–132. [CrossRef]
48. *Water Ecosystem Services: A Global Perspective*; Martin-Ortega, J., Ferrier, R.C., Gordon, I.J., Khan, S., Eds.; International Hydrology Series; Cambridge University Press: Cambridge, UK, 2015; ISBN 978-92-3-100068-3.
49. Porras, I.; Asquith, N. *Ecosystems, Poverty Alleviation and Conditional Transfers*; International Institute for Environment and Development: London, UK, 2018.
50. Wunder, S. Revisiting the Concept of Payments for Environmental Services. *Ecol. Econ.* **2015**, *117*, 234–243. [CrossRef]
51. Rissman, A.R.; Lozier, L.; Comendant, T.; Kareiva, P.; Kiesecker, J.M.; Shaw, M.R.; Merenlender, A.M. Conservation Easements: Biodiversity Protection and Private Use. *Conserv. Biol.* **2007**, *21*, 709–718. [CrossRef] [PubMed]
52. Merenlender, A.M.; Huntsinger, L.; Guthey, G.; Fairfax, S.K. Land Trusts and Conservation Easements: Who Is Conserving What for Whom? *Conserv. Biol.* **2004**, *18*, 65–76. [CrossRef]
53. Loomis, J.; Kent, P.; Strange, L.; Fausch, K.; Covich, A. Measuring the Total Economic Value of Restoring Ecosystem Services in an Impaired River Basin: Results from a Contingent Valuation Survey. *Ecol. Econ.* **2000**, *33*, 103–117. [CrossRef]
54. Rissman, A.R.; Merenlender, A.M. The Conservation Contributions of Conservation Easements: Analysis of the San Francisco Bay Area Protected Lands Spatial Database. *Ecol. Soc.* **2008**, *13*, 40. [CrossRef]
55. Smith, F.; Smillie, K.; Fitzsimons, J.; Lindsay, B.; Wells, G.; Marles, V.; Hutchinson, J.; O'Hara, B.; Perrigo, T.; Atkinson, I. Reforms Required to the Australian Tax System to Improve Biodiversity Conservation on Private Land. *Environ. Plan. Law J.* **2016**, *33*, 443–450.
56. Schuster, R.; Law, E.A.; Rodewald, A.D.; Martin, T.G.; Wilson, K.A.; Watts, M.; Possingham, H.P.; Arcese, P. Tax Shifting and Incentives for Biodiversity Conservation on Private Lands. *Conserv. Lett.* **2018**, *11*, e12377. [CrossRef]
57. Asbjornsen, H.; Wang, Y.; Ellison, D.; Ashcraft, C.M.; Atallah, S.S.; Jones, K.; Mayer, A.; Altamirano, M.; Yu, P. Multi-Targeted Payments for the Balanced Management of Hydrological and Other Forest Ecosystem Services. *For. Ecol. Manag.* **2022**, *522*, 120482. [CrossRef]
58. Schuster, R.; Arcese, P. Efficient Routes to Land Conservation given Risk of Covenant Failure. *PeerJ PrePrints* **2015**, *3*, e1033v1.
59. Farmer, J.R.; Ma, Z.; Drescher, M.; Knackmuhs, E.G.; Dickinson, S.L. Private Landowners, Voluntary Conservation Programs, and Implementation of Conservation Friendly Land Management Practices. *Conserv. Lett.* **2017**, *10*, 58–66. [CrossRef]
60. Paladini, L. *In Sintesi: Il Piano Strategico della PAC dell'ITALIA*; Agriculture and Rural Development–European Commission: Brussels, Belgium, 2023.
61. Ding, Z.; Zheng, H.; Wang, J.; O'Connor, P.; Li, C.; Chen, X.; Li, R.; Ouyang, Z. Integrating Top-Down and Bottom-Up Approaches Improves Practicality and Efficiency of Large-Scale Ecological Restoration Planning: Insights from a Social–Ecological System. *Engineering* **2023**, *31*, 50–58. [CrossRef]
62. Masiero, M.; Franceschinis, C.; Mattea, S.; Thiene, M.; Pettenella, D.; Scarpa, R. Ecosystem Services' Values and Improved Revenue Collection for Regional Protected Areas. *Ecosyst. Serv.* **2018**, *34*, 136–153. [CrossRef]
63. Daily, G.C.; Polasky, S.; Goldstein, J.; Kareiva, P.M.; Mooney, H.A.; Pejchar, L.; Ricketts, T.H.; Salzman, J.; Shallenberger, R. Ecosystem Services in Decision Making: Time to Deliver. *Front. Ecol. Environ.* **2009**, *7*, 21–28. [CrossRef]
64. Assennato, F.; Braca, G.; Calzolari, C.; Capriolo, A.; di Leginio, M.; Marchetti, M.; Marino, D.; Mascolo, R.; Morri, E.; Pettenella, D.; et al. Mappatura e valutazione dell'impatto del consumo di suolo sui servizi ecosistemici: Proposte metodologiche per il Rapporto sul consumo di suolo. *Soil Consum. Territ. Dyn. Ecosyst. Serv.* **2018**, 1–44.
65. Morya, C.P.; Punia, M. Impact of Urbanization Processes on Availability of Ecosystem Services in National Capital Region of Delhi (1992–2010). *Environ. Dev. Sustain.* **2022**, *24*, 7324–7348. [CrossRef] [PubMed]
66. La Notte, A. Mapping and Valuing Habitat Services: Two Applications at Local Scale. *Int. J. Biodivers. Sci. Ecosyst. Serv. Manag.* **2012**, *8*, 80–92. [CrossRef]
67. Häyhä, T.; Franzese, P.P.; Paletto, A.; Fath, B.D. Assessing, Valuing, and Mapping Ecosystem Services in Alpine Forests. *Ecosyst. Serv.* **2015**, *14*, 12–23. [CrossRef]

68. Manes, F.; Marando, F.; Capotorti, G.; Blasi, C.; Salvatori, E.; Fusaro, L.; Ciancarella, L.; Mircea, M.; Marchetti, M.; Chirici, G.; et al. Regulating Ecosystem Services of Forests in Ten Italian Metropolitan Cities: Air Quality Improvement by $PM_{10}$ and $O_3$ Removal. *Ecol. Indic.* **2016**, *67*, 425–440. [CrossRef]
69. Marino, D.; Palmieri, M.; Marucci, A.; Tufano, M. Comparison between Demand and Supply of Some Ecosystem Services in National Parks: A Spatial Analysis Conducted Using Italian Case Studies. *Conservation* **2021**, *1*, 36–57. [CrossRef]
70. Sebastiani, A.; Marando, F.; Manes, F. Mismatch of Regulating Ecosystem Services for Sustainable Urban Planning: PM10 Removal and Urban Heat Island Effect Mitigation in the Municipality of Rome (Italy). *Urban For. Urban Green.* **2021**, *57*, 126938. [CrossRef]
71. Schirpke, U.; Leitinger, G.; Tappeiner, U.; Tasser, E. SPA-LUCC: Developing Land-Use/Cover Scenarios in Mountain Landscapes. *Ecol. Inform.* **2012**, *12*, 68–76. [CrossRef]
72. Petrosillo, I.; Semeraro, T.; Zaccarelli, N.; Aretano, R.; Zurlini, G. The Possible Combined Effects of Land-Use Changes and Climate Conditions on the Spatial–Temporal Patterns of Primary Production in a Natural Protected Area. *Ecol. Indic.* **2013**, *29*, 367–375. [CrossRef]
73. Ferrari, M.; Geneletti, D. Mapping and Assessing Multiple Ecosystem Services in an Alpine Region: A Study in Trentino, Italy. *Ann. Bot.* **2014**, *4*, 65–71. [CrossRef]
74. Rova, S.; Pranovi, F.; Müller, F. Provision of Ecosystem Services in the Lagoon of Venice (Italy): An Initial Spatial Assessment. *Ecohydrol. Hydrobiol.* **2015**, *15*, 13–25. [CrossRef]
75. Arcidiacono, A.; Ronchi, S.; Salata, S. Managing Multiple Ecosystem Services for Landscape Conservation: A Green Infrastructure in Lombardy Region. *Procedia Eng.* **2016**, *161*, 2297–2303. [CrossRef]
76. Franzese, P.P.; Buonocore, E.; Donnarumma, L.; Russo, G.F. Natural Capital Accounting in Marine Protected Areas: The Case of the Islands of Ventotene and S. Stefano (Central Italy). *Ecol. Model.* **2017**, *360*, 290–299. [CrossRef]
77. Picone, F.; Buonocore, E.; D'Agostaro, R.; Donati, S.; Chemello, R.; Franzese, P.P. Integrating Natural Capital Assessment and Marine Spatial Planning: A Case Study in the Mediterranean Sea. *Ecol. Model.* **2017**, *361*, 1–13. [CrossRef]
78. Mancini, L.D.; Corona, P.; Salvati, L. Ranking the Importance of Wildfires' Human Drivers through a Multi-Model Regression Approach. *Environ. Impact Assess. Rev.* **2018**, *72*, 177–186. [CrossRef]
79. Salata, S.; Giaimo, C.; Barbieri, C.A.; Garnero, G. The Utilization of Ecosystem Services Mapping in Land Use Planning: The Experience of LIFE SAM4CP Project. *J. Environ. Plan. Manag.* **2019**, *63*, 523–545. [CrossRef]
80. Sacchelli, S.; Bernetti, I. Integrated Management of Forest Ecosystem Services: An Optimization Model Based on Multi-Objective Analysis and Metaheuristic Approach. *Nat. Resour. Res.* **2019**, *28*, 5–14. [CrossRef]
81. Salizzoni, E.; Allocco, M.; Murgese, D.; Quaglio, G. From Ecosystem Service Evaluation to Landscape Design: The Project of a Rural Peri-Urban Park in Chieri (Italy). In *Values and Functions for Future Cities*; Mondini, G., Oppio, A., Stanghellini, S., Bottero, M., Abastante, F., Eds.; Green Energy and Technology; Springer International Publishing: Cham, Switzerland, 2020; pp. 267–283. ISBN 978-3-030-23784-4.
82. Capriolo, A.; Boschetto, R.G.; Mascolo, R.A.; Balbi, S.; Villa, F. Biophysical and Economic Assessment of Four Ecosystem Services for Natural Capital Accounting in Italy. *Ecosyst. Serv.* **2020**, *46*, 101207. [CrossRef]
83. Buonocore, E.; Donnarumma, L.; Appolloni, L.; Miccio, A.; Russo, G.F.; Franzese, P.P. Marine Natural Capital and Ecosystem Services: An Environmental Accounting Model. *Ecol. Model.* **2020**, *424*, 109029. [CrossRef]
84. Buonocore, E.; Appolloni, L.; Russo, G.F.; Franzese, P.P. Assessing Natural Capital Value in Marine Ecosystems through an Environmental Accounting Model: A Case Study in Southern Italy. *Ecol. Model.* **2020**, *419*, 108958. [CrossRef]
85. Pacetti, T.; Castelli, G.; Bresci, E.; Caporali, E. Water Values: Participatory Water Ecosystem Services Assessment in the Arno River Basin, Italy. *Water Resour. Manag.* **2020**, *34*, 4527–4544. [CrossRef]
86. Masiero, M.; Biasin, A.; Amato, G.; Malaggi, F.; Pettenella, D.; Nastasio, P.; Anelli, S. Urban Forests and Green Areas as Nature-Based Solutions for Brownfield Redevelopment: A Case Study from Brescia Municipal Area (Italy). *Forests* **2022**, *13*, 444. [CrossRef]
87. Di Pirro, E.; Sallustio, L.; Capotorti, G.; Marchetti, M.; Lasserre, B. A Scenario-Based Approach to Tackle Trade-Offs between Biodiversity Conservation and Land Use Pressure in Central Italy. *Ecol. Model.* **2021**, *448*, 109533. [CrossRef]
88. Zulian, G.; La Notte, A. How to Account for Nature-Based Tourism in Europe. An Operational Proposal. *One Ecosyst.* **2022**, *7*, e89312. [CrossRef]
89. Boschetto, R.G.; Capriolo, A.; Mascolo, R.A.; Arrigotti, J.; Racevich, S.; Bulckaen, A.; Balbi, S.; Villa, F. Analysis of Changes over Time in Four Provisioning Ecosystem Services in Italy. *J. Environ. Account. Manag.* **2023**, *11*, 91–114. [CrossRef]
90. Marino, D.; Barone, A.; Marucci, A.; Pili, S.; Palmieri, M. Impact of Land Use Changes on Ecosystem Services Supply: A Meta Analysis of the Italian Context. *Land* **2023**, *12*, 2173. [CrossRef]

91. Catucci, E.; Buonocore, E.; Franzese, P.P.; Scardi, M. Assessing the Natural Capital Value of *Posidonia oceanica* Meadows in the Italian Seas by Integrating Habitat Suitability and Environmental Accounting Models. *ICES J. Mar. Sci.* **2023**, *80*, 739–750. [CrossRef]
92. Pignatti, S.; Carfora, M.F.; Coluzzi, R.; D'Amato, L.; De Feis, I.; Mora, D.F.; Laneve, G.; Imbrenda, V.; Lanfredi, M.; Mirzaei, S.; et al. Detection of Critical Areas Prone to Land Degradation Using Prisma: The Metaponto Coastal Area in South Italy Test Case. In Proceedings of the IGARSS 2024—2024 IEEE International Geoscience and Remote Sensing Symposium, Athens, Greece, 7–12 July 2024; pp. 1063–1066.
93. Pelorosso, R.; Gobattoni, F.; Geri, F.; Monaco, R.; Leone, A. Evaluation of Ecosystem Services Related to Bio-Energy Landscape Connectivity (BELC) for Land Use Decision Making across Different Planning Scales. *Ecol. Indic.* **2016**, *61*, 114–129. [CrossRef]
94. Schirpke, U.; Scolozzi, R.; Concetti, B.; Comini, B.; Tappeiner, U. Supporting the Management of Ecosystem Services in Protected Areas: Trade-Offs Between Effort and Accuracy in Evaluation. *J. Environ. Assess. Policy Manag.* **2017**, *19*, 1750007. [CrossRef]
95. Appolloni, L.; Sandulli, R.; Vetrano, G.; Russo, G.F. A New Approach to Assess Marine Opportunity Costs and Monetary Values-in-Use for Spatial Planning and Conservation; the Case Study of Gulf of Naples, Mediterranean Sea, Italy. *Ocean Coast. Manag.* **2018**, *152*, 135–144. [CrossRef]
96. Lai, S.; Motroni, A.; Santona, L.; Schirru, M. One Place, Different Communities' Perceptions. Mapping Cultural Ecosystem Services in the Asinara National Park (Italy). In *Computational Science and Its Applications–ICCSA 2021*; Gervasi, O., Murgante, B., Misra, S., Garau, C., Blečić, I., Taniar, D., Apduhan, B.O., Rocha, A.M.A.C., Tarantino, E., Torre, C.M., Eds.; Lecture Notes in Computer Science; Springer International Publishing: Cham, Switzerland, 2021; Volume 12955, pp. 675–691; ISBN 978-3-030-87006-5.
97. Gebre, T.; Gebremedhin, B. The Mutual Benefits of Promoting Rural-Urban Interdependence through Linked Ecosystem Services. *Glob. Ecol. Conserv.* **2019**, *20*, e00707. [CrossRef]
98. Schirpke, U.; Scolozzi, R.; De Marco, C. *Modello Dimostrativo di Valutazione Qualitativa e Quantitativa dei Servizi Ecosistemici nei siti Pilota*; EURAC Research: Bolzano, Italy, 2014; p. 75.
99. Regione Lombardia. Strategic Environmental Assessment of PTR 2021. 2021. Available online: https://www.regione.lombardia.it/wps/portal/istituzionale/HP/DettaglioPubblicazione/servizi-e-informazioni/Enti-e-Operatori/territorio/pianificazione-regionale/conferenza-vas-revisione-ptr-pvp/conferenza-vas-revisione-ptr-pvp (accessed on 18 January 2025).
100. Arcidiacono, A.; Ronchi, S.; Salata, S. Ecosystem Services Assessment Using InVEST as a Tool to Support Decision Making Process: Critical Issues and Opportunities. In *Computational Science and Its Applications–ICCSA 2015*; Gervasi, O., Murgante, B., Misra, S., Gavrilova, M.L., Rocha, A.M.A.C., Torre, C., Taniar, D., Apduhan, B.O., Eds.; Lecture Notes in Computer Science; Springer International Publishing: Cham, Switzerland, 2015; Volume 9158, pp. 35–49; ISBN 978-3-319-21409-2.
101. Gaglioppa, P.; Guadagno, R.; Marino, D.; Marucci, A.; Palmieri, M.; Pellegrino, D.; Schirpke, U.; Caracausi, C. Forest management based on ecosystem services and payments for ecosystem services: Considerations after the project LIFE+ Making Good Natura. *Forest@* **2017**, *14*, 99–106. [CrossRef]
102. Leonhardt, S.D.; Gallai, N.; Garibaldi, L.A.; Kuhlmann, M.; Klein, A.-M. Economic Gain, Stability of Pollination and Bee Diversity Decrease from Southern to Northern Europe. *Basic Appl. Ecol.* **2013**, *14*, 461–471. [CrossRef]
103. Alias, C.; Benassi, L.; Bertazzi, L.; Sorlini, S.; Volta, M.; Gelatti, U. Environmental Exposure and Health Effects in a Highly Polluted Area of Northern Italy: A Narrative Review. *Environ. Sci. Pollut. Res.* **2019**, *26*, 4555–4569. [CrossRef]
104. Pezzagno, M.; Richiedei, A.; Tira, M. Spatial Planning Policy for Sustainability: Analysis Connecting Land Use and GHG Emission in Rural Areas. *Sustainability* **2020**, *12*, 947. [CrossRef]
105. Richiedei, A.; Giuliani, M.; Pezzagno, M. Unveiling the Soil beyond Definitions: A Holistic Framework for Sub-Regional Soil Quality Assessment and Spatial Planning. *Sustainability* **2024**, *16*, 6075. [CrossRef]
106. *Climate Change 2007: The Physical Science Basis*; Intergouvernemental Panel on Climate Change, Ed.; Cambridge University Press: Cambridge, UK, 2007; ISBN 978-0-521-88009-1.
107. Eggleston, H.S.; Buendia, L.; Miwa, K.; Ngara, T.; Tanabe, K. *2006 IPCC Guidelines for National Greenhouse Gas Inventories*; IPCC: Geneva, Switzerland, 2006.
108. *Natural Capital Committee Annual Report 2018*; England's Economic Affairs Committee: London, UK, 2018.
109. Sallustio, L.; Quatrini, V.; Geneletti, D.; Corona, P.; Marchetti, M. Assessing Land Take by Urban Development and Its Impact on Carbon Storage: Findings from Two Case Studies in Italy. *Environ. Impact Assess. Rev.* **2015**, *54*, 80–90. [CrossRef]
110. IPCC. *Good Practice Guidance for Land Use, Land-Use Change and Forestry*; Penman, J., Ed.; The Intergovernmental Panel on Climate Change: Hayama, Japan, 2003; ISBN 978-4-88788-003-0.
111. Marchetti, M.; Sallustio, L.; Ottaviano, M.; Barbati, A.; Corona, P.; Tognetti, R.; Zavattero, L.; Capotorti, G. Carbon Sequestration by Forests in the National Parks of Italy. *Plant Biosyst.-Int. J. Deal. All Asp. Plant Biol.* **2012**, *146*, 1001–1011. [CrossRef]
112. Interagency Working Group on Social Cost of Greenhouse Gases, United States Government. *Technical Update of the Social Cost of Carbon for Regulatory Impact Analysis Under Executive Order 12866*; United States Government: Washington, DC, USA, 2016.

113. Maluccio, S.; Grassi, T.; Romano, R. *Progetti Forestali di Sostenibilità in Italia 2021–2022*; Nucleo Monitoraggio del Carbonio, CREA: Roma, Italy, 2024.
114. *Guidelines on Best Practice to Limit, Mitigate or Compensate Soil Sealing*; European Commission, Ed.; Publications Office: Luxembourg, 2012; ISBN 978-92-79-26210-4.
115. Provincia di Brescia. *Documento di Scoping del Piano Territoriale di Coordinamento Provinciale della Provincia di Brescia*, Versione 2.0. 2013. Available online: https://territorioweb.provincia.brescia.it/ptcp-luglio-2016/05_Rapporto%20ambientale%20e%20studio%20di%20incidenza/Documento%20di%20Scoping.pdf (accessed on 18 January 2025).
116. Terrado, M.; Sabater, S.; Chaplin-Kramer, B.; Mandle, L.; Ziv, G.; Acuña, V. Model Development for the Assessment of Terrestrial and Aquatic Habitat Quality in Conservation Planning. *Sci. Total Environ.* **2016**, *540*, 63–70. [CrossRef] [PubMed]
117. Sallustio, L.; De Toni, A.; Strollo, A.; Di Febbraro, M.; Gissi, E.; Casella, L.; Geneletti, D.; Munafò, M.; Vizzarri, M.; Marchetti, M. Assessing Habitat Quality in Relation to the Spatial Distribution of Protected Areas in Italy. *J. Environ. Manag.* **2017**, *201*, 129–137. [CrossRef] [PubMed]
118. Nieto, A.; Roberts, S.P.M.; Kemp, J.; Rasmont, P.; Kuhlmann, M.; García Criado, M.; Biesmeijer, J.C. *European Red List of Bees*; Publications Office of the European Union: Luxembourg, 2014; ISBN 978-92-79-44512-5.
119. *ESTIMAP: Ecosystem Services Mapping at European Scale*; Zulian, G., Paracchini, M.L., Maes, J., Liquete, C., European Commission, Eds.; Publications Office: Luxembourg, 2013; ISBN 978-92-79-35274-4.
120. Klein, A.-M.; Vaissière, B.E.; Cane, J.H.; Steffan-Dewenter, I.; Cunningham, S.A.; Kremen, C.; Tscharntke, T. Importance of Pollinators in Changing Landscapes for World Crops. *Proc. R. Soc. B* **2007**, *274*, 303–313. [CrossRef]
121. European Environment Agency. *Costs of Air Pollution from European Industrial Facilities 2008–2012: An Updated Assessment*; Publications Office: Luxembourg, 2014.
122. *Air Quality in Europe: 2015 Report*; Guerreiro, C., de Leeuw, F., Foltescu, V., González Ortiz, A., Horálek, J., European Environment Agency, Eds.; Publications Office: Luxembourg, 2015; ISBN 978-92-9213-701-4.
123. Nedkov, S.; Burkhard, B. Flood Regulating Ecosystem Services—Mapping Supply and Demand, in the Etropole Municipality, Bulgaria. *Ecol. Indic.* **2012**, *21*, 67–79. [CrossRef]
124. Tate, K.W. Interception on Rangeland Watersheds. *Rangel. Watershed Program Fact Sheet* **1995**, *36*, 1–4. Available online: http://ucanr.edu/sites/UCCE_LR/files/180588.pdf (accessed on 18 January 2025).
125. De Groot, R.; Brander, L.; Van Der Ploeg, S.; Costanza, R.; Bernard, F.; Braat, L.; Christie, M.; Crossman, N.; Ghermandi, A.; Hein, L.; et al. Global Estimates of the Value of Ecosystems and Their Services in Monetary Units. *Ecosyst. Serv.* **2012**, *1*, 50–61. [CrossRef]
126. Stoker, G. Public Value Management: A New Narrative for Networked Governance? *Am. Rev. Public Adm.* **2006**, *36*, 41–57. [CrossRef]
127. Bagstad, K.J.; Villa, F.; Batker, D.; Harrison-Cox, J.; Voigt, B.; Johnson, G.W. From Theoretical to Actual Ecosystem Services: Mapping Beneficiaries and Spatial Flows in Ecosystem Service Assessments. *Ecol. Soc.* **2014**, *19*, art64. [CrossRef]
128. Aryal, K.; Maraseni, T.; Apan, A. How Much Do We Know about Trade-Offs in Ecosystem Services? A Systematic Review of Empirical Research Observations. *Sci. Total Environ.* **2022**, *806*, 151229. [CrossRef] [PubMed]
129. Burkhard, B.; Kandziora, M.; Hou, Y.; Müller, F. Ecosystem Service Potentials, Flows and Demands-Concepts for Spatial Localisation, Indication and Quantification. *Landsc. Online* **2014**, *34*, 1–32. [CrossRef]
130. Tao, Y.; Wang, H.; Ou, W.; Guo, J. A Land-Cover-Based Approach to Assessing Ecosystem Services Supply and Demand Dynamics in the Rapidly Urbanizing Yangtze River Delta Region. *Land Use Policy* **2018**, *72*, 250–258. [CrossRef]
131. Stoll, S.; Frenzel, M.; Burkhard, B.; Loke, R. Assessment of Ecosystem Integrity and Service Gradients across Europe Using the LTER Europe Network. *Ecol. Model.* **2015**, *295*, 75–87. [CrossRef]
132. Cai, W.; Gibbs, D.; Zhang, L.; Ferrier, G.; Cai, Y. Identifying Hotspots and Management of Critical Ecosystem Services in Rapidly Urbanizing Yangtze River Delta Region, China. *J. Environ. Manag.* **2017**, *191*, 258–267. [CrossRef]
133. Hou, Y.; Burkhard, B.; Müller, F. Uncertainties in Landscape Analysis and Ecosystem Service Assessment. *J. Environ. Manag.* **2013**, *127*, S117–S131. [CrossRef]
134. Ruhl, J.B.; Kraft, S.E.; Lant, C.L. *The Law and Policy of Ecosystem Services*; Island Press: Washington, DC, USA, 2007; ISBN 978-1-55963-094-8.
135. Tallis, H.; Kareiva, P.; Marvier, M.; Chang, A. An Ecosystem Services Framework to Support Both Practical Conservation and Economic Development. *Proc. Natl. Acad. Sci. USA* **2008**, *105*, 9457–9464. [CrossRef] [PubMed]
136. Fisher, B.; Turner, R.K.; Morling, P. Defining and Classifying Ecosystem Services for Decision Making. *Ecol. Econ.* **2009**, *68*, 643–653. [CrossRef]
137. Johnson, G.W.; Bagstad, K.J.; Snapp, R.R.; Villa, F. Service Path Attribution Networks (SPANs): A Network Flow Approach to Ecosystem Service Assessment. *Int. J. Agric. Environ. Inf. Syst.* **2012**, *3*, 54–71. [CrossRef]

138. Bagstad, K.J.; Johnson, G.W.; Voigt, B.; Villa, F. Spatial Dynamics of Ecosystem Service Flows: A Comprehensive Approach to Quantifying Actual Services. *Ecosyst. Serv.* **2013**, *4*, 117–125. [CrossRef]

*land*

*Article*

# Habitat Quality Dynamics in Urumqi over the Last Two Decades: Evidence of Land Use and Land Cover Changes

**Siying Chen [1,2], Ümüt Halik [1,2,*], Lei Shi [3,*], Wentao Fu [1,2], Lu Gan [4] and Martin Welp [5]**

1 College of Ecology and Environment, Xinjiang University, Urumqi 830046, China; csy@stu.xju.edu.cn (S.C.); fwindow@stu.xju.edu.cn (W.F.)
2 Key Laboratory of Oasis Ecology, Ministry of Education, Xinjiang University, Urumqi 830046, China
3 School of Architecture and Urban Planning, Shandong Jianzhu University, Jinan 250101, China
4 Postdoctoral Research Station of Theoretical Economics, School of Economics and Management, Xinjiang University, Urumqi 830046, China; lugan@xju.edu.cn
5 Faculty of Forest and Environment, Eberswalde University for Sustainable Development, 16225 Eberswalde, Germany; martin.welp@hnee.de
* Correspondence: halik@xju.edu.cn (Ü.H.); shilei15@mails.ucas.ac.cn (L.S.)

**Abstract:** The integrity of habitat quality is a pivotal cornerstone for the sustainable advancement of local ecological systems. Rapid urbanization has led to habitat degradation and loss of biodiversity, posing severe threats to regional sustainability, particularly in extremely vulnerable arid zones. However, systematic research on the assessment indicators, limiting factors, and driving mechanisms of habitat quality in arid regions is notably lacking. This study takes Urumqi, an oasis city in China's arid region, as a case study and employs the InVEST and PLUS models to conduct a dynamic evaluation of habitat quality in Urumqi from 2000 to 2022 against the backdrop of land use changes. It also simulates habitat quality under different scenarios for the year 2035, exploring the temporal and spatial dynamics of habitat quality and its driving mechanisms. The results indicate a decline in habitat quality. The habitat quality in the southern mountainous areas is significantly superior to that surrounding the northern Gurbantunggut Desert, and it exhibits greater stability. The simulation and prediction results suggest that from 2020 to 2035, habitat degradation will be mitigated under Ecological Protection scenarios, while the decline in habitat quality will be most pronounced under Business-As-Usual scenarios. The spatial distribution of habitat quality changes in Urumqi exhibits significant autocorrelation and clustering, with these patterns intensifying over time. The observed decline in habitat quality in Urumqi is primarily driven by anthropogenic activities, urban expansion, and climate change. These factors have collectively contributed to significant alterations in the landscape, leading to the degradation of ecological conditions. To mitigate further habitat quality loss and support sustainable development, it is essential to implement rigorous ecological protection policies, adopt effective ecological risk management strategies, and promote the expansion of ecological land use. These actions are crucial for stabilizing and improving regional habitat quality in the long term.

**Keywords:** Urumqi; habitat quality; PLUS-InVEST; land use/land cover changes; scenario simulation

## 1. Introduction

Land use change is a key factor influencing regional habitat quality [1]. Habitat quality refers to the capacity of the natural environment to provide the essential conditions necessary for species' survival [2]. Land use change can reflect the status of regional biodiversity and ecosystem health to a certain extent, serving as a crucial guarantee and fundamental prerequisite for the survival and reproduction of organisms [3]. Therefore, the level of habitat quality is directly related to the maintenance of biodiversity and the function of ecosystem services. With the intensification of human activities, especially the rapid advancement of urbanization, the dramatic expansion of impermeable surfaces has had a significant impact on the quality of regional habitats, which in turn poses a great challenge to biodiversity conservation [4], and coupled with global climate change, the quality of habitats and biodiversity poses a serious threat to ecosystems [5]. In particular, urban expansion over the past two decades has encroached upon extensive areas of arable and natural land with high ecological value, resulting in the loss of over 80% of biologically important habitats [6]. This problem is exacerbated in extreme arid regions, where limited water resources and land desertification intensify habitat degradation under human-induced disturbances, making habitats more vulnerable [7]. Without timely intervention, habitat quality will be subjected to substantial threats in the future [8].

To enhance regional habitat quality, numerous scholars have conducted extensive research. Amphibian biologists were among the first to focus on this topic, primarily assessing specific wildlife species or communities through field surveys [9,10]. Swiss scholar Glenz et al. proposed an application of a predictive wolf habitat model to address the issue of large-scale migration and the establishment of small populations in *Canis lupus* populations [9]. With the intensification of urbanization's impact on habitat quality since the Anthropocene, current research primarily focuses on two aspects: First is quantifying habitat quality to identify priority conservation areas [11,12]. Ikaunece et al. conducted a forest key habitat inventory and established a forest network of endangered tree species in order to establish priority protected areas for Nordic endangered tree species [11]. To effectively protect the biodiversity of coastal areas, Zhu et al. utilized geomorphological models, species habitat models, and other methodologies to determine the priority sequence for conservation in response to sea-level rise and land use changes within the Matanzas River Basin [12]. On the other hand, current research explores the impacts of land use changes on habitat quality from both micro and macro perspectives during the urbanization process. At the microlevel, biodiversity is regarded as the best indicator of habitat quality. Researchers have primarily analyzed the effects of land use changes on biodiversity, focusing on plant and animal communities [13,14]. Urban growth's conversion of ecological land to construction land significantly fragments habitats, disrupts ecological flows, and diminishes biodiversity [14]. At the macrolevel, most studies combine InVEST (Integrated Valuation of Ecosystem Services and Trade-offs)and predictive models to assess the impacts of historical and future land use changes on habitat quality [15,16]. Habitat quality is closely related to changes in land use types, with some researchers evaluating the relationship between land use and habitat quality through changes in habitat quality caused by land use transitions [17,18]. Chinese researchers, utilizing Geographically and Temporally Weighted Regression (GTWR) and Multiscale Geographically Weighted Regression (MGWR) models, have identified urban expansion as a primary driver of shifts in habitat quality [19]. Habitat quality is closely related to changes in land use types, and some researchers have evaluated the relationship between land use and habitat quality through changes in habitat quality caused by land use conversion. Wu et al. [20] and Zhang et al. [21] developed frameworks to assess future land use impacts on habitat quality. However, it is important to note that the contribution of land use conversion to changes in habitat quality has rarely been measured across different historical and future scenarios. This is particularly true in varying developmental contexts, where the changes in habitat quality and their driving mechanisms remain poorly understood. Consequently, there is a limited understanding of the long-term dynamic changes in regional habitat quality, as well as inadequate support for regional ecological quality enhancement and sustainable economic development.

Despite current research addressing habitat quality changes across various temporal and spatial scales—such as provincial, metropolitan, protected areas, and river basins—there is relatively little focus on arid regions, and there is a lack of systematic assessment indicators. Arid regions are among the most vulnerable natural landscapes in the world, covering approximately 40% of the Earth's total area. These regions are characterized by scarce precipitation, high evaporation rates, low runoff, and poor soil quality. Their ecological environments are exceptionally fragile and highly sensitive to both climate change and human activities [22]. Since the 21st century, with the advancement of the national Western Development Strategy and the implementation of the Belt and Road Initiative, Urumqi has experienced rapid economic development, population growth, and intensified the development and utilization of water and soil resources. As a result, the city's ecological environment issues have become increasingly prominent [23]. The dynamics and future trends of habitat quality in the region are critical for the country's ecological security and biodiversity conservation. In response, this study focused on Urumqi, an oasis city in Northwest China's arid region. First, the land use transfer matrix from 2000 to 2022 was used to identify the direction and extent of land use changes, as well as analyze their temporal and spatial characteristics. Next, the InVEST-PLUS model simulated land use changes under four scenarios for 2035, measuring historical and future habitat quality to analyze its temporal and spatial evolution: Business-As-Usual (BAU), cropland protection (CP), ecological protection (EP), and economic development (ED). Finally, this study quantitatively assessed the impact of land use changes on habitat quality. The findings provide a theoretical basis for stakeholders to optimize land use structures and improve habitat quality.

## 2. Materials and Methods

### 2.1. Study Area

Urumqi, located in the arid northwest region of China, holds a pivotal position within the core area of the Silk Road Economic Belt. The municipality governs seven districts and one county, covering an area of approximately $1.42 \times 10^4$ km$^2$, with 545.1 km$^2$ developed. The permanent population is 4.085 million. The climate is classified as a mid-temperate continental arid climate, with an average annual temperature of 25.7 °C and annual precipitation of 286.3 mm. Urumqi is surrounded by mountains on three sides, with the Gurbantunggut Desert to the north (Figure 1), which is a typical composite system of "Mountains-Rivers-Forests-Fields-Lakes-Grass-Sand, and Ice" [24]. As a result, it is rich in flora and fauna. Urumqi is home to approximately 1100 species of wild plants and more than 300 species of vertebrates. As the capital city of Xinjiang, Urumqi, driven by the Urumqi metropolitan area, has cultivated a favorable environment for economic growth. However, nature faces numerous challenges. Research on the spatiotemporal evolution and driving mechanisms of habitat quality in oasis cities under the impact of urbanization can provide valuable theoretical guidance for regional sustainable development and ecological balance.

### 2.2. Data Description

The types of data used in this study mainly included spatial data and statistical data. The spatial data were mainly land use data and land use simulations of factors affecting land use. The data on land use and land cover were sourced from the Geospatial Data Cloud. Land use data were obtained through the interpretation of remote sensing imagery. To ensure optimal data accuracy, a series of preprocessing tasks were conducted. Each image must have had less than 10% cloud cover, and the imaging period was selected from July to September. Missing images were replaced with those from adjacent years, with a spatial resolution of 30 m. The main preprocessing steps included geometric correction, radiometric calibration, atmospheric correction, image mosaicking, and cropping. The ENVI software was primarily used for radiometric calibration, with the purpose of eliminating errors inherent in the sensor during imaging and converting image brightness grayscale values into radiance values and other physical quantities. Atmospheric correction used the model tool "Atmospheric Correction Module"-"FLAASH Atmospheric Correction". Subsequently, image mosaicking was performed using the "Mosaicking"-"Seamless Mosaic" tool. Finally, cropping was performed using the "Regions of Interest"-"Subset Data from ROIs" tool with the administrative boundary vector data of the study area, thereby obtaining the regional image data for Urumqi in the years 2000, 2010, 2020, and 2022. All image data were unified using the World Geodetic System 1984 geodetic coordinate system. The PLUS (Patch-level Land Use Simulation Model, PLUS) model's influencing factor data included a total of 13 categories, such as temperature and precipitation. Details of the data are provided in Table 1. Before using the PLUS model for simulation and prediction, it is necessary to rasterize the driving factors used in this study. First, in ArcGIS 10.8 software, the projection tool was used to ensure that the projection coordinates of the road network, railway network, residential points, and river system vector data were consistent with the land use data of the study area Next, the Euclidean distance tool was used to rasterize the four accessibility factors. Finally, all raster data were unified in terms of spatial extent and resolution using the resampling and cropping raster tools, with the projection coordinates set to Albers Conic Equal Area and the spatial resolution standardized to 30 m.

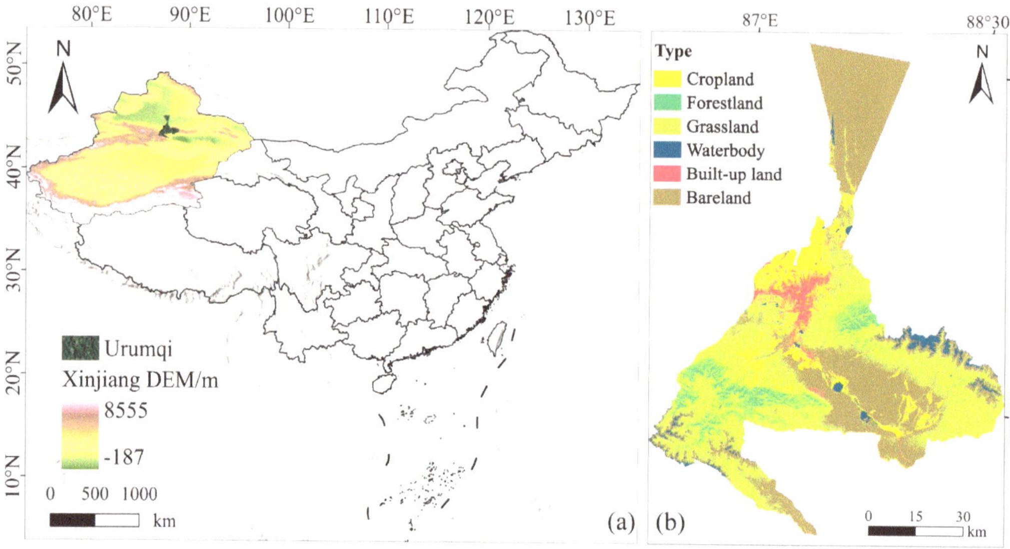


**Figure 1.** Sketch map of study area. (**a**) illustrates the geographical position of Urumqi, Xinjiang, within China; (**b**) presents the spatial distribution of different land use categories in Urumqi in 2022.

The methodology of this article consists of four steps, as illustrated in Figure 2. The first step involves preprocessing land use data and land use impact factor data. The second step defines four scenarios based on regional planning policies, including ecological red lines, the Urumqi 14th Five-Year Plan, and permanent basic farmland: BAU, CP, EP, and ED. The Markov chain model is used to predict land use demand for 2035, and the PLUS model simulates future land use changes under these four scenarios. The third step involves using the InVEST model to evaluate and compare the spatiotemporal distribution of habitat quality for the years 2000, 2010, 2022, and 2035. The fourth step is to analyze the research results and propose optimization recommendations.

**Table 1.** Data sources of the land use/land cover changes and its impact factors.

| Sub-Data | Year(s) | Resolution | Database Sources | Access Date |
|---|---|---|---|---|
| Land use/Land cover | 2000\2010\2020\2022 | 30 m | https://www.gscloud.cn/ | 20 December 2023 |
| DEM/Slope/Elevation | 2020 | 30 m | https://www.gscloud.cn/ | 20 December 2023 |
| NDVI/Soil | 2020 | 1000 m | https://www.resdc.cn/ | 25 April 2024 |
| Precipitation/Temperature | 2020 | 30 m | http://data.cma.cn/ | 22 April 2024 |
| GDP/Population | 2020 | 1000 m | https://www.resdc.cn/ | 22 April 2024 |
| Railway/Highway/Road/ Settlement/Water | 2020 | 1:1,000,000 | https://www.webmap.cn/ | 20 December 2023 |

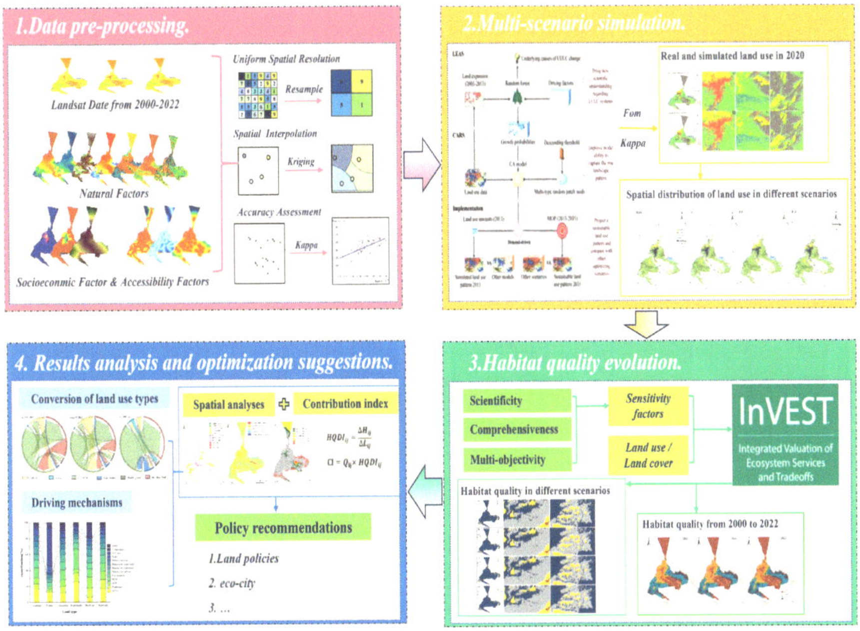


**Figure 2.** Technology road map for land use modeling and habitat quality evaluation.

### 2.3. Habitat Quality Evaluation Indicator System

The InVEST model is a comprehensive valuation model of ecosystem services and trade-offs that provides users with a variety of options for evaluating ecosystem functioning. It consists of three main modules, each containing lots of valuation items: terrestrial, marine, and freshwater ecosystem valuation. The main purpose of this study was to evaluate the habitat quality in Urumqi using the habitat quality module of the InVEST model. The habitat quality module assumes that the higher the quality of the habitat, the greater the biodiversity in the area, and vice versa [25]. This model correlates the land use type with the threat factors of the ecosystem and derives the degree of the habitat degradation of the ecosystem based on the sensitivity of the ecosystem to the threat sources, the distance of influence of the threat sources, the degree of interaction between the threat sources, etc. (taking the range of values from 0 to 1), and the value is directly proportional to the level of habitat quality, as specified in the following formulas:

$$Q_{xj} = H_j\left[1 - \left(\frac{D_{xj}^Z}{D_{xj}^Z + k^Z}\right)\right] \tag{1}$$

where $Q_{xj}$: habitat quality of individual raster $x$ at habitat or land use type $j$; $H_j$: habitat suitability at habitat or land use type $j$; $D_{xj}$: weighted mean of threat levels of raster cell $x$ at habitat or land use type $j$; $k$: half-saturation parameter, taken as $1/2$ of the maximum value of $D_{xj}$; and $Z$: normalization constant, taken as 2.5.

$$D_{xj} = \sum_{r=1}^{R}\sum_{y=1}^{Y_r}\left(w_r \div \sum_{r=1}^{R} w_r\right) \times r_y \times i_{rxy} \times \beta_x \times S_{jr} \tag{2}$$

$$i_{rxy} = 1 - \left(\frac{dxy}{dr_{max}}\right) \tag{3}$$

where $D_{xj}$: weighted average of the total threat level to which raster $x$ is exposed, a certain threat factor specifically; $R$: all grids on the raster layer of threat factor $r$; $Y_r$: phase element set on the raster layer of threat factor $r$; $W_r$: normalized threat weight, ranging from 0 to 1; $r_y$: used to determine whether a raster $y$ is the source of the threat factor $r$; $i_{rxy}$: distance function between the habitat class and threat factor; $\beta_x$: under the relevant environmental protection state; $\beta_x$: the accessibility level of the threat source to grid $x$ under the relevant environmental protection status; and $S_{jr}$: the sensitivity to threat factor $r$ when the habitat or land use type is $j$.

Based on the land use data, the model estimates habitat quality by considering the distance and intensity of threat factors, as well as the sensitivity of different habitat types to these factors. In this study, we refer to the InVEST model manual, consider the actual conditions in the study area, and draw on relevant research to define the threat source factors (Table 2) and sensitivity factors (Table 3).

**Table 2.** Threat factor parameter.

| Threat Factors | Weight | Influence Distance (km) | Spatial Decay Type |
|---|---|---|---|
| Cropland | 0.7 | 8 | linear |
| Built-up land | 1 | 10 | exponential |
| Bareland | 0.2 | 3 | exponential |

**Table 3.** Sensitivity of different land use types to habitat threats.

| Land Use Type | Habitat Suitability | Threats | | |
|---|---|---|---|---|
| | | Cropland | Built-Up Land | Bareland |
| Cropland | 0.5 | 0 | 0.5 | 0.4 |
| Forest land | 1 | 0.8 | 0.9 | 0.5 |
| Grassland | 0.7 | 0.5 | 0.6 | 0.5 |
| Water body | 0.9 | 0.7 | 0.8 | 0.2 |
| Built-up land | 0 | 0 | 0 | 0 |
| Bareland | 0.1 | 0.1 | 0.2 | 0 |

In the geographic information map model, cropland, forest land, grassland, water body, built-up land, and bareland are assigned values from 1 to 6. In ArcGIS 10.8, the land use conversion map cells were algebraically overlaid to integrate spatial information from the map-coded values [15]. Based on the geographic information map model, the map algebra overlay operation was performed on the Urumqi 2022 and future land use atlas grid cells, and the habitat quality changes were analyzed based on the results of the operation. The formula is

$$W = 10B + Q \tag{4}$$

where $W$ is the map cell grid map of land use class changes; $B$ is the 2022 land use map cell grid value; and $Q$ is the attribute value of the future land use mapping unitary grid.

The relationship between land use change and habitat quality was determined through the Habitat Quality Dynamic index. The formula is as follows [15]:

$$HQDI_{ij} = \frac{\Delta H_{ij}}{\Delta L_{ij}} \tag{5}$$

where $HQDI_{ij}$ is the dynamic index of habitat quality; positive values indicate positive impacts of land use change on habitats; negative values indicate negative impacts; and the larger the absolute value, the larger the impacts. $\Delta H_{ij}$ indicates the change in habitat quality caused by land use type conversion, and $\Delta L_{ij}$ indicates the area of land transformation in the region during the same period.

We used the contribution index ($CI$) to characterize the extent to which land use type conversion contributes to changes in habitat quality, calculated as follows [15]:

$$CI = Q_{ij} \times HQDI_{ij} \tag{6}$$

where $CI$ is the contribution index, and $Q_{ij}$ is the proportion of the area converted from land use type $i$ to land use type $j$ to the total converted area. Positive and negative values of the $CI$ represent positive and negative contributions, respectively. The higher the absolute value of the $CI$, the greater the impact of land use conversion on habitat quality.

### 2.4. PLUS Model

The Patch-generating Land Use Simulation (PLUS) model is an advanced version of the Future Land Use Simulation (FLUS) model. It integrates the Land Expansion Analysis Strategy (LEAS) and Cellular Automata model based on multi-type random patch seeds (CARS). This model not only enables a better exploration of the triggers behind various types of land use changes but also improves the simulation of changes among multiple types of land use patches [4]. We used the PLUS model to simulate future land use, which integrates a Markov chain. The advantage of this model is its ability to better explain the mechanisms driving various land use changes and more accurately simulate future land use changes under different policy scenarios [26]. Based on the land use data of Urumqi in 2015 and 2020, the development probability of the six land use types was determined, and then based on the actual situation, 13 factors were selected from natural, socio-economic, and accessibility aspects, including elevation, slope, slope direction, temperature, precipitation, soil type, Normalized Vegetation Index (NDVI), population density, and Gross Domestic Product (GDP), as well as distance to highway, distance to railway, distance to river, and distance to settlements; however, future development plans for the region were not considered when setting land use simulation scenarios. In the LEAS, the probability of development for each land use type was determined using a random forest algorithm (Figure 3). The current land use map of Urumqi in 2020 was simulated based on the

2015 land use data, combined with the development probability of each type of land use incorporating the development probability for each land use type, the projected future demand for each land use, the transfer cost matrix, and the weights of domain factors. The simulated land use data for 2020 were then compared with the actual 2020 data through spatial overlay analysis. The Kappa statistic and Figure of Merit (FoM) were calculated, and the parameters were iteratively adjusted to improve the simulation accuracy. Among them, the Kappa coefficient is an indicator for measuring classification accuracy. Usually, Kappa ranges from 0 to 1 and can be divided into five groups to indicate different degrees of consistency: 0.0 to 0.20 indicates very low consistency (mild), 0.21 to 0.40 indicates general consistency, 0.41 to 0.60 indicates moderate consistency (moderate), 0.61 to 0.80 indicates basic consistency, and 0.81 to 1 indicates almost complete consistency [26]. The FoM coefficient is concentrated on the grid where land use change occurs and is usually used to measure the goodness of fit of changes in land use composition. The FoM value is usually between 1% and 59%, and larger FoM values often match higher model accuracy [26]. Finally, based on the 2020 land use data, the model parameters were repeatedly debugged in the PLUS model to simulate the land use data of Urumqi in 2035 under the B-A-U, CP, EP, and ED scenarios.

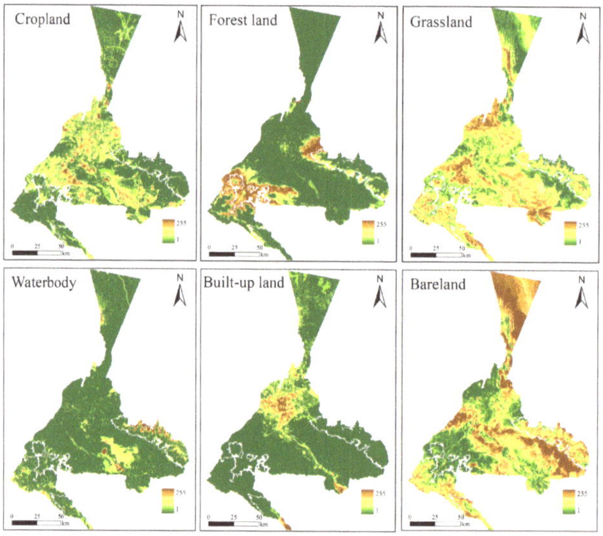


**Figure 3.** Development probability of various types of land.

### 2.5. Multi-Scenario Simulation

The Business-As-Usual (B-A-U) scenario refers to a situation where historical land use changes continue without any changes or interventions. In this scenario, land use transition probabilities, transition cost matrices, and constraints on conversion areas remain unchanged. Croplands in farming areas are crucial for food production, agricultural supply, and rural development. Therefore, this study introduces the CP scenario, which focuses on ensuring food security and preserving permanent basic agricultural land. In the CP scenario, the likelihood of cropland converting to urban areas is reduced by 60%, following the Business-As-Usual (BAU) probabilities, while the transition probabilities for other land uses remain unchanged [27]. The EP scenario focuses on the conservation of land types important for ecological restoration, such as forests, grasslands, and water body, with the aim of ensuring ecological security and maintaining ecosystem service functions [28]. The

ED scenario requires the rational development of urban boundaries within the constraints of cropland protection and ecological protection lines, with the aim of creating a compact and intensive urban spatial pattern [29]. The proportions of land use transition probabilities are detailed in Table 4.

**Table 4.** Transition probability matrix of each land use type in the EP and ED scenarios.

| Scenarios | Land Use Type | Cropland | Forest | Grassland | Built-Up Land |
|---|---|---|---|---|---|
| Business-As-Usual | not adjusted | | | | |
| Cropland Protection | Cropland | | | | −60% |
| Ecological Protection | Cropland | | +30% | +60% | −50% |
| | Forest land | −80% | −80% | | −90% |
| | Grassland | +20% | | | −80% |
| | Built-up land | +20% | +20% | +50% | |
| Economic Development | Cropland | | | | +60% |
| | Forest land | | | | +40% |
| | Grassland | | | | +40% |
| | Water body | | | | +20% |
| | Bareland | | | | +10% |

Notes: Based on BAU scenario: "+" and "−" indicates "increasing" and "decrease".

In this study, based on the land use data of Urumqi in 2015 and 2020, the development probability (Figure 3) of each land use type was generated using the LEAS module in the PLUS model. This was then further simulated to generate the 2020 land use map by combining the 2015 land use data with the development probabilities in the CARS module. After comparing the actual 2020 land use with the simulated 2020 land use, the following results were obtained: Kappa = 0.8917, FoM = 11.74% (Figure 4).

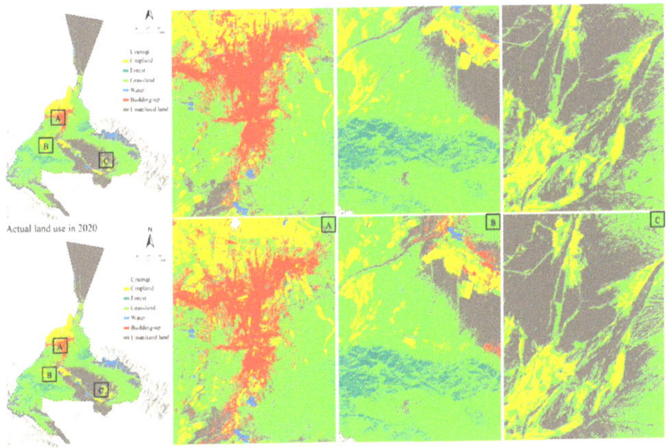


**Figure 4.** Comparing real and simulated land use in 2020. The simulated spatial distribution of land use exhibits discrepancies primarily in the central urban area (**A**) and two ecological transition zones (**B**,**C**). Consequently, (**A**) depicts an enlarged view of the central urban area, (**B**) provides an enlarged view of the southwestern region, and (**C**) presents an enlarged view of the southeastern region.

Consequently, the outcomes of the PLUS Model are closely aligned with the actual land use patterns, exhibiting a high degree of precision, and thus are suitable for forecasting land use in Urumqi by the year 2035.

# 3. Results

## 3.1. Characteristics of Historical and Future Land Use Spatiotemporal Changes

Based on the land use transfer matrix of Urumqi from 2000 to 2020 (Figure 5), the land types in Urumqi are ranked by area proportion from largest to smallest as follows: grassland > bareland > cropland > forest land > built-up land > water body. Notably, the area of arable land first increased and then decreased, exhibiting land use dynamic indices of 0.42% and −24.74%, respectively. The main reason is that large-scale reclamation after 2000 led to an increase in arable land area, and after 2010, the policy of converting farmland back to forest and grassland caused most of the arable land to be transformed into forest and grassland; the area of grassland showed a decreasing trend over 22 years, possibly due to overgrazing; the area of water body first decreased and then increased, with land use dynamics of -1.16% and 14.28%. The area of construction land showed a rapid growth trend, with new construction primarily emerging from arable land and grassland. Overall, the study area is characterized by a high coverage of grassland and desert but exhibits poor habitat quality. This trend is compounded by the rapid expansion of construction land, leading to the loss of ecological areas such as forest land, grassland, and arable land.

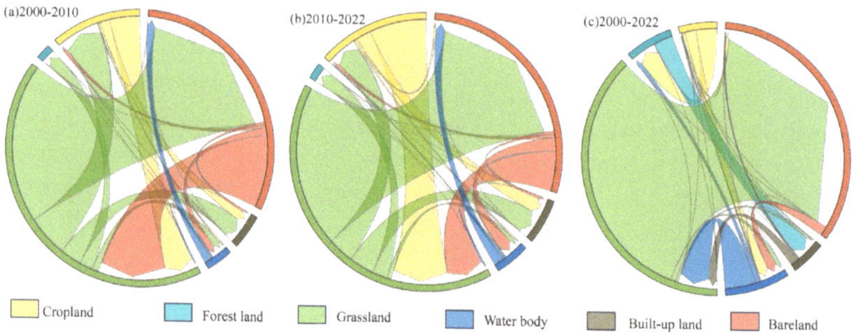


**Figure 5.** Conversion of land use types, 2000–2022.

The model results of land use area changes in Urumqi for the year 2035 under different scenarios are as follows (Figure 6). Under the BAU scenario, land use types that saw an increase in area include built-up land, bare land, and forest land, with built-up land experiencing the most significant growth at 27.55%. In contrast, cropland, water bodies, and grassland areas decreased, with cropland showing the largest reduction at 25.30%. Under the CP scenario, compared to the 2020 land use areas, cropland decreased by 159.63 km$^2$, grassland by 563.48 km$^2$, and water body by 31.4 km$^2$. Forest land, built-up land, and bareland areas increased by 37.61 km$^2$, 112.48 km$^2$, and 604.41 km$^2$, respectively. The CP scenario's land use evolution pattern is consistent with the BAU scenario. The decline in cropland is attributed to the lack of control over the conversion of permanent cropland areas. Under the EP scenario, forest land, grassland, and water body increased, while cropland decreased by 27.04%, and built-up land and bareland decreased by 1.49% and 3.36%, respectively. Under the ED scenario, built-up land, bareland, and forest land increased, with built-up land exhibiting the highest growth rate of 39.15%. This increase is closely linked to factors such as population density and GDP. In contrast, cropland, water bodies, and grassland decreased, with cropland experiencing the most substantial reduction of 20.10%.

A comparative analysis of land use data for Urumqi in 2022 and 2035, based on a geographic information system model (Figure 7), shows that unchanged land (coded as 11, 22, 33, 44, 55, 66) accounted for 91.02%, 90.35%, 94.39%, and 90.64% across the four time periods. From 2022 to 2035, the BAU scenario's total land use change area is 1367.38 $km^2$, with the largest changes occurring from grassland to bareland (36) and from bareland to grassland (63). In the CP scenario, the total land use change area is 1326.38 $km^2$, with the greatest change from grassland to bareland, followed by the conversion of cropland to built-up land. The ED scenario shows a total land use change area of 1273.48 $km^2$, with the most significant change from grassland to unused land (667.05 $km^2$), followed by cropland to built-up land (148.87 $km^2$). In the EP scenario, the total land use change area is 795.04 $km^2$, with the largest change from bareland to grassland (253.17 $km^2$), followed by the conversion of cropland to grassland (181.62 $km^2$). This reduction in the conversion of water body, forest land, and grassland to arable and built-up land under the EP scenario controls excessive growth in construction and cropland, resulting in significant ecological improvements. In summary, the simulation of Urumqi's land use situation in 2035 indicates that future changes in land use will primarily focus on the oasis transition zones and ecological intersection areas.

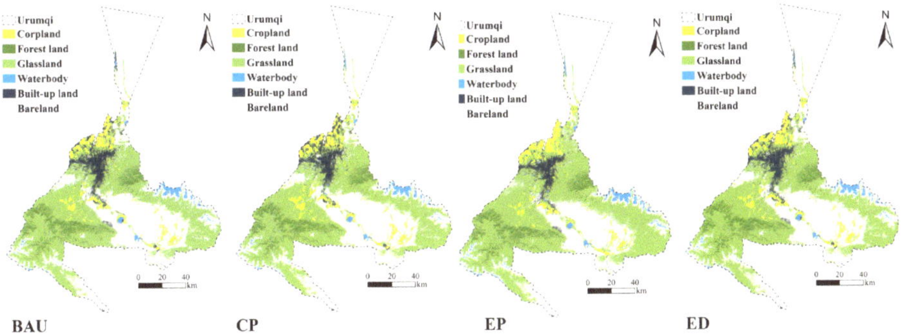


**Figure 6.** Spatial distribution of land use in different scenarios for 2035.

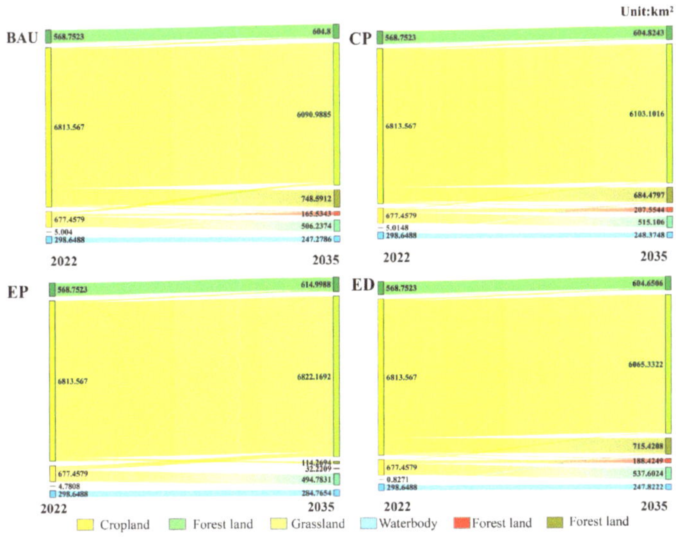


**Figure 7.** Transformation of land use in 2022–2035.

### 3.2. Spatiotemporal Changes in Habitat Quality in the Past and Future

From 2000 to 2022, the habitat quality scores in Urumqi showed a declining trend, with scores of 0.4180, 0.3937, and 0.3838, reflecting change rates of −5.8% and −2.5% (Table 5). During this period, the implementation of the Western Development Strategy and rapid urbanization significantly contributed to the continuous decline in habitat quality. After 2010, the continuous advancement of the "Grain-to-Green" policy slowed the rate of decrease in arable land and grassland areas, thus mitigating the downward trend in habitat quality. During the study period, areas with excellent habitat quality consistently constituted 1% of the total area. The proportion of good habitat quality areas decreased from 37% in 2000 to 32% in 2010, remaining stable thereafter. The proportion of areas with moderate habitat quality initially increased and then decreased, rising from 26% in 2000 to 28% in 2010, before returning to 26% in 2022. The proportion of areas with poor habitat quality gradually increased from 36% in 2000 to 41% in 2022. Overall, habitat quality in Urumqi displayed a trend of deterioration (Figure 8).

**Table 5.** Average habitat quality (HQ) for 2000–2022 and different scenarios (BAU, CP, EP, and ED).

| HQ | 2000 | 2010 | 2022 | BAU | CP | EP | ED |
| --- | --- | --- | --- | --- | --- | --- | --- |
| Mean | 0.4180 | 0.3937 | 0.3838 | 0.0898 | 0.0893 | 0.0906 | 0.0892 |

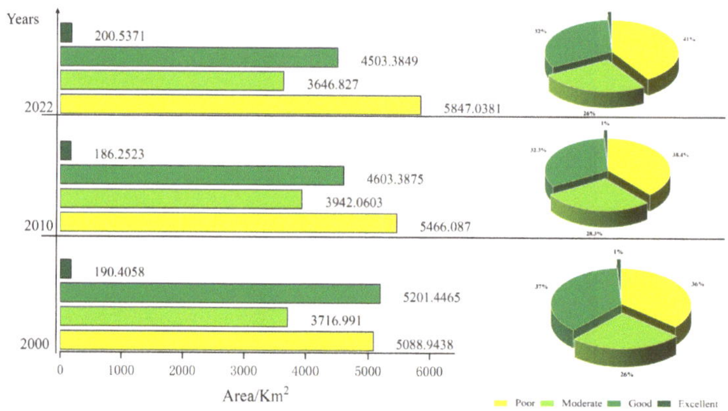


**Figure 8.** Quantitative changes in habitat quality level.

The spatial distribution of habitat quality in Urumqi exhibits significant regional heterogeneity. The distribution pattern is characterized by a high-density zone along the northern side of the Tianshan Mountains, a low-density zone in the Gurbantunggut Desert, and transitional distribution features in other areas (Figure 9). High-quality habitats are concentrated on the northern slopes of the Tianshan Mountains and in the northwestern part of Tianchi, primarily due to the dominance of forest land use types in these areas, which have high forest coverage and thus ensure a higher level of habitat quality. Other regions exhibit sporadic distributions of high-quality habitat. Relatively good habitat quality areas are primarily found within the Tianshan Mountains, where grassland vegetation dominates, and urban development and population density remain relatively low. These areas maintain better ecological conditions, contributing to higher habitat quality compared to urbanized zones. The effective implementation of ecological protection policies has further stabilized grassland landscapes, promoting the improvement of habitat quality. Moderate habitat quality regions are primarily found within the central oasis area. Poor habitat quality regions are predominantly distributed in the northern desert zone and adjacent areas to Turpan, where land use types are primarily bareland and low-vegetation grasslands, with limited potential for ecological improvement. Overall, the spatial distribution pattern of habitat quality in Urumqi is predominantly influenced by land cover types, showing significant spatial variability. Areas with natural vegetation, such as grasslands and forests, tend to have higher habitat quality, while urbanized areas and regions with high human activity, like built-up lands and agricultural zones, typically exhibit lower habitat quality. This spatial variability underscores the impact of land use and land cover on the ecological balance and habitat quality within the region.

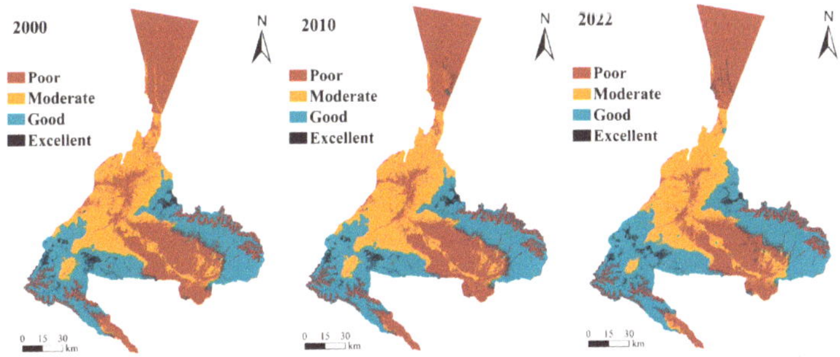


**Figure 9.** Spatial distribution of habitat quality from 2000 to 2022.

Based on the changes in habitat quality over four categories, the habitat changes from 2000 to 2022 are categorized into five types (Figure 10a). The majority, 86.01%, are stable zones with minimal anthropogenic disruption. Significant degradation and significant improvement areas account for 2.62% and 0.81%, respectively, while minor degradation and minor improvement areas account for 8.52% and 2.03% (Figure 10b). These changes are largely concentrated in regions of heightened human activity, reflecting the dual influence of human actions on habitat quality. The hot-spot analysis of habitat quality was employed to study the spatial variation in habitat quality. With little spatial fluctuation in habitat quality from 2000 to 2022, the year 2022 serves as an illustrative case (Figure 10c).

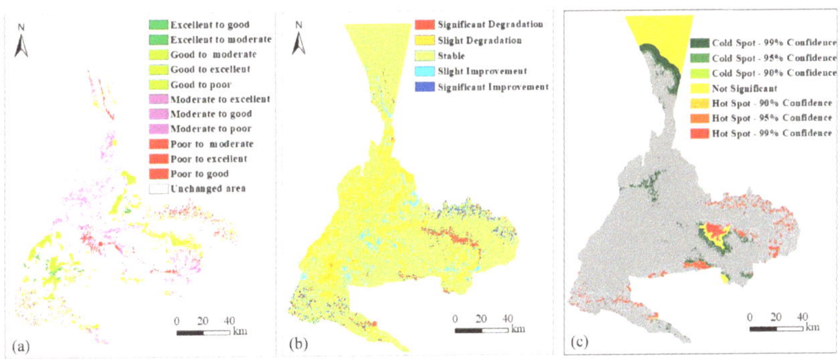


**Figure 10.** Spatial variation in habitat quality from 2000 to 2022. (**a**) Spatial pattern of habitat quality grade shift. (**b**) Spatial changes in habitat quality. (**c**) Spatiotemporal characterization of HQ cold spots and hot spots.

The results indicate that hot-spot areas are primarily found in the Tianshan Ecological Protection Zone and the Salt Lakes, with a focus on forest–grassland mosaic zones. These areas have higher habitat quality due to their natural vegetation and relatively low human disturbance. In contrast, cold-spot areas are concentrated in the oasis aggregation development zones, the periphery of the Gurbantunggut Desert, and the outskirts of the Salt Lakes. These cold spots are dominated by agricultural lands, with grasslands occupying secondary positions. The largest proportion of cold spots is located at the edges of oases, characterized by bareland, which indicates a significant loss of ecological quality in these regions due to urban expansion and agricultural development.

Different scenarios (BAU, CP, EP, and ED) for 2035 show a different spatial distribution of habitat quality (Figure 11). Among them, the EP scenario demonstrates the highest habitat quality, attributed to significant increases in grasslands, forest lands, and water bodies. Spatially, all scenarios show that the Tianshan Ecological Protection Zone has high habitat quality, while other areas have lower quality.

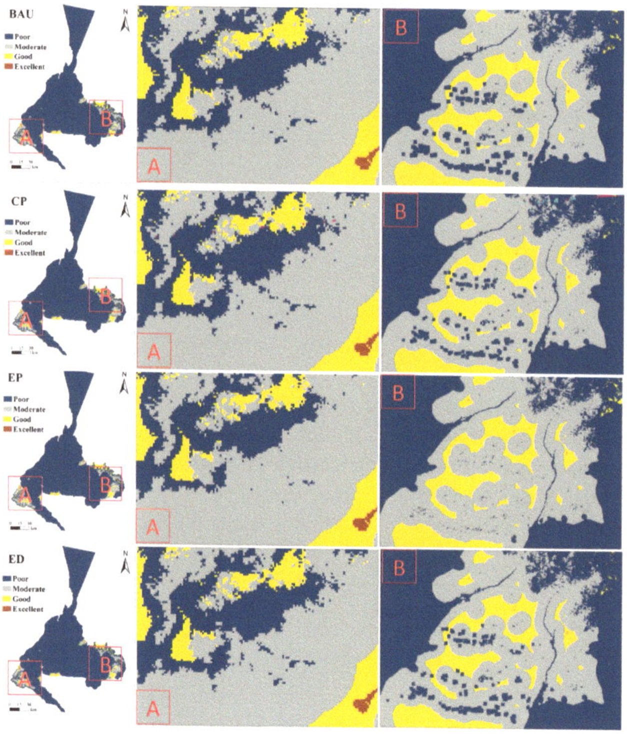


**Figure 11.** Habitat quality in different scenarios for 2035. (**A**) represents the magnified view of the southwestern area, while (**B**) depicts the magnified view of the southeastern area.

In comparing habitat quality ratings across scenarios, the BAU scenario has good and excellent habitat quality areas totaling 458.15 km$^2$, or 3.23% of the total area. In contrast, the EP scenario shows an increase in good and excellent habitat quality areas by 20.72 km$^2$, representing 3.38% of the total area. The ED scenario integrates elements of both the CP and BAU scenarios, aiming to balance ecological and economic development. Overall, the average habitat quality in Urumqi is significantly declining (Table 5). Reasons include various factors such as land use transformation, urbanization and economic development, water resource scarcity and pollution, climate change, overgrazing and irrational development, and insufficient ecological sensitivity and protection [5,23,24,30]. These factors, including urbanization, agricultural expansion, and land degradation, are intertwined and contribute to the continuous deterioration of Urumqi's ecological environment quality. The rapid urban sprawl, especially in the oasis transition zones, leads to the loss of critical habitats, including grasslands and forest areas. This, combined with the pressures of intensive farming and the expansion of infrastructure, has significantly impacted the overall ecological health of the region. Moreover, the development of areas near deserts and salt lakes exacerbates soil degradation and reduces the capacity of ecosystems to support biodiversity. The ongoing decline in habitat quality underscores the need for effective land use planning and sustainable development practices to protect the region's natural resources and biodiversity. Although the ED scenario may better reflect the actual conditions of the city, the EP scenario offers the best habitat quality in Urumqi. To improve this situation, it is necessary to adopt comprehensive ecological governance measures, optimize the sustainable use of land resources, and strengthen ecological and environmental protection.

### *3.3. The Influence of Land Use Alterations on Habitat Quality Dynamics*

The desert–oasis transition zone is the area where environmental quality changes are most pronounced. The interconversion between farmland, grassland, desert, and construction land is the main factor affecting ecological environment quality. Land use changes have both positive and negative impacts on habitat quality. Land use changes contribute both positively and negatively to habitat quality changes. Under the BAU scenario from 2022 to 2035, the positive contribution index to habitat quality change is 0.43, while the negative contribution index is −41.46, yielding a total contribution index of −41.03 (Table 6).

**Table 6.** Contribution index for the impact of land use conversions on HQ-CA from 2022 to 2035.

| Habitat Quality Change Type | Land Use Transfer | BAU | | CP | | EP | | ED | |
|---|---|---|---|---|---|---|---|---|---|
| | | CA (km$^2$) | CI | CA (km$^2$) | CI | CA (km$^2$) | CI | CA (km$^2$) | CI |
| Reduction | 12 | 0.22 | −0.01 | 0.22 | −0.04 | 0.22 | −0.01 | 0.22 | −0.01 |
| | 13 | 98.94 | −3.16 | 34.04 | −1.12 | 33.71 | −1.16 | 33.71 | −1.16 |
| | 14 | 2.27 | −0.08 | 0.48 | −0.02 | 0.49 | −0.02 | 0.49 | −0.02 |
| | 15 | 109.07 | −3.3 | 173.46 | −5.48 | 148.87 | −4.87 | 148.87 | −4.87 |
| | 16 | 6.49 | −0.2 | 2.41 | −0.07 | 1.50 | −0.05 | 1.50 | −0.05 |
| | 21 | 0.21 | −0.01 | 0.22 | −0.01 | 0.23 | −0.01 | 0.23 | −0.01 |
| | 23 | 27.35 | −1.43 | 27.42 | −1.47 | 27.39 | −1.53 | 27.39 | −1.53 |
| | 26 | 0.17 | −0.01 | 0.16 | −0.01 | 0.16 | −0.01 | 0.16 | −0.01 |
| | 31 | 39.28 | −1.09 | 42.35 | −1.22 | 43.24 | −1.30 | 43.24 | −1.30 |
| | 32 | 63.58 | −2.62 | 63.67 | −2.70 | 63.48 | −2.81 | 63.48 | −2.81 |
| | 34 | 2.68 | −0.02 | 2.60 | −0.03 | 2.59 | −0.03 | 2.59 | −0.03 |
| | 35 | 36.12 | −0.89 | 33.13 | −0.85 | 38.52 | −1.02 | 38.52 | −1.02 |
| | 36 | 712.82 | −25.70 | 635.72 | −23.49 | 667.05 | −25.78 | 667.05 | −25.78 |
| | 41 | 1.73 | −0.04 | 0.67 | −0.02 | 0.65 | −0.02 | 0.65 | −0.02 |
| | 43 | 5.62 | −0.23 | 5.54 | −0.23 | 5.54 | −0.24 | 5.54 | −0.24 |
| | 45 | 20.32 | −0.48 | 0.95 | −0.02 | 1.01 | −0.02 | 1.01 | −0.02 |
| | 46 | 29.11 | −1.63 | 46.20 | −2.66 | 46.71 | −2.80 | 46.71 | −2.80 |
| | 53 | 27.14 | −0.02 | 26.23 | −0.02 | 18.92 | −0.01 | 18.92 | −0.01 |
| | 61 | 1.23 | −0.01 | 2.07 | −0.01 | 5.25 | −0.02 | 5.25 | −0.02 |
| | 63 | 140.40 | −0.49 | 134.26 | −0.47 | 138.26 | −0.51 | 138.26 | −0.51 |
| | 65 | 8.70 | −0.06 | 6.50 | 0.05 | 13.37 | −0.10 | 13.37 | −0.10 |
| Improvement | 56 | | 73.14 | 0.02 | | 46.71 | −2.80 | | |
| | 64 | 10.45 | 0.43 | 9.16 | 0.38 | 9.22 | 0.40 | 9.22 | 0.40 |
| Total | | 1333.88 | −41.03 | 1238.27 | −39.54 | 766.46 | -28.08 | 1257.16 | −42.92 |

Notes: HQ-CA means Habitat Quality Change Area; numbers 1 to 6 in the table represent cropland, forest land, grassland, water body, built-up land, and bareland, respectively.

This indicates that land use changes will severely degrade habitat quality during this period. The main cause of habitat quality degradation is the transfer of grassland and cropland to other land types, with a cumulative change area of 1071.91 km$^2$, contributing 89.4% of the total impact. In the CP scenario, the positive contribution index to habitat quality change is 0.4, and the negative contribution index is −39.94, resulting in a total contribution index of −39.54. This suggests that land use changes will lead to habitat quality degradation, with the main cause being the transfer of grassland to other land types, with a cumulative change area of 777.46 km$^2$, contributing 70.81% of the total impact. Under the EP scenario, the positive contribution index to habitat quality change is 1.26, while the negative contribution index is −29.34, resulting in a total contribution index of −28.08. This indicates that habitat quality will still decline, though less severely compared to other scenarios. The reduction in the transfer area of grassland to other land types is significant, with a cumulative change area of only 206.17 km$^2$. In the ED scenario from 2022 to 2035, the positive contribution index is 0.4, the negative contribution index is −42.31, and the total contribution index is −41.92. This indicates that land use changes will lead to habitat quality degradation, with the main cause being the transfer of grassland to bareland and cropland to built-up land, with a cumulative change area of 815.92 km$^2$, contributing 72.44% to the total impact.

## 4. Discussion

### 4.1. Spatial and Temporal Variations in Habitat Quality and Driving Mechanisms in the Arid Zone

This study provides a comprehensive analysis of land use spatiotemporal changes in Urumqi from 2000 to 2022 and used the PLUS model to simulate land use under four scenarios for the year 2035. The Urumqi oasis ecosystem encompasses a diverse range of environments, including mountains, water bodies, forests, fields, lakes, grasslands, and deserts. Throughout the study period, both natural factors and human activities have jointly influenced changes in land use, thereby impacting habitat quality. With the acceleration of urbanization, the area of built-up land has consistently increased, while grassland areas have steadily declined. Spatially, urban expansion has exhibited a trend of moving from south to north and from west to east [31]. The deterioration in habitat quality is closely related to the reduction in grassland area and the increase in built-up land. The deterioration in habitat quality is closely related to the reduction in grassland area and the increase in built-up land.

From the perspective of natural factors, elevation, precipitation, and the NDVI impose significant constraints on ecological land use over long-term scales, thereby influencing overall habitat quality. Elevation affects factors such as slope and annual precipitation, which in turn impact land use types and habitat quality. Generally, habitat quality is higher in mid- to low-elevation areas and lower in high-elevation areas. Previous studies have indicated that high elevation and steep slopes provide favorable conditions for the growth of natural vegetation such as forests and grasslands, resulting in higher vegetation cover and habitat quality, which aligns closely with the findings of this study [21]. Temperature and precipitation affect species' habitat suitability, with significant impacts on species composition, ecological functions, processes, and surface vegetation growth. Adequate rainfall promotes habitat suitability [32]; however, excessive rainfall can trigger landslides, mudslides, and other disasters, exacerbating soil erosion and negatively impacting habitat quality. This study finds that ecological land use in the region is more susceptible to precipitation factors. Since 1980, the northwestern arid region has experienced a continuous trend of warming and increased moisture [33], though precipitation remains unevenly distributed spatially. For desert plants in arid areas, water is a critical limiting factor for survival. Compared to studies in the eastern regions, precipitation is crucial for habitat quality in Urumqi. The NDVI serves as a key indicator of vegetation growth, with elevation and precipitation affecting vegetation growth and types in arid regions [30]. Soil type is a determinant of terrestrial ecosystem habitat quality, with soil pH and organic matter showing negative and positive correlations with habitat quality, respectively. A reduction in soil organic matter may lead to decreased soil fertility and quality, thereby impairing habitat quality [5]. In arid regions, where most soils are aeolian, soil fertility is low and soil types are homogeneous, which limits their impact on desert plants.

From a socio-economic perspective, the rapid urbanization in Urumqi over the past two decades has led to the expansion of urban built-up areas and transportation networks, which have encroached upon large areas of grassland, thereby disrupting grassland ecosystems [34]. Additionally, the implementation of the Western Development policy has resulted in population growth in the region. Noteworthy are the negative impacts of population migration on receiving areas, such as increased cropland, urban expansion, rising household waste, and greater human disturbance of natural environments. Although various ecological protection policies have been implemented, restoration efforts remain challenging. In summary, the spatial variation in habitat quality in Urumqi is directly influenced by factors such as elevation, precipitation, NDVI, construction land, and population density, while other factors exert indirect effects. From 2015 to 2020, the contribution of various factors to land use changes highlights that all considered driving factors impact land category conversion, though to varying degrees across different land category conversions (Figure 12). Future research should explore the driving forces behind land cover changes and habitat quality variations in Urumqi in greater depth [17]. This will help elucidate the mechanisms of habitat quality evolution and provide a solid foundation for improving habitat quality and promoting sustainable regional development.

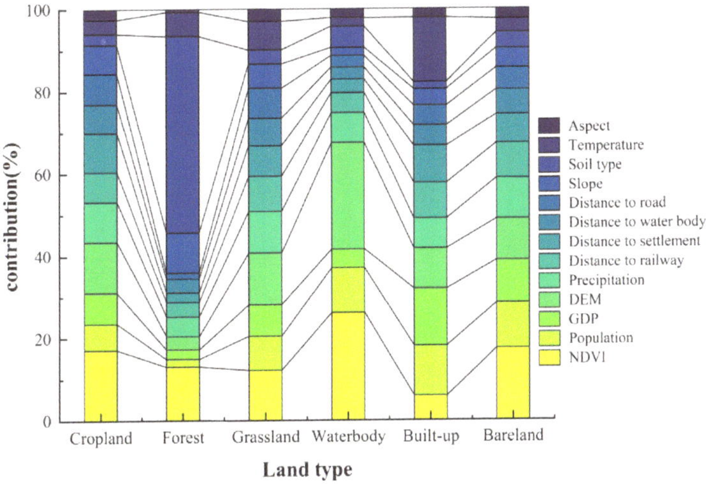


**Figure 12.** Contribution of driving factors by land use type.

Urumqi plays an important role in the northern sand control zone of China's "two screens and three belts" ecological security pattern, and the habitat in this region has a positive impact on China's ecological civilization construction [35]. We analyze the changes in various land use types in Urumqi, with the main purpose of better understanding the spatial and temporal distribution patterns of the land use status and habitat quality of different land use types in the region, and providing a theoretical basis for the urban development and ecological protection of Urumqi. The expansion of construction land in Urumqi is based on the premise of reducing the area of grassland and arable land. This is obviously not in line with the concept of sustainable development. Urumqi should firmly grasp the two development opportunities of the "Western Development Strategy" and "the Belt and Road Initiative", make full use of the preferential policies of the country and the autonomous region, adhere to the concept of ecological civilization construction, correctly guide all efforts, and achieve a win–win situation in regional economy and ecological protection.

### *4.2. Multi-Scenario Projections of Habitat Quality in Arid Zones and Policy Recommendations*

The PLUS model effectively simulates the spatial changes in land use structure [4]. In this study, the LEAS module of the PLUS model was employed to identify 13 selected driving factors. The default setting for the number of decision trees in the random forest was set to 20, with a sampling rate of 1% and 13 features for training the random forest, thereby generating probability distribution maps for different land use types. Parameters such as the transition cost matrix and neighborhood weights were set in the CARS module of the model. Following relevant studies, domain range parameters were set to 3, and the attenuation coefficient of the decay threshold, probability of random patch seeds, and other adjustments were made, with maximum proportions of random seeds for 2020 simulations set to 0.8, 0.1, and 0.0001, respectively, and parallel thread numbers set to 12 as initial assumptions [25].

Additionally, the InVEST model was used to assess habitat quality, with spatial comparisons between assessed habitat quality and landscape patterns revealing consistency, further validating the feasibility of the model. Through ecological process modeling, we observed that under the ED scenario, the area of built-up land experienced the most significant increase, by 40.64%. In the EP scenario, the area of forest and grassland increased substantially by 334.35 km$^2$, aligning with the land use conversion probabilities set for this scenario. The simulation results for the four scenarios met the expected outcomes, further confirming the reasonableness of the PLUS model's parameter settings. Additionally, significant spatial variation in habitat quality was found. The Tianshan Ecological Protection Zone exhibited higher habitat quality, characterized by forest and grassland dominance, low population density, and minimal human disturbance. Furthermore, the influence of ecological protection policies enhanced the stability of forest and grassland ecosystems, allowing these areas to maintain high habitat quality. Low habitat quality was predominantly observed in the northern Gurbantunggut Desert Conservation and Restoration Area, where land use types are primarily bareland, with low vegetation cover and exposed surfaces, making it highly sensitive to environmental changes and limiting habitat quality. The spatial distribution of historical and future habitat quality was consistent with ecosystem patterns, validating the model's evaluation. Finally, under the B-A-U scenario for 2022–2035, the greatest changes were observed in grassland to bareland (36) and bareland to grassland (63), with the most severe habitat quality degradation. This is attributed to the historical overgrazing and other human activities that have damaged grasslands, putting pressure and risks on the ecosystem, leading to habitat quality degradation in some areas [36]. At the same time, certain unused lands with favorable soil conditions have allowed resilient desert plants to thrive and reproduce, or have even accelerated the spread of these plants through human intervention [37]. In the EP scenario for 2022–2035, the most significant change was observed in cropland to grassland (13), primarily due to policies promoting cropland conversion to grassland.

Habitat quality is a crucial indicator of regional ecology [2]. Therefore, forecasting habitat quality under various future scenarios is crucial for providing scientific evidence to support government policies aimed at ecological environmental protection and preventing regional habitat quality degradation. By integrating the InVEST and PLUS models, future habitat quality was predicted under the B-A-U, CP, EP, and ED scenarios. Under the B-A-U scenario, the average habitat quality is expected to decline by 0.3039 compared to 2022, with a contribution index of −41.03 for land use change on habitat quality, which is consistent with previous research findings [15]. Without adjustments to land use structures, allowing them to evolve freely with natural trends, Urumqi's ecological habitat quality will continue to deteriorate. Compared to in the B-A-U scenario, it is projected that by 2035, the area of habitat with good and excellent quality in the EP scenario will expand by 20.72 km$^2$, yet the overall average habitat quality will still deteriorate. It is a fact that, in this scenario, only the conversion area from cropland to forest and grassland has increased, while the conversion amount from forest and grassland to other types has decreased, with the transition probability between other types remaining unchanged. Moreover, certain areas under the EP scenario maintained higher habitat quality, which is consistent with increased forest and grassland areas, indicating the effective mitigation of habitat quality degradation.

To promote the sustainable development of ecological economy in Urumqi, it is crucial to integrate land use regulation policies and actions. Therefore, this study proposes three measures to enhance habitat quality in arid regions: (1) Control the amount of land converted between different land types. In formulating land use policies, it is important to consider the trend and extent of transfers between different land use types, as well as the impact of land use conversions on habitat quality, with the aim of striking a balance between ED and EP. (2) Strictly control urban expansion. During urbanization, it is essential to tightly regulate the construction land within the urban development boundary to prevent outward expansion and encroachment on cropland. In particular, attention should be given to protecting existing forest, grassland, and water resources from unreasonable human activities, preventing arable land abandonment, and construction land encroaching on arable land to ensure stable arable land areas [38]. (3) Plan urban green space construction comprehensively. Urban green spaces are potential green areas. By establishing ecological corridors to connect various green patches across the region, we can provide refuges for oasis flora and fauna, migratory birds, and more. Especially through vertical greening and green roofs, creating green spaces as much as possible will significantly contribute to biodiversity conservation, ecosystem stability, and regional ecological security.

### *4.3. Restriction and Uncertainty*

Due to data limitations, the spatial resolution for most driving factors used in the PLUS model simulations of future land use is 30 m, which precludes the acquisition of higher-resolution data. Consequently, some data were resampled during the simulation process, which **introduced** potential uncertainties. The InVEST model, with its advantages over traditional methods, is widely used for visualization and dynamic studies [34]. During the habitat quality assessment using the InVEST model, parameters for the habitat quality evaluation model were set based on the model manual and literature. The resulting spatial distribution of habitat quality was consistent with ecosystem patterns, indicating good simulation results. However, there remain uncertainties in model performance evaluation and parameter settings, and further exploration of model performance evaluation methods and appropriate parameters is needed. Additionally, despite considering four different simulation scenarios for land use in 2035, there is still some uncertainty in the simulation results. First, although 13 driving factors were selected for the PLUS model simulations, actual land use driving factors are more complex. On the other hand, future development plans for the region were not considered when setting land use simulation scenarios. Future research should place greater emphasis on the complexity of land use driving factors, conduct more detailed investigations, and develop land use simulation scenarios that align with regional development plans to support medium- and long-term growth.

## 5. Conclusions

This study examines landscape dynamics in Urumqi under various scenarios and their potential impacts on habitat quality. Using the PLUS model, we analyzed multiple driving factors, including natural, social, and proximity influences, to simulate land use evolution. From 2022 to 2035, Markov chain analysis was employed to derive land use transition probabilities and simulate land use changes under different future scenarios. The InVEST model was subsequently used to quantify habitat quality across diverse landscape patterns. The results indicate that the ongoing loss of natural landscapes, such as forests, grasslands, and water bodies, is the primary driver of habitat quality decline in Urumqi. The effects of built-up land changes on habitat quality exhibit both quantitative and spatial variations across different scenarios. Under the BAU scenario from 2022 to 2035, habitat quality is expected to significantly decrease due to extensive grassland degradation. Areas around the Gurbantunggut Desert should focus on ecological protection to mitigate this trend. In contrast, the EP scenario shows a marked improvement in habitat quality, with only 95.45 $km^2$ of grassland transitioning to bareland. Overall, while land use changes under all scenarios from 2022 to 2035 will result in varying degrees of habitat quality degradation, the EP scenario effectively minimizes the extent of this decline.

**Author Contributions:** S.C.: conceptualization, writing—original draft preparation, and methodology; Ü.H.: conceptualization, writing—review and editing, supervision, and project administration; L.S.: visualization, software, and writing—review and editing; W.F.: visualization; L.G.: technical support on the methodology; M.W.: writing—review and editing. All authors have read and agreed to the published version of the manuscript.

**Funding:** This work was supported by the Regional Collaborative Innovation Program-Shanghai Cooperation Organization Science and Technology Partnership and International Cooperation Project of the Science & Technology Department of Xinjiang Uygur Autonomous Region (No. 2023E01026), Graduate Education Innovation Project of Xinjiang Uygur Autonomous Region (No. XJ2023G037).

**Data Availability Statement:** The original contributions presented in the study are included in the article, further inquiries can be directed to the corresponding authors.

**Acknowledgments:** We thank Aishajiang Aili (Xinjiang Institute of Ecology and Geography, CAS) for language polishing and improving this manuscript. We would also like to express our gratitude to the editors and reviewers for their helpful comments and careful revision of this manuscript.

**Conflicts of Interest:** The authors declare no conflicts of interest.

## References

1. Li, Y.; Duo, L.; Zhang, M.; Wu, Z.; Guan, Y. Assessment and Estimation of the Spatial and Temporal Evolution of Landscape Patterns and Their Impact on Habitat Quality in Nanchang, China. *Land* **2021**, *10*, 10. [CrossRef]
2. Gomes, E.; Inácio, M.; Bogdzevič, K.; Kalinauskas, M.; Karnauskaitė, D.; Pereira, P. Future scenarios impact on land use change and habitat quality in Lithuania. *Environ. Res.* **2021**, *197*, 111101. [CrossRef] [PubMed]
3. Qin, X.; Yang, Q.; Wang, L. The evolution of habitat quality and its response to land use change in the coastal China, 1985–2020. *Sci. Total Environ.* **2024**, *952*, 175930. [CrossRef] [PubMed]

4. Zhang, X.; Song, W.; Lang, Y.; Feng, X.; Yuan, Q.; Wang, J. Land use changes in the coastal zone of China's Hebei Province and the corresponding impacts on habitat quality. *Land Use Policy* **2020**, *99*, 104957. [CrossRef]
5. Pei, X.; Zhao, X.; Liu, J.; Liu, W.; Zhang, H.; Jiao, J. Habitat degradation changes and disturbance factors in the Tibetan plateau in the 21st century. *Environ. Res.* **2024**, *260*, 119616. [CrossRef]
6. Li, S.; Dong, B.; Gao, X.; Xu, H.; Ren, C.; Liu, Y.; Peng, L. Study on spatio-temporal evolution of habitat quality based on land-use change in Chongming Dongtan, China. *Environ. Earth Sci.* **2022**, *81*, 220. [CrossRef]
7. Wang, J.; Zhou, W.; Guan, Y. Optimization of management by analyzing ecosystem service value variations in different watersheds in the Three-River Headwaters Basin. *J. Environ. Manag.* **2022**, *321*, 115956. [CrossRef]
8. Janus, J.; Bozek, P. Land abandonment in Poland after the collapse of socialism: Over a quarter of a century of increasing tree cover on agricultural land. *Ecol. Eng.* **2019**, *138*, 106–117. [CrossRef]
9. Glenz, C.; Massolo, A.; Kuonen, D.; Schlaepfer, R. A wolf habitat suitability prediction study in Valais (Switzerland). *Landsc. Urban Plan.* **2001**, *55*, 55–65. [CrossRef]
10. Bryant, M.D.; Edwards, R.T.; Woodsmith, R.D. An approach to effectiveness monitoring of floodplain channel aquatic habitat: Salmonid relationships. *Landsc. Urban Plan.* **2005**, *72*, 157–176. [CrossRef]
11. Ikauniece, S.; Brūmelis, G.; Zariņš, J. Linking woodland key habitat inventory and forest inventory data to prioritize districts needing conservation efforts. *Ecol. Indic.* **2012**, *14*, 18–26. [CrossRef]
12. Zhu, M.; Hoctor, T.S.; Volk, M.; Frank, K.I.; Zwick, P.D.; Carr, M.H.; Linhoss, A.C. Spatial conservation prioritization to conserve biodiversity in response to sea level rise and land use change in the Matanzas River Basin, Northeast Florida. *Landsc. Urban Plan.* **2015**, *144*, 103–118. [CrossRef]
13. Yang, Y.; Tian, Y.; Zhang, Q.; Tao, J.; Huang, Y.; Gao, C.; Lin, J.; Wang, D. Impact of current and future land use change on biodiversity in Nanliu River Basin, Beibu Gulf of South China. *Ecol. Indic.* **2022**, *141*, 109093. [CrossRef]
14. Venter, O.; Sanderson, E.W.; Magrach, A.; Allan, J.R.; Beher, J.; Jones, K.R.; Possingham, H.P.; Laurance, W.F.; Wood, P.; Fekete, B.M.; et al. Sixteen years of change in the global terrestrial human footprint and implications for biodiversity conservation. *Nat. Commun.* **2016**, *7*, 12558. [CrossRef]
15. Li, X.; Zhang, X.; Feng, H.; Li, Y.; Yu, J.; Liu, Y.; Du, H. Dynamic evolution and simulation of habitat quality in arid regions: A case study of the Hexi region, China. *Ecol. Model.* **2024**, *493*, 110726. [CrossRef]
16. Dang, H.; Lü, Y.; Guo, J.; Wu, X. Multi-scenario simulation can contribute to identify priorities for regional ecological corridors conservation. *Ecol. Indic.* **2024**, *165*, 112166. [CrossRef]
17. Chen, S.; Liu, X. Spatio-temporal variations of habitat quality and its driving factors in the Yangtze River Delta region of China. *Glob. Ecol. Conserv.* **2024**, *52*, e02978. [CrossRef]
18. Wang, B.; Oguchi, T.; Liang, X. Evaluating future habitat quality responding to land use change under different city compaction scenarios in Southern China. *Cities* **2023**, *140*, 104410. [CrossRef]
19. Hu, J.; Zhang, J.; Li, Y. Exploring the spatial and temporal driving mechanisms of landscape patterns on habitat quality in a city undergoing rapid urbanization based on GTWR and MGWR: The case of Nanjing, China. *Ecol. Indic.* **2022**, *143*, 109333. [CrossRef]
20. Wu, J.; Luo, J.; Zhang, H.; Qin, S.; Yu, M. Projections of land use change and habitat quality assessment by coupling climate change and development patterns. *Sci. Total Environ.* **2022**, *847*, 157491. [CrossRef]
21. Zhang, S.; Lei, J.; Zhang, X.; Tong, Y.; Lu, D.; Fan, L.; Duan, Z. Assessment and optimization of urban spatial resilience from the perspective of life circle: A case study of Urumqi, NW China. *Sustain. Cities Soc.* **2024**, *109*, 105527. [CrossRef]
22. Chen, Y.N.; Li, Z.; Fang, G.H.; Li, Y.P. Global drought variation and its adaptation. *Sci. Technol. Rev.* **2024**, 1–6. [CrossRef]
23. Li, P.; Zhang, R.; Xu, L. Three-dimensional ecological footprint based on ecosystem service value and their drivers: A case study of Urumqi. *Ecol. Indic.* **2021**, *131*, 108117. [CrossRef]
24. Shi, L.; Halik, Ü.; Mamat, Z.; Aishan, T.; Abliz, A.; Welp, M. Spatiotemporal investigation of the interactive coercing relationship between urbanization and ecosystem services in arid northwestern China. *Land Degrad. Dev.* **2021**, *32*, 4105–4120. [CrossRef]
25. Aishan, T.; Song, J.; Halik, Ü.; Betz, F.; Yusup, A. Predicting Land-Use Change Trends and Habitat Quality in the Tarim River Basin: A Perspective with Climate Change Scenarios and Multiple Scales. *Land* **2024**, *13*, 1146. [CrossRef]
26. Liang, X.; Guan, Q.; Clarke, K.C.; Liu, S.; Wang, B.; Yao, Y. Understanding the drivers of sustainable land expansion using a patch-generating land use simulation (PLUS) model: A case study in Wuhan, China. *Comput. Environ. Urban Syst.* **2021**, *85*, 101569. [CrossRef]
27. Xu, X.; Kong, W.; Wang, L.; Wang, T.; Luo, P.; Cui, J. A novel and dynamic land use/cover change research framework based on an improved PLUS model and a fuzzy multiobjective programming model. *Ecol. Inform.* **2024**, *80*, 102460. [CrossRef]
28. Zhou, Y.D.; Wang, J.; Zhou, F. Evaluation of land use change and ecosystem service value based on multi-scenario analysis of PLUS model. *J. Gansu Agric. Univ.* **2023**, *58*, 198–209.

29. Deng, H.; Shao, J.A.; Wang, J.L.; Wei, C.F. Modelling of future land use scenarios in the Three Gorges reservoir area under multi-factor coupling. *Acta Geogr. Sin.* **2016**, *71*, 1979–1997. [CrossRef]
30. Tian, J.; Zhang, Z.; Zhao, T.; Tao, H.; Zhu, B. Warmer and wetter climate induced by the continual increase in atmospheric temperature and precipitable water vapor over the arid and semi–arid regions of Northwest China. *J. Hydrol. Reg. Stud.* **2022**, *42*, 101151. [CrossRef]
31. Zhang, Z.; Hu, B.; Jiang, W.; Qiu, H. Spatial and temporal variation and prediction of ecological carrying capacity based on machine learning and PLUS model. *Ecol. Indic.* **2023**, *154*, 110611. [CrossRef]
32. You, Y.; Jiang, W.; Yi, L.; Zhang, G.; Peng, Z.; Chang, S.; Hou, F. Seeding alpine grasses in low altitude region increases global warming potential during early seedling growth. *J. Environ. Manag.* **2024**, *356*, 120679. [CrossRef] [PubMed]
33. James, C.; Iverson, L.; Woodall, C.W.; Allen, C.D.; Bell, D.M.; Bragg, D.C.; D'Amato, A.W.; Davis, F.W.; Hersh, M.H.; Ibanez, I.; et al. The impacts of increasing drought on forest dynamics, structure, and biodiversity in the United States. *Glob. Change Biol.* **2016**, *22*, 2329–2352. [CrossRef]
34. Zhang, J.; Wu, X.; Shi, Y.; Jin, C.; Yang, Y.; Wei, X.; Mu, C.; Wang, J. A slight increase in soil pH benefits soil organic carbon and nitrogen storage in a semi-arid grassland. *Ecol. Indic.* **2021**, *130*, 108037. [CrossRef]
35. Zhao, Y.; Kasimu, A.; Liang, H.; Reheman, R. Construction and Restoration of Landscape Ecological Network in Urumqi City Based on Landscape Ecological Risk Assessment. *Sustainability* **2022**, *14*, 8154. [CrossRef]
36. Fang, C.L.; Huang, J.C.; Bu, W.N. Theoretical study on urbanization process and ecological effect with the restriction of water resource in arid area of northwest China. *Arid. Land Geogr.* **2004**, *27*, 1–7.
37. Xiang, Q.; Yu, H.; Huang, H.; Yan, D.; Yu, C.; Wang, Y.; Xiong, Z. The impact of grazing activities and environmental conditions on the stability of alpine grassland ecosystems. *J. Environ. Manag.* **2024**, *360*, 121176. [CrossRef]
38. Yi, Z.; Zhou, W.; Razzaq, A.; Yang, Y. Land resource management and sustainable development: Evidence from China's regional data. *Resour. Policy* **2023**, *84*, 103732. [CrossRef]

*land*

MDPI

*Article*

# Dynamic Simulation of the Supply and Demand of Ecosystem Windbreak and Sand Fixation Service in the Wuding River Basin

**Xuanchen Xiong [1], Zixiang Zhou [1,*], Xiaofeng Wang [2] and Jing Li [3]**

1 College of Geomatics, Xi'an University of Science and Technology, Xi'an 710054, China; xiongxuanchen@stu.xust.edu.cn
2 School of Land Engineering, Chang'an University, Xi'an 710054, China; wangxf@chd.edu.cn
3 School of Geography and Tourism, Shaanxi Normal University, Xi'an 710119, China; lijing@snnu.edu.cn
* Correspondence: zhouzixiang@xust.edu.cn

**Abstract:** Wind erosion can cause land degradation and other harmful effects. Examining the ecosystem windbreak and sand fixation service (WSFS) from the perspectives of supply and demand plays a crucial role in the continuous regulation of regional wind erosion. Through the enhancement of the revised wind erosion equation (RWEQ) model, integrated with uncertainty analysis, scenario simulation, and environmental factors calculation, the dynamic simulation of the supply of ecosystem windbreak and sand fixation service (WSFSS) and the demand of ecosystem windbreak and sand fixation service (WSFSD) in the Wuding River Basin in China was achieved, and specifically, a simulation framework for WSFSD and WSFSS was constructed. The results show that: (1) the uncertainty analysis can calculate the upper and lower limits of the range of parameter x (downwind distance) in the RWEQ model, and changes in the parameter x can make the simulation results of WSFSS and WSFSD more reasonable; (2) In the past 20 years, the WSFSS has shown a spatial distribution pattern of high in the northwest and low in the southeast. In terms of time, the annual WSFSS has shown a fluctuating growth trend with a growth rate of 8.06 t/a. The monthly WSFSS has shown a rising-fluctuating-declining trend; (3) The rationality of WSFSD was indirectly verified through the setting of scenario simulation. In terms of time, across the 252 months under study (January 2000–December 2020), 85% of the months witnessed WSFSD within the range of 1.0–1.4 kg/m$^2$ in the Wuding River Basin. At the same time, the WSFSD also presented seasonal variation patterns. The WSFSD was relatively high in spring (March–May) and relatively low in summer (July–September) each year.

**Keywords:** ecosystem windbreak and sand fixation service; the supply and demand of ecosystem windbreak and sand fixation service; RWEQ model; Wuding River Basin

## 1. Introduction

The ecosystem windbreak and sand fixation service (WSFS) [1] is a kind of ecosystem service that mitigates the hazards of wind erosion by means of surface vegetation. The supply of ecosystem windbreak and sand fixation service (WSFSS) [2] refers to the rate of reduction in soil wind erosion resulting from the existence of vegetation. The demand of ecosystem windbreak and sand fixation service (WSFSD) [3] refers to the maximum soil wind erosion rate that humans allow to occur in a certain region within a given period. At present, under the influence of climate change and human activities, there is an increasing risk of global soil erosion [4]. Wind erosion can lead to soil loss, land degradation, and land desertification, among other harmful effects [5]. The sand and dust after wind erosion can also cause severe air pollution, sandstorms, and other disasters [6]. In arid, semi-arid, and semi-humid regions around the world, WFSF can significantly reduce the hazards of wind erosion [7,8]. Therefore, how to enhance the WFSF has become the focus of many scholars. Examining the WSFS from the perspectives of WSFSS and WSFSD plays a crucial role in maintaining ecosystem functional stability [9], quantifying management for wind erosion control [10], and establishing ecological compensation policies [11]. It also effectively maintains the sustainable development of WSFS within the region.

The simulation methods for WSFSS and WSFSD often involve the soil wind erosion rate obtained from wind erosion simulations. The revised wind erosion equation model (RWEQ) is easy to expand and has accurate simulation results [12], and thus it is widely used to assess the soil wind erosion rate. Many scholars also adjust the input parameters of the model according to actual conditions. For example, Gong et al. [13] conducted soil particle size conversion when calculating soil factor with the model. Guo et al. [14] modified the wind factor in the model using limited wind speed data. Zhang et al. [15] added soil compaction data to calculate the soil erodibility factor in the model. However, there have been fewer studies on modifying model parameters from the perspectives of WSFSS and WSFSD. The parameter x (downwind distance) is a key parameter in the RWEQ model for calculating soil wind erosion. Changing its range of values can avoid resulting redundancies [16]. Meanwhile, uncertainty analysis is a fast and effective parametric analysis method that can reveal the impact of parameter changes on simulation results [17]. Based on the definitions of WSFSS and WSFSD and in combination with uncertainty analysis to modify the parameter x in RWEQ model, fundamental data support can be provided for the simulation of WSFSS and WSFSD.

In terms of the simulation methods of WSFSS and WSFSD, the simulation method of WSFSD has not yet been standardized. For example, Xu et al. [18] considered that WSFSD represents the wind erosion that affects human life, and used the actual wind erosion rate in the region as the WSFSD. Zhao et al. [19] used a fixed value (allowable soil loss rate) to measure WSFSD. Mi et al. [20] generated weighting operators for different slope land use types and calculated WSFSD by fitting the actual wind erosion rate with it. Environmental factors [21] are used to quantify the current environment's ability to resist wind erosion, which requires consideration of factors such as wind speed, topography, vegetation, soil moisture, and relevant data. Based on previous research, incorporating environmental factors into the simulation of WSFSD can help to explore the sensitivity of the environment to wind erosion changes, endowing the results with practical significance and temporal variation characteristics. Meanwhile, scenario simulation is a method of observing the characteristics of result changes by setting up scenarios with controlled variables [22]. Establishing reasonable scenarios and observing whether the WSFSD still complies with its definition under different scenarios can offer indirect validation for the rationality of the WSFSD.

The Wuding River Basin is a coarse-sand area located at the junction of the northern part of the Loess Plateau and the southern part of the Mu Us Sandy Land in China. It is also an environmentally sensitive area (ESA) for the implementation of the afforestation and grassland restoration project on the Loess Plateau [23]. Constant exposure to wind erosion has made this area susceptible to phenomena such as land degradation and drought. In this study, by integrating and developing methods such as uncertainty analysis, the RWEQ model, the calculation of environmental factors, and scenario simulation, a simulation framework for WSFSS and WSFSD is constructed to simulate the spatiotemporal variations in WSFSS and WSFSD in the Wuding River Basin from 2000 to 2020. The research findings have theoretical supplementary value and methodological reference significance for regulating wind erosion and maintaining WSFS in the Wuding River Basin.

## 2. Materials and Methods

### *2.1. Overview of the Study Area*

The Wuding River is a first-order tributary of the Yellow River and originates from the north slope of the Baiyushan Mountains in Dingbian County, Shaanxi Province, with a total length of 491 km. The geographical scope of the Wuding River Basin (as shown in Figure 1) is between 37°03′ and 39°04′ N latitude and 108°03′ and 110°57′ E longitude, with an area of approximately $2.8 \times 10^{-4}$ km$^2$. The terrain of the Wuding River Basin is south-high-north-low, with an elevation range of 581–1824 m. The terrain is hilly and undulating, with numerous ravines and sparse vegetation. The dominant vegetation types are shrubs and grasslands. Due to the frequent occurrence of strong winds in spring, with most wind speeds far exceeding the wind speed for erosion (the maximum wind speed can reach 17.5 m/s), large-scale wind erosion and land degradation is prone to occur in the Wuding River Basin. Consequently, it is highly indispensable to enhance the WSFS in this area.

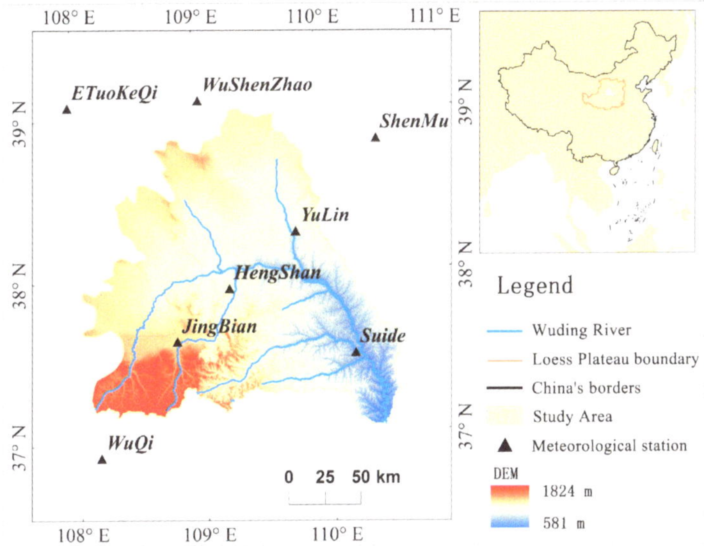


**Figure 1.** The geographical scope of the Wuding River Basin.

### *2.2. Data Sources and Processing*

The main data included meteorological, soil, vegetation, and topographic data. Meteorological data were derived from the final operational global analysis (FNL) data and were simulated by the weather research and forecasting (WRF) model [24], including hourly wind speed, precipitation, temperature, and solar radiation data. The soil data were sourced from the harmonized world soil database version 2.0 (HWSD2.0), which contains soil data at a depth of seven layers [25]. Since wind erosion primarily affects the surface, the first layer was chosen as the data for calculation. The vegetation data came from the China regional 250 m vegetation coverage dataset, which was then modified to remove lakes, rivers, and other areas, thus improving the accuracy of wind erosion simulation results. Other data sources are shown in Table 1, and the detailed information regarding the usage of the datasets is elucidated in Appendix A. Before inputting into the model, all data were resampled to 1 km resolution and projected into the lambert conformal conic projection with a central meridian of 105°E and standard parallels of 25° N and 47° N.

**Table 1.** Data sources.

| Data Name | Time Resolution | Spatial Resolution | Data Download Platform | Website |
|---|---|---|---|---|
| Vector boundary data | N/A | N/A | China Resources and Environmental Science Data Platform | https://www.resdc.cn/ (accessed on 1 December 2024) |
| China regional fractional vegetation cover dataset | monthly | 250 m | Tibetan Plateau Scientific Data Center | https://data.tpdc.ac.cn/ (accessed on 5 December 2024) |
| Long-term series of daily snow depth dataset | daily | 0.25° | | |
| Temperature data<br>Wind speed data<br>Precipitation data | hourly<br>hourly<br>hourly | 0.05°<br>0.05°<br>0.05° | National Center for Atmospheric Research | https://rda.ucar.edu/ (accessed on 3 December 2024) |
| Harmonized world soil database | N/A | 1 km | United Nations Food and Agriculture Organization | https://gaez.fao.org/pages/hwsd (accessed on 11 December 2024) |
| SRTMDEM | N/A | 90 m | Geospatial Data Cloud | https://www.gscloud.cn/ (accessed on 12 December 2024) |

### 2.3. *Wind Erosion Rate Simulation Model*

The RWEQ (revised wind erosion equation) is an empirical model for wind erosion simulation [26]. This model can simulate the soil erosion amount in accordance with the wind in a field (from the surface to a height of 2 m). Based on the input time period, the simulated soil wind erosion amount can be converted to soil wind erosion rate. According to the definition of model, the influence factors of wind erosion are divided into weather factor (WF), snow depth factor (SD), soil moisture factor (SW), soil erodibility factor (EF), soil crust factor (SCF), surface roughness factor (K′), and combined vegetation factor (COG), and the detailed information regarding the calculation of factors is elucidated in Appendix A. The environmental factors in this study include both actual environmental factors (Factor5) and potential environmental factors (Factor4).

Based on actual environmental factors, the actual maximum sediment transport capacity, actual critical field length, and actual wind erosion rate can be calculated. The calculation formula [26] is as follows:

$$\mathrm{Factor5} = \mathrm{WF} \times \mathrm{SCF} \times \mathrm{SCF} \times \mathrm{EF} \times \mathrm{K'} \times \mathrm{COG} \quad (1)$$

$$\mathrm{Qmax} = 109.8 \times \mathrm{Factor5} \quad (2)$$

$$\mathrm{S} = 150.71 \times (\mathrm{Factor5} + 0.001)^{-0.3711} \quad (3)$$

$$\mathrm{SL} = \frac{2\mathrm{x}}{\mathrm{S}^2} \times \mathrm{Qmax} \times \mathrm{e}^{-(\mathrm{x}/\mathrm{S})^2} \quad (4)$$

In the formula, Factor5 represents the actual environmental factors; Qmax represents the actual maximum sediment transport capacity (kg/m); S represents the actual critical field length (m), which is the distance required to achieve a maximum sediment transport capacity of 63.2%; SL represents the actual soil wind erosion rate (kg/m$^2$); and the parameter x represents the downwind distance (m).

### 2.4. The Calculation of WSFSS and WSFSD

#### 2.4.1. The Calculation of WSFSS

The calculation of WSFSS is expressed as the difference between the potential wind erosion rate and the actual wind erosion rate. The formula [2] for calculation is as follows:

$$\text{Factor4} = \text{WF} \times \text{SCF} \times \text{SCF} \times \text{EF} \times \text{K}' \tag{5}$$

$$\text{Qmax\_r} = 109.8 \times \text{Factor4} \tag{6}$$

$$\text{S\_r} = 150.71 \times (\text{Factor4} + 0.001)^{-0.3711} \tag{7}$$

$$\text{SL\_r} = \frac{2\text{x}}{\text{S\_r}^2} \times \text{Qmax\_r} \times \text{e}^{-(\text{x}/\text{S\_r})^2} \tag{8}$$

$$\text{WSFSS} = \text{SL\_r} - \text{SL} \tag{9}$$

In the formula, Factor4 represents the potential environmental factors (kg/m); Qmax_r represents the potential maximum sediment transport capacity (kg/m); S_r represents the potential critical field length (m); SL_r represents the potential wind erosion rate (kg/m$^2$); and WSFSS represents the supply of ecosystem windbreak and sand fixation service (kg/m$^2$).

#### 2.4.2. The Calculation of WSFSD

In this study, the WSFSD was defined as the soil wind erosion rate that could be allowed to occur, in other words, the soil wind erosion rate that humans define as a limit that should not be exceeded. Furthermore, on the basis of the studies of previous researchers [18,20], in this study, we attempt to improve the calculation method of WSFSD by incorporating environmental factors into the RWEQ model.

Its calculation is expressed as follows: constructing the functional relationship between actual wind erosion rate and actual environmental factors using the RWEQ model, and finding the point where actual environmental factors have the greatest influence on the actual wind erosion rate (the most sensitive environmental areas within the region to wind erosion changes), which is also the point of demand. The method has the following advantages:

- It takes into account the wind erosion rate that actually affects human beings;
- The results may have temporal variation characteristics;
- It also takes into account all the factors that affect wind erosion; and
- the RWEQ model used for calculation can also have a good connection with the WSFSS.

The calculation formula is as follows:

$$\text{SL} = \frac{2\text{x}}{\left(150.71 \times (\text{Factor5} + 0.001)^{-0.3711}\right)^2} \times 109.8 \times \text{Factor5} \times \text{e}^{-\left(\text{x}/150.71 \times (\text{Factor5}+0.001)^{-0.3711}\right)^2} \tag{10}$$

$$f(\text{Factor5}) = \frac{2\text{x}}{\left(150.71 \times (\text{Factor5} + 0.001)^{-0.3711}\right)^2} \times 109.8 \times \text{Factor5} \times \text{e}^{-\left(\text{x}/150.71 \times (\text{Factor5}+0.001)^{-0.3711}\right)^2} \tag{11}$$

$$f(\text{Factor5})'' = \frac{2x}{\left(150.71 \times (\text{Factor5} + 0.001)^{-0.3711}\right)^2} \times 109.8 \times \text{Factor5} \times e^{-\left(x/150.71 \times (\text{Factor5}+0.001)^{-0.3711}\right)^2} \quad (12)$$

We set Formula (12) as equal to zero and solved for the demand point (Factor5_d, WSFSD). WSFSD represents the demand of ecosystem windbreak and sand fixation service ($kg/m^2$). The second derivative equals zero, and can be solved using analytical methods or Newton iteration methods, etc. Since a large number of sample pixels are used in this study (for example, the Factor5 range varies within a time period from 0 to 15 kg/m, with a total of 27,843 sampled pixels and an average interval of $5.387 \times 10^{-4}$ kg/m), a method using the maximum value of the first derivative was used to solve it, which can reduce the computational load. The reference formula is:

$$\text{dy\_dx}_i = \frac{dy}{dx} = \frac{\text{SL}_i - \text{SL}_{i-1}}{\text{Factor5}_i - \text{Factor5}_{i-1}} (i \in 2, \ldots, n-1, n) \quad (13)$$

$$\text{dy\_dx}_s = \max(\text{dy\_dx}_i) \quad (14)$$

$$\text{WSFSD} = f(\text{Factor5}_s) \quad (15)$$

In this formula, i represents the index of the Factor5 raster pixels after sorting; $SL_i$ represents the actual wind erosion rate obtained from $\text{Factor5}_i$. When i = s, $\text{dy\_dx}_i$ reaches the maximum value.

### 2.5. Uncertainty Analysis

The parameter x is a key constant parameter for calculating the wind erosion rate in the RWEQ model, as shown in Formulas (4) and (8). The physical meaning of the parameter in the model is the downwind distance. The parameter x directly affects the wind erosion rate simulation results. Therefore, the value of the parameter x should be determined before calculating the WSFSS and WSFSD. However, there is no clear method for calculating the numerical range of parameter x. Based on the understanding of the RWEQ model, the definition of WSFSS and WSFSD, and the principles of wind and sand physics [27], this study attempts to develop the following algorithm to determine the upper and lower limits of the parameter x.

The lower limit calculation formula is:

$$\text{SL\_rx} = \frac{2x}{\text{S\_r}^2} \times \text{Qmax\_r} \times e^{-(x/S)^2} \quad (16)$$

$$\text{lst(Qmax\_r)}, \text{lst(S\_r)}, \text{lst(SL\_r)} = \text{geo\_lst(Qmax\_r, S\_r, SL\_rx)} \quad (17)$$

$$\text{lst(Qmax\_sort)}, \text{lst(S\_sort)}, \text{lst(SL\_sort)} = \text{sort(lst(Qmax\_r), lst(S\_r), lst(SL\_r), para} = 1) \quad (18)$$

$$\min(\mu x - \text{max\_arg(lst(SL\_sort))}) \geq 0 \quad (19)$$

In the formula SL_rx represents the potential wind erosion rate raster data calculated at a downwind distance of x; Formula (17) indicates that pixel lists are generated based on the geographic location corresponding to Qmax_r, S_r, and SL_rx raster pixels; Formula (18) indicates that lst(Qmax_r) will be sorted in ascending order, and lst(S_r) and lst(SL_rx) will be reordered according to the corresponding values of lst(Qmax_r) from before; Formula (19) indicates that the position of the maximum pixel in the SL_sort list is subtracted from μx. If the minimum condition is met, we obtain the optimal SL_rx, where x is the desired lower limit; μx represents the number of raster pixels, used to control the threshold values of Qmax_r and S_r outliers.

The upper limit calculation formula is:

$$SL_x = f_1(\text{Factor5}) = \frac{2x}{\left(150.71 \times (\text{Factor5} + 0.001)^{-0.3711}\right)^2} \times 109.8 \times \text{Factor5} \times e^{-\left(x/150.71 \times (\text{Factor5}+0.001)^{-0.3711}\right)^2} \quad (20)$$

$$SL_{x+1} = f_2(\text{Factor5}) = \frac{2(x+1)}{\left(150.71 \times (\text{Factor5} + 0.001)^{-0.3711}\right)^2} \times 109.8 \times \text{Factor5} \times e^{-\left(x+1/150.71 \times (\text{Factor5}+0.001)^{-0.3711}\right)^2} \quad (21)$$

$$\forall \text{Factor5}\ \epsilon\ (0, \max(\text{Factor5})), \exists f_1(\text{Factor5})'' = 0 \quad (22)$$

$$\forall \text{Factor5}\ \epsilon\ (0, \max(\text{Factor5})), \nexists f_2(\text{Factor5})'' = 0 \quad (23)$$

In this formula, SLx represents the actual wind erosion rate calculated at a downwind distance of x, and x is the desired upper limit when function 1 and function 2 satisfy Formulas (22) and (23).

### *2.6. Scenarios Setting*

To investigate the impact of WSFSS changes on WSFSD and indirectly verify the rationality of the WSFSD, the scenarios of magnified FVC (Fractional Vegetation Cover) within the entire zone and partial zone were established. Here are the specific steps:

We chose two soil wind erosion rates from different years but within the same month, and made sure they had a significant difference in WSFSS (e.g., April 2000 and April 2020). After that, two scenarios were set up:

1. Magnified FVC within the entire zone:

   We took the selected data of April 2000 as an example, scaled the FVC (Fractional Vegetation Cover) proportionally across the entire zone (zone below/above WSFSD), and re-simulated the WSFSS, WSFSD, and the area of the zone above the WSFSD.

2. Magnified FVC within a partial zone:

   We took the selected data of April 2000 as an example, scaled the FVC proportionally in a partial zone (zone above WSFSD), and re-simulated the WSFSS, WSFSD, and the area of the zone above the WSFSD.

Because vegetation is one of the important factors that affect wind erosion rate changes [28], we chose to enlarge the FVC as the driving force to increase WSFSS. The results of the scenario simulation can reveal the impact of WSFSS changes on WSFSD.

Following the change in FVC, the areas of the zones above WSFSD in both scenarios were compared with the areas of the zones above WSFSD in April 2020. Once the area was observed to be smaller than the area of the zone above WSFSD in April 2020, the magnification ratio of the FVC was recorded. If the proportion of records is similar in both scenarios, it indicates that the zone above the WSFSD can play a more critical role in wind and sand control than the entire zone, which is consistent with the definition of WSFSD itself and indirectly proves the rationality of the WSFSD. If the ratio of the recorded values is too different, it indicates that the WSFSD is not reasonable.

### *2.7. The Simulate Framework for WSFSS and WSFSD*

The simulation framework for WSFSS and WSFSD consists of two parts (as shown in Figure 2). The first part aims to calculate the WSFSS and WSFSD, and then actual wind erosion rate and WSFSD are selected as input for the second part. The second part aims to indirectly verify the reasonableness of the WSFSD through scenario simulation setting, and supports the results of part one.

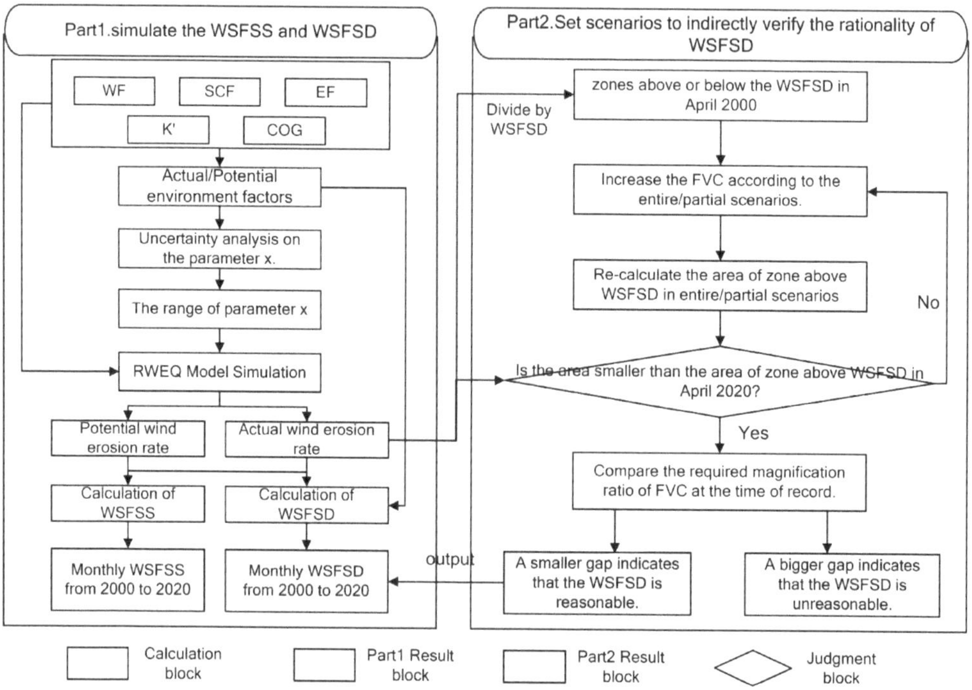


**Figure 2.** The simulation framework for WSFSS and WSFSD.

## 3. Results

### 3.1. Uncertainty Analysis on the Parameter x

We took the calculation of the potential soil wind erosion rate in the Wuding River Basin in January 2000 as an example to demonstrate the calculation process and results of the upper and lower limits of parameter x. Before calculating the lower limit of parameter x, the output results of the RWEQ model were sorted by geographical location and Qmax_r (potential maximum sediment transport capacity) according to Formulas (17) and (18). The result is presented in Figure 3a. During the process of the increase in the value of Qmax_r, the value of S_r (the potential critical field length) corresponding to its geographical location was constantly decreasing, which was in accordance with the definition of the model. Nevertheless, the SL_r (the potential wind erosion rate) corresponding to the geographical location increased initially and then decreased. This indicates that a greater Qmax_r does not necessarily imply a greater SL_r, which is not in line with the definition of the model [29,30].

Without altering the parameters in Qmax_r and S_r, only parameter x is modified. As depicted in Figure 3b, Qmax_r and S_r remain unchanged. This indicates that changing x does not result in data redundancy, only SL_r changes. With the decrease in x, the maximum pixel point of the SL_r curve continuously increases. When x is less than or equal to 98, the maximum pixel point of the SL_r curve does not change (equal to the number of raster pixel). At this juncture, SL_r increases along with the augmentation of Qmax_r, which is in accordance with the expected change; 98 was recorded as the lower limit of the parameter x during the simulation period of January 2000.

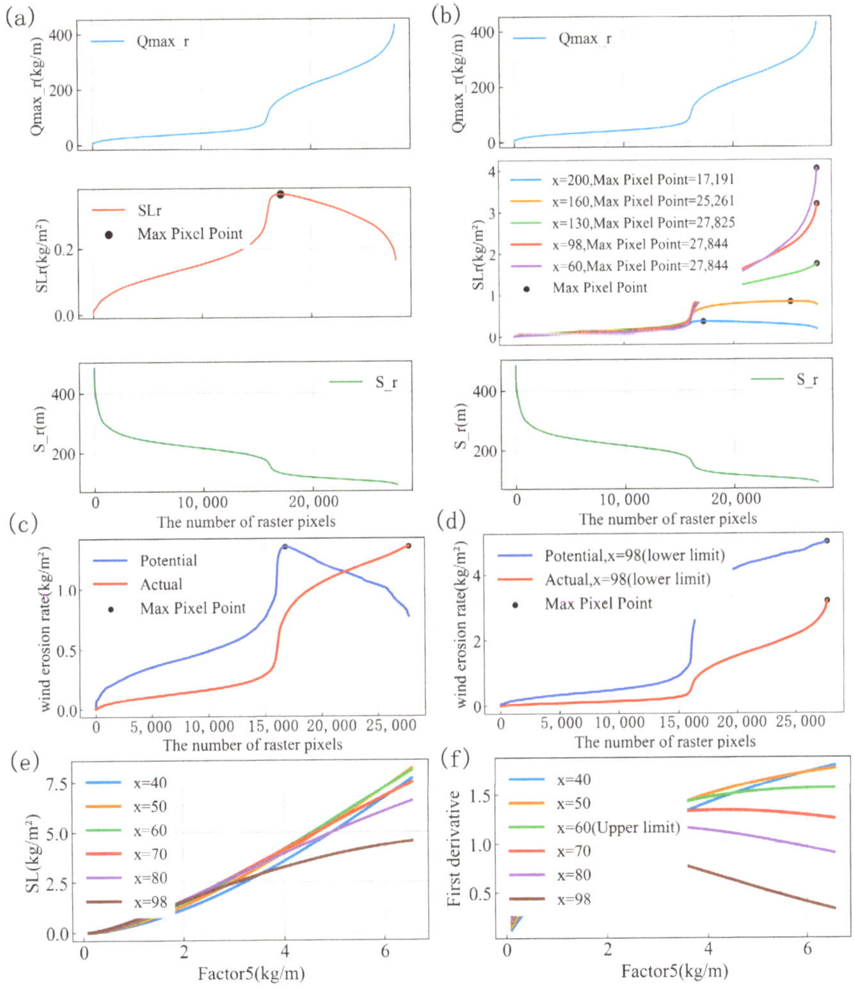

**Figure 3.** (**a**) The output of the RWEQ model before calculating the lower limit of parameter x; (**b**) The calculation of the lower limit of parameter x using uncertainty analysis; (**c**) The output of potential and actual wind erosion rate before calculating the lower limit of parameter x; (**d**) The output of potential and actual wind erosion rate after calculating the lower limit of parameter x; (**e**) The calculation of the upper limit of parameter x using uncertainty analysis; (**f**) The calculation of the upper limit of parameter x using uncertainty analysis (the first derivative of SL).

Meanwhile, wind erosion rate can directly affect changes in the WSFSS. As shown in Figure 3c, with the actual wind erosion rate increasing, there is a situation where the potential wind erosion rate is smaller than the actual wind erosion rate, which obviously does not meet the requirements of WSFSS. We set x as a number that is less than or equal to 98. As depicted in Figure 3d, this can satisfy the requirements of the WSFSS, thereby making the simulated WSFSS more rational. However, as x continuous decreases, WSFSD will also be affected, as shown in Figure 3d, where SL (actual wind erosion rate) keeps increasing as Factor5 (actual environment factors) grows. Taking the first-order derivatives of these curves, as illustrated in Figure 3e, the variation in the parameter x is capable of influencing whether there is a demand point in the first-order derivative curve of SL. When x is less than 60, the first derivative curve does not have a maximum value, and the point of demand does not exist. To avoid this situation, 60 was recorded as the upper limit of the parameter x during the simulation period of January 2000.

In summary, using the uncertainty analysis to determine the upper and lower limits of parameter x can not only make the output of the RWEQ model more rational, but also make the simulation results of the WSFSS and WSFSD models more reasonable.

### 3.2. The Variation of the WSFSS

After determining the parameter x, the WSFSS in the Wuding River Basin from 2000 to 2020 was simulated. As shown in Figure 4a, from the perspective of annual scale spatial variation, the high WSFSS in the Wuding River Basin has been concentrated mainly in the northern and western regions over the past 20 years, with a maximum value of up to 71.8 kg/m$^2$. The WSFSS in the southern and eastern regions has been persistently low for many years, so the overall spatial distribution pattern of WSFSS in the study area is characterized as high in the northwest and low in the southeast. As shown in Figure 4b, from the annual scale temporal variation, there is a fluctuating trend with large amplitude in the annual WSFSS, but the overall trend continues to rise, with a growth rate of 8.06 t/a. The highest value was reached in 2018 at 719.5 t, and the lowest value was reached in 2003 at 550.2 t. As shown in Figure 4c, the trend changes of WSFSS are related to the month. January to April is the ascending stage, May to October is the fluctuating stage, and November to December is the descending stage. Additionally, the standard deviation (sd) of WSFSS for each month also varies greatly, with the highest variance in April and the lowest in October. Overall, despite the occurrence of fluctuations, it is demonstrated that the WSFS has been effectively sustained in this area.

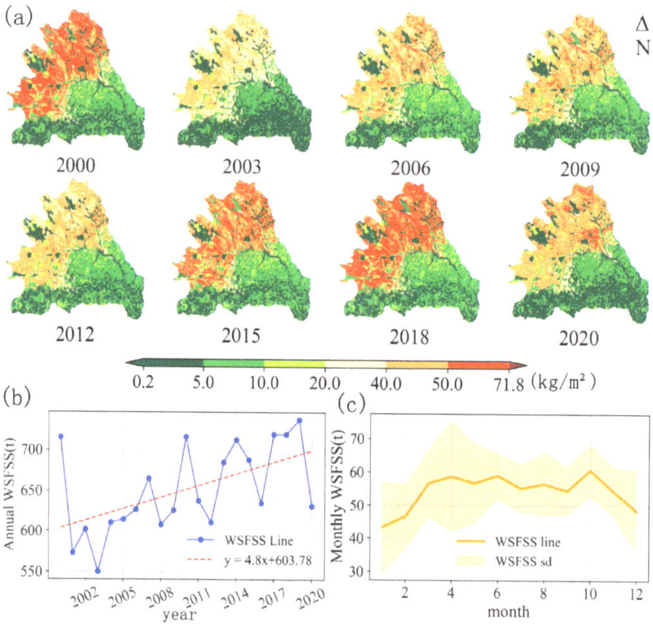


**Figure 4.** (**a**) Spatial distribution of annual WSFSS; (**b**) Temporal variation in the annual WSFSS; (**c**) Temporal variation in the monthly WSFSS.

### 3.3. The Variation of the WSFSD

We simulated the WSFSD based on the SL (actual wind erosion rate) function, with Factor5 (actual environmental factors) as the independent variable. As shown in Figure 5a, with an increase in Factor5, the SL keeps increasing, and the first derivative of the SL function initially increases and then decreases. The WSFSD in January 2000 was calculated to be 1.29 kg/m$^2$. Similarly, as shown in Figure 5b, the WSFSD was calculated to be 2.86 kg/m$^2$ in May 2000. As shown in Figure 5c, during the period from 2000 to 2020, 85% of the monthly WSFSD was within the range of 1.0–1.4 kg/m$^2$. The overall range of WSFSD in the study area fluctuated little. The maximum WSFSD beyond the range reached 3.46 kg/m$^2$, and the minimum WSFSD reached 0.09 kg/m$^2$. From the perspective of year, the WSFSD in different years showed a gradual decrease over time. For example, the WSFSD was 3.46 kg/m$^2$ in April 2000, but it had decreased to just 1.1 kg/m$^2$ by April 2020. From the perspective of month, WSFSD across different years changed significantly with month. Low WSFSD was more likely to occur in July to September (Autumn), while high WSFSD was more likely to occur in March to May (Spring).

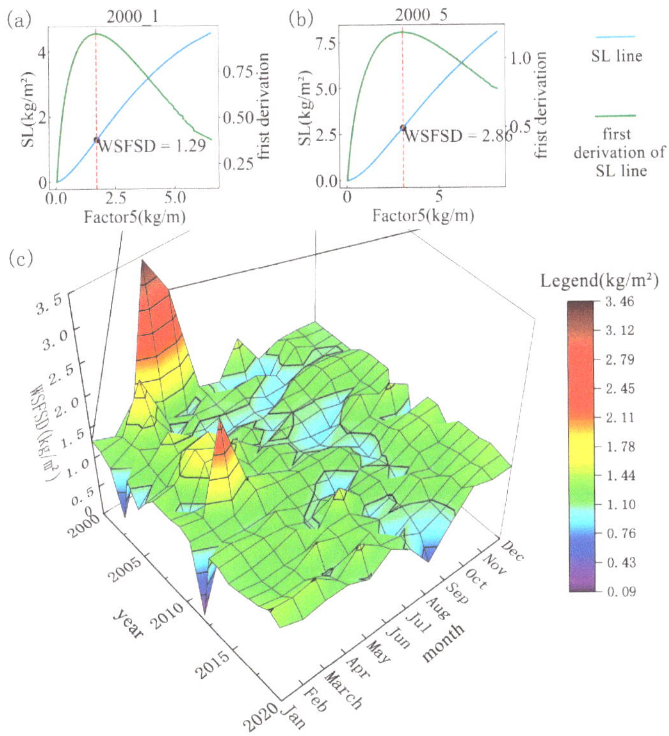


**Figure 5.** (**a**) WSFSD in January 2000 (**b**) WSFSD in May 2000 (**c**) Monthly WSFSD from 2000 to 2020.

### 3.4. Indirect Verification of WSFSD

According to the results shown in Section 3.1, the month with the greatest variation in WSFSS across different years was April, so the actual wind erosion rates in April 2000 and April 2020 were chosen as the input data for scenario simulation. From the change in WSFSD (as shown in Figure 6a), as the FVC in both partial and entire zones was enlarged, both WSFSS values were increasing, and the gap between them was also widening, with the largest gap reaching 8.57 t. Additionally, the WSFSD in both simulation scenarios showed a downward trend, and the magnitude of the change was consistent. However, the decrease in WSFSD was not significant, with the maximum value of difference being only 0.08 kg/m$^2$. From the perspective of the area of the zone above WSFSD (as shown in Figure 6b), the area in both simulation scenarios showed a downward trend, with the maximum reduction areas being 8073 and 8067 km$^2$, and exhibiting a high degree of consistency in the change amplitude. Therefore, under certain conditions, the growth of WSFSS in entire and partial zones will not affect WSFSD, but both can significantly reduce the area of the zone above WSFSD.

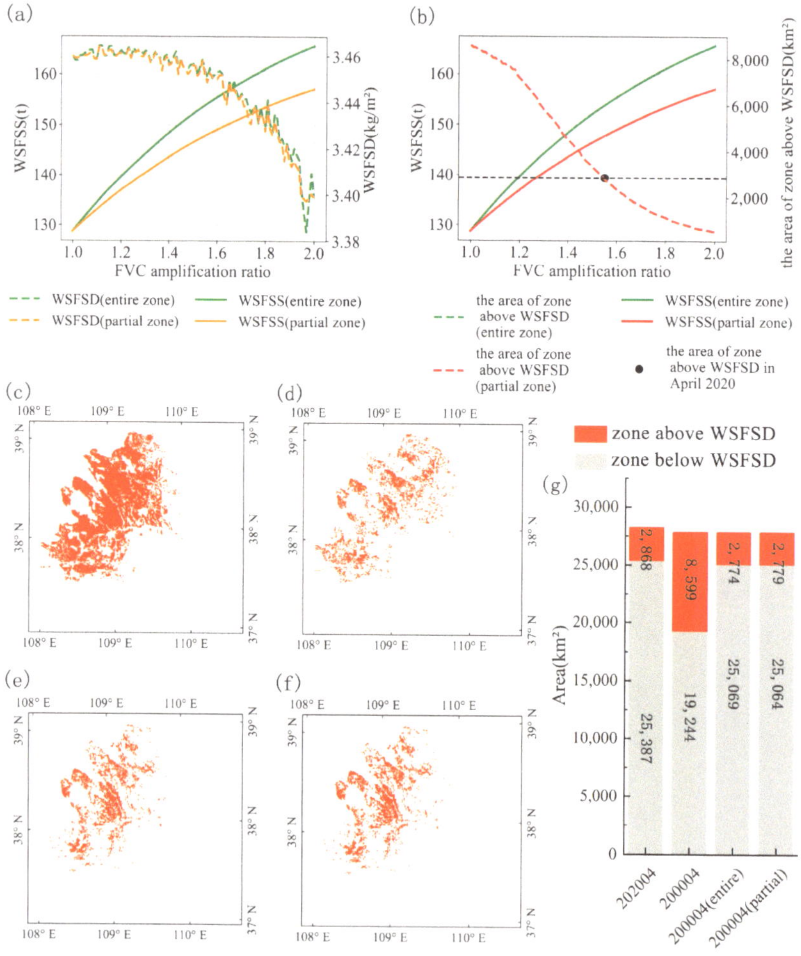


**Figure 6.** (**a**) The variations in WSFSS and WSFSD after magnifying the FVC proportion within the entire or partial zone; (**b**) The variations in WSFSS and the area of the zone above WSFSD after magnifying the FVC proportion within the entire or partial zone; (**c**) The spatial distribution of the zone above WSFSD in April 2000; (**d**) The spatial distribution of the zone above WSFSD in April 2020; (**e**) The spatial distribution of the zone above WSFSD at the time of record (entire zone); (**f**) The spatial distribution of the zone above WSFSD at the time of record (partial zone); (**g**) Comparison of the area of zone above WSFSD at the time of record in different scenarios.

The simulated data were divided into zones above or below the WSFSD. As shown in Figure 6c,d,g, the spatial distribution of the zone above the WSFSD in April 2000 and

April 2020 was quite different, with the areas being 8599 km$^2$ and 2868 km$^2$, respectively. We magnified the FVC proportion according to the scenario settings until the recording time. From the perspective of the change in FVC, in the entire zone and the partial zone, the required amplification ratio to meet the area of the zone above the WSFSD was 1.51 and 1.53 (as shown in Figure 6b). The areas of the zone above the WSFSD decreased by 5825 km$^2$ and 5820 km$^2$, respectively (as shown in Figure 6e,f,g). Therefore, in the zone above the WSFSD, it is necessary to prioritize increasing WSFSS. This is consistent with the definition of WSFSD, thereby demonstrating the rationality of the WSFSD.

## 4. Discussion

### *4.1. The Variation of Parameter x*

The results show that the upper and lower limits of the parameter x calculated in all time periods were all within the range of 40–110, which is consistent with the research results of Zhang [31] and Cao [32] et al. This indicates that the uncertainty analysis method developed in this study possesses a certain degree of reliability when dealing with long-term data, thereby ameliorating the situation where the x-values are set as fixed values. Regarding the factors influencing the variation in parameter x, the crucial factors are Factor5 (the actual environmental factors) or Factor4 (the potential environmental factors), followed by Qmax (the maximum sediment transport capacity), Qmax_r (the potential maximum sediment transport capacity), S (the critical field distance), and S_r (the potential critical field distance). This aligns with the idea of external factors affecting the x parameter, as proposed by Pelt et al. [33]. Additionally, we found the simulation time period is also a factor that affects the change in parameter x; for example, the value of parameter x in April was smaller than that in August. This is because wind erosion is more prone to occur in April than in August, increasing the soil wind erosion rate within the region and relatively reducing the downwind distance. This is consistent with a study by Borrelli et al. [34], which showed that the soil wind erosion rate is greater closer to the downwind edge. Therefore, when simulating WSFSS and WSFSD, it is advisable to consider adjusting the value of the x parameter to cope with the variations in environmental factors, thereby making the results more rational.

### *4.2. The Range of WSFSD*

From the simulation results of WSFSD, most WSFSD are within a small range. This aligns with the idea proposed by Zhao et al. [19] and Xie et al. [35] of using a fixed value (allowable soil loss rate) to measure WSFSD. Because Xie et al. [35] determined the WSFSD through years of field experiments and observations, the WSFSD in this study has practical significance. In terms of the number of pixels with values above or below the WSFSD, although the WSFSD in the month within the range is not significantly different, it can still effectively distinguish pixels with a value above the WSFSD from those with a value below the WSFSD. As shown in Figure 7a,b, although the WSFSD is only 0.12 kg/m$^2$ different, there were more pixels with values above the WSFSD in February 2000, while there were fewer pixels with values above the WSFSD in September 2000. In months outside the range, there may be either too small or too large WSFSD values, but it can also effectively complete the task of partitioning. As shown in Figure 7c,d, the WSFSD in January 2003 and April 2000 were 0.38 kg/m$^2$ and 3.46 kg/m$^2$, respectively, but the number of pixels above the WSFSD still fits the characteristics of the month within the range. Throughout the partitioning situation of WSFSD in all time periods (as depicted in Figure 7e), the variation curve of the area of zone above the WSFSD has the characteristic of changing along with the months and presents a fluctuating upward trend. This indicates that the area of zone above the WSFSD is shrinking over time, which is in line with the situation where the environment of the study area has improved. Therefore, combining the results in Section 3.4 shows that the WSFSD is capable of indicating the regions within the study area where an increase in WSFSS is necessary and the wind erosion rate that needs to be reduced (shown as the red dashed line in the Figure 7), which also aligns with the requirements for WSFSD put forward by Xia et al. [36].

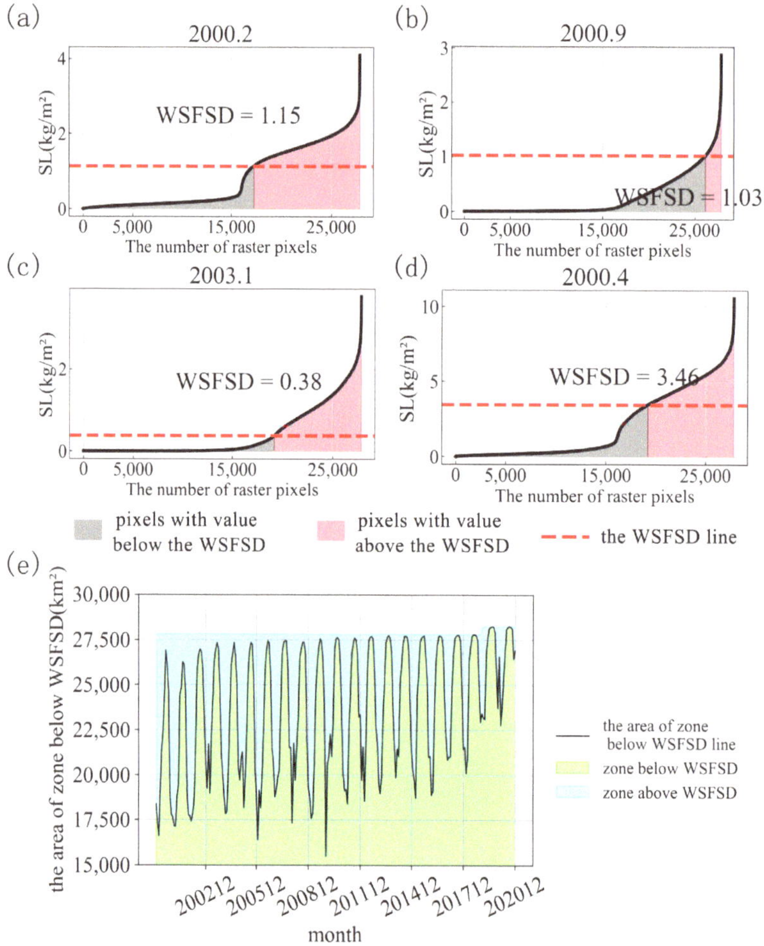


**Figure 7.** (**a**) Zoning by WSFSD in February 2000; (**b**) Zoning by WSFSD in September 2000; (**c**) Zoning by WSFSD in January 2003; (**d**) Zoning by WSFSD in April 2000; (**e**) partitioning situation of WSFSD from 2000 to 2020.

## 4.3. *Influencing Factors of WSFSS and WSFSD*

The spatial distribution characteristics of the WSFSS over multiple years are distinct and can be classified into high-value and low-value areas. The reason for the high-value area may be that the high value of the actual wind erosion rate is also distributed here, which is consistent with the geographical location of the wind sand area studied by Nong et al. [37] in the Wuding River Basin. In terms of annual variation, the fluctuating growth of WSFSS is consistent with the findings of Wang et al. [38]. The possible cause for this situation might be that within the study area, projects such as the Three-North Shelter Forest have been continuously implemented in the past two decades, leading to a continuous decrease in the wind erosion rate [39]. Furthermore, changes in WSFSS are also influenced by a variety of factors. For example, the WSFSS in July and August, when vegetation is lush, is actually slightly lower than that in March and April, when vegetation is sparse. It is likely that the relatively high wind speeds and the drier soil in spring lead to an excessive wind erosion rate, which enlarges the wind erosion rate that vegetation can reduce, thereby passively increasing the WSFSS. This is consistent with the study on the driving force of wind erosion rate by Wei et al. [40].

In the scenario simulation, as the scale of vegetation coverage was increased proportionally, the WSFSD did not drop significantly. This is because the key parameters in the WSFSD calculation process are the parameter x and Factor5 (the actual environmental factors), the parameter x is limited by the change in Factor4 (the potential environmental factors), and according to the scenario simulation, the change in FVC only leads to a change in Factor5. In actual circumstances, the FVC will influence the variations in other factors in the model, thereby affecting the changes in WSFSD. For example, vegetation restoration may result in a temporary decrease in soil moisture, but in the long run, it can increase soil moisture [41]. This is not taken into account in the scenario simulation, so the interactions between factors may also have an impact on changes in wind erosion rate and WSFSD. Interaction factors should be taken into account in future studies, such as improving factor calculations in the RWEQ model, which is consistent with the study on wind erosion simulation model construction by Zou et al. [42].

### *4.4. Limitations*

The uncertainty analysis developed in this study is currently unable to solve the initial value problem. Instead, an initial value must be set based on experience and the iteration process continued until the best result is achieved. The simulation framework constructed in this study was only tested in the Wuding River Basin and not tested in other regions. There may be errors due to differences in the environment. The calculation method of WSFSD used in this study only judges the rationality of the method through comparisons of simulation results, reference to previous research results, etc. There may be random errors. In the future, we will increase the accuracy of the results by conducting field surveys and social experiments.

## 5. Conclusions

To simulate the dynamic changes in the WSFSS and WSFSD in Wuding River Basin, this study constructed a simulation method framework for WSFSS and WSFSD by combining uncertainty analysis, the RWEQ model, and calculation of environmental factors and scenario simulation, etc. The results show:

(1) In the simulation framework, uncertainty analysis can calculate the variation range of parameter x in the RWEQ model. Changing parameter x can make the simulation results of WSFSS and WSFSD more reasonable. The introduction of environmental factors for calculating the WSFSD makes it more practically significant. The results of the scenario simulation can demonstrate the rationality of the WSFSD. It is demonstrated that this simulation framework has certain application value for the research of WSFS.

(2) In the past 20 years, the WSFSS in the Wuding River Basin has shown a spatial distribution pattern of high in the northwest and low in the southeast. In terms of time, the annual WSFSS has shown a fluctuating growth trend, with a growth rate of

8.06 t/a. The monthly WSFSS has shown a "rising-fluctuating-declining" trend. This indicates that the WSFS has been effectively maintained in the area.

(3) Although 85% of the months witnessed WSFSD within the range of 1.0–1.4 kg/m$^2$ in the Wuding River Basin, the WSFSD also presented seasonal variation patterns. The WSFSD was relatively high in spring (March–May) and relatively low in summer (July–September) each year. The results of the scenario simulation also show WSFSD is capable of indicating the regions within the study area where an increase in WSFSS is necessary, i.e., where the wind erosion rate needs to be reduced.

The simulation framework and the WSFSS and WSFSD in this study can offer methodological references and data support for studies on soil wind erosion, the maintenance of WSFS, and the establishment of ecological compensation policies in environmentally sensitive areas (such as the Wuding River Basin).

**Author Contributions:** Conceptualization, Z.Z.; methodology, X.X. and Z.Z.; formal analysis, X.W.; writing—original draft preparation, X.X.; writing—review and editing, Z.Z. and J.L.; funding acquisition, Z.Z. and J.L. All authors have read and agreed to the published version of the manuscript.

**Funding:** This work was supported by the National Natural Science Foundation of China (42071285, 42371297, 41771576), the Key R & D Program Projects in Shaanxi Province of China (2022SF-382), and the Fundamental Research Funds for the Central Universities (GK202302002).

**Data Availability Statement:** The original contributions presented in this study are included in the article. Further inquiries can be directed to the corresponding author.

**Conflicts of Interest:** The authors declare no conflicts of interest.

## Appendix A

Appendix A mainly contains an introduction to the method of calculating the factor parameters in the RWEQ model and details on how to use the datasets.

The factor parameters in the RWEQ model mainly include WF, SCF, EF, K′, and COG. WF refers to weather factor. The calculation formula of WF is as follows:

$$WF = \frac{\sum_{i=1}^{N} WS_2 \times (WS_2 - WS_t)^2 \times N_d \times \rho}{N \times g} \times SW \times SD \quad (A1)$$

$$\rho = 348.0\left(\frac{1.013 - 0.1183EL + 0.0048EL^2}{T}\right) \quad (A2)$$

$$SW = \frac{ET_p - (R + I)\frac{R_d}{N_d}}{ET_p} \quad (A3)$$

$$ET_p = 0.0162 \times \left(\frac{SR}{58.5}\right) \times (DT + 17.8) \quad (A4)$$

$$SD = 1 - P(sd > 25.4mm) \quad (A5)$$

In the formula, $WS_2$ represents the wind speed at a height of 2 m (m/s); $WS_t$ is the critical wind speed for the movement of soil particles by wind (assumed to be 5 m/s); $N_d$ represents the number of days of the experiment (d); ρ is the density of air (kg/m$^3$); N is the number of wind speed observations; g is the acceleration of gravity (m/s$^2$); SW is the soil moisture factor (dimensionless); $ET_p$ is the potential evapotranspiration (mm); R is the precipitation (mm); I is the irrigation (mm); $R_d$ is the number of rainy days (d); SR is the total solar radiation (cal/cm$^2$); and DT is the average temperature (°C). SD is the snow depth factor, where P() is the probability that the snow depth exceeds 25.4 mm during the calculation period. R, SR, DT, WS, and sd are derived from the final operational global analysis (FNL) data and simulated by the weather research and forecasting (WRF) model.

EF refers to soil erodibility factor, and SCF refers to soil crusting factor. The calculation formulas of EF and SCF are as follows [43]:

$$EF = \frac{29.09 + 0.31Sa + 0.17Si + 0.33Sa/Cl - 2.59OM - 0.95CaCO_3}{100} \quad (A6)$$

$$SCF = \frac{1}{1 + 0.0066(Cl)^2 + 0.021(OM)^2} \quad (A7)$$

In the formula, Sa represents the percentage of sand particles in the soil (dimensionless); Si represents the percentage of fine sand in the soil (dimensionless); Sa/Cl represents the ratio of sand and clay content in the soil (dimensionless); Cl represents the percentage of clay content (dimensionless); OM represents the percentage of organic matter content (dimensionless); and $CaCO_3$ represents the percentage of calcium carbonate content (dimensionless). Sa, Cl, Si, OM, and $CaCO_3$ values were sourced from the harmonized world soil database version 2.0 (HWSD2.0).

$K'$ refer to surface roughness factor. The calculation formula of $K'$ is as follows:

$$K' = e^{1.86Kr - 2.41Kr^{0.934} - 0.127Crr} \quad (A8)$$

$$Kr = 0.2 \times \left(H^2/L\right) \quad (A9)$$

$$RR = 0.025 + 2.464 \times FVC^{3.56} \quad (A10)$$

$$Crr = 17.46 \times RR^{0.738} \quad (A11)$$

In the formula, Kr represents the terrain roughness (dimensionless); Crr represents the chain random roughness (dimensionless); RR represents the random roughness (dimensionless); and FVC represents the fractional vegetation cover. FVC was derived from the China regional 250 m vegetation coverage dataset.

COG refers to combined vegetation factor. The calculation formula of COG is as follows:

$$COG = SLR_f \times SLR_c \times SLR_s \quad (A12)$$

$$SLR_f = e^{-0.0483sc} \quad (A13)$$

$$SLR_s = e^{-0.0344(SA^{0.6413})} \quad (A14)$$

$$SLR_c = e^{-5.614} \times \left(FVC^{0.7366}\right) \quad (A15)$$

In the formula, SLRf represents the factor of fallen plants (dimensionless); SLRc represents the factor of growing plant canopy (dimensionless); SLRs represents the factor of standing plants (dimensionless); sc represents fractional fallen vegetation cover (dimensionless); and SA is calculated by multiplying the number of standing stems per 1 m$^2$ area by the average stem diameter (cm) and average stem height (cm) (dimensionless). Zhang et al. [15] found that in Inner Mongolia grassland (adjacent to the Wuding River Basin), sc and FVC have a corresponding relationship, which is expressed as:

$$sc = 0.703FVC_{max} - 0.052 \quad (A16)$$

$FVC_{max}$ represents the maximum synthesized FVC.

## References


1. Han, Y.; Ma, X.; Yan, W.; Wang, Y. Windbreak and Sand Fixation Service Flow Simulation in the Terminal Lake Basin of Inland Rivers in Arid Regions: A Case Study of the Aral Sea Basin. *Sci. Total Environ.* **2024**, *945*, 174047. [CrossRef]
2. Xu, J.; Xiao, Y.; Xie, G.; Wang, Y.; Lei, G. Assessment of the Benefit Diffusion of Windbreak and Sand Fixation Service in National Key Ecological Function Areas in China. *Aeolian Res.* **2021**, *52*, 100728. [CrossRef]
3. Li, P. Study on Supply and Demand Relationship of Key Ecosystem Services in Shiyang River Basin. Master's Thesis, Northwest Normal University, Lanzhou, China, 2021. (In Chinese)
4. Wall, D.H.; Six, J. Give Soils Their Due. *Science* **2015**, *347*, 695. [CrossRef] [PubMed]
5. Borrelli, P.; Panagos, P.; Alewell, C.; Ballabio, C.; de Oliveira Fagundes, H.; Haregeweyn, N.; Lugato, E.; Maerker, M.; Poesen, J.; Vanmaercke, M.; et al. Policy Implications of Multiple Concurrent Soil Erosion Processes in European Farmland. *Nat. Sustain.* **2023**, *6*, 103–112. [CrossRef]
6. Zhang, H.; Song, H.; Wang, X.; Wang, Y.; Min, R.; Qi, M.; Ru, X.; Bai, T.; Xue, H. Effect of Agricultural Soil Wind Erosion on Urban PM2.5 Concentrations Simulated by WRF-Chem and WEPS: A Case Study in Kaifeng, China. *Chemosphere* **2023**, *323*, 138250. [CrossRef] [PubMed]
7. Li, C.; Fu, B.; Wang, S.; Stringer, L.C.; Wang, Y.; Li, Z.; Liu, Y.; Zhou, W. Drivers and Impacts of Changes in China's Drylands. *Nat. Rev. Earth Environ.* **2021**, *2*, 858–873. [CrossRef]
8. Yang, G.; Sun, R.; Jing, Y.; Xiong, M.; Li, J.; Chen, L. Global Assessment of Wind Erosion Based on a Spatially Distributed RWEQ Model. *Prog. Phys. Geogr. Earth Environ.* **2022**, *46*, 28–42. [CrossRef]
9. Wang, Q.; Cao, W.; Huang, L. Evolution Characteristics of Ecosystem Functional Stability and Ecosystem Functional Zoning on the Qinghai-Tibet Plateau. *J. Geogr. Sci.* **2023**, *33*, 2193–2210. [CrossRef]
10. Lyu, X.; Li, X.; Wang, H.; Gong, J.; Li, S.; Dou, H.; Dang, D. Soil Wind Erosion Evaluation and Sustainable Management of Typical Steppe in Inner Mongolia, China. *J. Environ. Manag.* **2021**, *277*, 111488. [CrossRef]
11. Xu, J.; Xiao, Y.; Xie, G.; Wang, Y.; Jiang, Y. Computing Payments for Wind Erosion Prevention Service Incorporating Ecosystem Services Flow and Regional Disparity in Yanchi County. *Sci. Total Environ.* **2019**, *674*, 563–579. [CrossRef]
12. Fryrear, D.W. Mechanics, Measurement and Modeling of Wind Erosion. *Adv. Geoecology* **1998**, *1*, 291–300.
13. Gong, G.; Liu, J.; Shao, Q.; Zhai, J. Sand-Fixing Function under the Change of Vegetation Coverage in a Wind Erosion Area in Northern China. *J. Resour. Ecol.* **2014**, *5*, 105–114. (In Chinese) [CrossRef]
14. Guo, Z.; Chang, C.; Wang, R. Calculate the Wind Factor for the RWEQ Model Using Limited Wind Speed Data. *Jiangsu Agric. Sci.* **2015**, *43*, 350–353. (In Chinese) [CrossRef]
15. Zhang, X.; Zhang, C.; Zuo, X.; Zou, X.; Wang, X.; Zhao, J.; Li, W.; Zhou, Z.; Zhang, Y. Extension of the Revised Wind Erosion Equation (RWEQ) to Calculate Grassland Wind Erosion Rates Based on the 137Cs Tracing Technique. *CATENA* **2024**, *234*, 107544. [CrossRef]
16. Fryrear, D.W.; Bilbro, J.D.; Saleh, A.; Schomberg, H.; Stout, J.E.; Zobeck, T.M. RWEQ: Improved Wind Erosion Technology. *J. Soil. Water Conserv.* **2000**, *55*, 183.
17. Huang, Q.; Badreldin, N.; Lobb, D.A.; Li, S.; Feng, G.; McConkey, B.G. Uncertainty and Sensitivity Analyses of the Modified Wind Erosion Equation for Application in Canada. *Land. Degrad. Dev.* **2017**, *28*, 2298–2307. [CrossRef]
18. Xu, M. Relationship between Supply and Demand of Ecosystem Services and Human Well-Being in the Arid Region of Northwest China. Master's Thesis, China University of Geosciences, Beijing, China, 2021. (In Chinese)
19. Zhao, W.; Liu, Y.; Daryanto, S.; Fu, B.; Wang, S.; Liu, Y. Metacoupling Supply and Demand for Soil Conservation Service. *Curr. Opin. Environ. Sustain.* **2018**, *33*, 136–141. [CrossRef]
20. Mi, Z. Simulation of Soil Conservation Services and Analysis of Supply and Demand in the Ecosystem of the Jinghe River Basin. Master's Thesis, Xi'an University of Science and Technology, Xi'an, China, 2022. (In Chinese)
21. Song, R.; Zhao, X.; Jing, Y.; Li, X.; Su, J.; Wang, X.; Zhao, D. Analysis of Ecosystem Protection and Sustainable Development Strategies—Evidence Based on the RWEQ Model on the Loess Plateau, China. *Sustainability* **2022**, *14*, 11502. [CrossRef]

22. ZHANG, H.; GAO, Y.; CUI, Y. Multi-Scenarios Simulation of Land Use for Reducing Wind Erosion in Typical Agro-Pastoral Ecotone. *Res. Environ. Sci.* **2019**, *32*, 1081–1089. (In Chinese) [CrossRef]
23. Han, J.; Gao, J.; Luo, H. Changes and Implications of the Relationship between Rainfall, Runoff and Sediment Load in the Wuding River Basin on the Chinese Loess Plateau. *CATENA* **2019**, *175*, 228–235. [CrossRef]
24. Zhang, S.; Zhou, Z.; Xiong, X.; Zhou, J. Extreme Climate Characteristics in the Wuding River Basin Based on WRF Model. *Arid. Land. Geogr.* **2024**, *47*, 1482–1495. (In Chinese) [CrossRef]
25. Freddy, N.; Harrij, V.; Luc, V.; Dave, W. *Harmonized World Soils Database Version 2.0*; the Food and Agriculture Organization of the United Nations: Rome, Italy; International Institute for Applied Systems Analysis: Laxenburg, Austria, 2023. [CrossRef]
26. Fryrear, D.W.; Saleh, A.; Bilbro, J.D.; Schromberg, H.M.; Stout, J.E.; Zobeck, T.M. Revised Wind Erosion Equation. USDA Technical Bulletin No. 1. 1998. Available online: https://soilerosion.net/old/sen/doc/report_6/RWEQmod.htm (accessed on 1 December 2024).
27. Ding, G.; Zhao, Y. *Aeolian Physics*, 3rd ed.; China Forestry Publishing House: Beijing, China, 2021; ISBN 978-7-5219-1011-7.
28. Talukdar, I.N.; Kowal, V.A.; Huang, B., Weil, C. Vegetation Drastically Reduces Wind Erosion: An Implementation of the RWEQ in the Mongolian Gobi Steppe. *Land* **2022**, *11*, 1204. [CrossRef]
29. Fryrear, D.W.; Saleh, A.; Bilbro, J.D. A Single Eventwind Erosion Model. *Trans. ASAE* **1998**, *41*, 1369–1374. [CrossRef]
30. Physics and Modelling of Wind Erosion. In *Atmospheric and Oceanographic Sciences Library*, 1st ed.; Shao, Y., Ed.; Springer: Dordrecht, The Netherlands, 2009; Volume 37, ISBN 978-1-4020-8894-0.
31. Zhang, H.; Fan, J.; Cao, W.; Harris, W.; Li, Y.; Chi, W.; Wang, S. Response of Wind Erosion Dynamics to Climate Change and Human Activity in Inner Mongolia, China during 1990 to 2015. *Sci. Total Environ.* **2018**, *639*, 1038–1050. [CrossRef]
32. Cao, T.; Yi, Y.; Liu, H.; Xu, Q.; Yang, Z. The Relationship between Ecosystem Service Supply and Demand in Plain Areas Undergoing Urbanization: A Case Study of China's Baiyangdian Basin. *J. Environ. Manag.* **2021**, *289*, 112492. [CrossRef] [PubMed]
33. Pelt, R.S.V.; Zobeck, T.M.; Potter, K.N.; Stout, J.E.; Popham, T.W. Validation of the Wind Erosion Stochastic Simulator (WESS) and the Revised Wind Erosion Equation (RWEQ) for Single Events. *Environ. Model. Softw.* **2004**, *19*, 191–198. [CrossRef]
34. Borrelli, P.; Lugato, E.; Montanarella, L.; Panagos, P. A New Assessment of Soil Loss Due to Wind Erosion in European Agricultural Soils Using a Quantitative Spatially Distributed Modelling Approach. *Land. Degrad. Dev.* **2017**, *28*, 335–344. [CrossRef]
35. Xie, Y.; Duan, X.; Liu, B.; Liu, G. Soil Loss Tolerance for Black Soil Species in Northeast China. *Acta Geogr. Sin.* **2011**, *66*, 940–952. (In Chinese) [CrossRef]
36. Xia, P.; Peng, J.; Xu, Z.; Gu, W. Conceptual connotation and quantification methods of ecosystem service flows. *Acta Geogr. Sin.* **2024**, *79*, 584–599. (In Chinese) [CrossRef]
37. Nong, X.; Nie, W.; Ma, X. Baseflow Variation and Driving Factors in the Blown-Sand Region of Wuding River Basin. *J. Soil. Water Conserv.* **2023**, *37*, 103–113. (In Chinese) [CrossRef]
38. Wang, G.; Li, J.; Zhang, Y. Simulation and Analysis of Supply, Demand and Flow of Soil Conservation Service in the Wuding River Basin, China. *Acta Ecol. Sin.* **2024**, *44*, 2323–2336. (In Chinese) [CrossRef]
39. Huang, L.; Zhu, P.; Xiao, T. The Sand Fixation Effects of Three-North Shelter Forest Program in Recent 35 Years. *Sci. Geogr. Sin.* **2018**, *38*, 600–609. (In Chinese) [CrossRef]
40. Wei, X.; Wu, X.; Wang, D.; Wu, T.; Li, R.; Hu, G.; Zou, D.; Bai, K.; Ma, X.; Liu, Y.; et al. Spatiotemporal Variations and Driving Factors for Potential Wind Erosion on the Mongolian Plateau. *Sci. Total Environ.* **2023**, *862*, 160829. [CrossRef] [PubMed]
41. Ma, Q.; Cheng, F.; Liu, Y.; Wang, F.; Zhang, D.; Jin, H. Spatial Heterogeneity of Soil Water Content in the Reversion Process of Desertification in Arid Areas: Spatial Heterogeneity of Soil Water Content in the Reversion Process of Desertification in Arid Areas. *J. Arid. Land.* **2011**, *3*, 268–277. [CrossRef]
42. Zou, X.; Zhang, C.; Kang, L. Classification and Representation of Factors Affecting Soil Wind Erosion in a Model. *Prog. Geogr.* **2014**, *29*, 875–889. (In Chinese) [CrossRef]
43. Saleh, A.; Fryrear, D.W. Soil Roughness for the Revised Wind Erosion Equation (RWEQ). *J. Soil. Water Conserv.* **1999**, *54*, 473–476.

land

MDPI

Article

# Spatiotemporal Variation in Wind Erosion in Tarim River Basin from 2010 to 2018

**Qinqin Zhang [1,2], Fang Gu [2], Sicong Zhang [2], Xuehua Chen [2], Xue Ding [1,3,4] and Zhonglin Xu [1,3,4,*]**

1 College of Ecology and Environment, Xinjiang University, Urumqi 830017, China; haozqqhao@163.com (Q.Z.); 188476dx@stu.xju.edu.cn (X.D.)
2 Xinjiang Academy of Surveying and Mapping, Urumqi 830002, China; scsfdxgf@126.com (F.G.); sczhang2023@126.com (S.Z.); xuehuachen2023@163.com (X.C.)
3 Technology Innovation Center for Ecological Monitoring and Restoration of Desert-Oasis, Ministry of Natural Resources, Urumqi 830002, China
4 Key Laboratory of Oasis Ecology, Ministry of Education, Xinjiang University, Urumqi 830017, China
* Correspondence: galinwa@gmail.com

**Abstract:** The Tarim River Basin, China's largest inland river basin, is renowned for its ecological fragility characterized by concurrent greening and desertification processes. Soil wind erosion emerges as a critical factor impacting the natural ecosystem of this region. This study employs a soil wind erosion model tailored to cultivated land, grassland, and desert terrains to analyze the multitemporal characteristics of and spatial variations in soil wind erosion across nine subbasins within the Tarim River Basin, utilizing observed data from 2010, 2015, and 2018. Additionally, this study investigates the influence of various factors, particularly wind speed, on the soil wind erosion dynamics. Following established standards of soil erosion classification, the intensity levels of soil erosion are assessed for each calculation grid within the study area alongside an analysis of the environmental factors influencing soil erosion. Findings indicate that approximately 38.79% of the total study area experiences soil wind erosion, with the Qarqan River Basin exhibiting the highest erosion modulus and the Aksu River Basin registering the lowest. Light and moderate erosion predominates in the Tarim River Basin, with an overall decreasing trend observed over the study period. Notably, the Qiemo River Basin, Dina River Basin, and Kaidu Kongque River Basin display relatively higher proportions of eroded area compared to their total subbasin area. Furthermore, this study underscores the substantial influence of the annual average wind speed on soil erosion within the study area, advocating for prioritizing soil and water conservation programs, particularly in the downstream regions of the Tarim River Basin, to mitigate future environmental degradation.

**Keywords:** Tarim River Basin; soil wind erosion; spatiotemporal variation; driving factors; monitoring

## 1. Introduction

Soil wind erosion is a natural process of soil transportation and deposition by the wind. This is a common phenomenon, mainly occurring in dry sandy soil or any loose, dry, and fine-grained soil. Wind erosion causes a wide range of ecological and environmental problems, such as land desertification [1], crop yield reduction [2], sandstorms [3], and the deterioration of human settlements [4] by removing and depositing soil from one place to another. Given its detrimental impacts and extensive repercussions, soil wind erosion represents a critical scientific issue within arid and semi-arid regions. Beginning in the 1920s, researchers from the United States and the former Soviet Union embarked on comprehensive studies aimed at unraveling the processes, mechanisms, and principal factors influencing soil wind erosion, particularly focusing on the Great Plains of the Midwest United States and the expansive steppes of central Asia. This body of research has laid a foundational understanding, guiding subsequent investigations into the complexities of soil wind erosion and its significant environmental implications. In 1965, the United States Department of Agriculture (USDA) established the wind erosion equation (WEQ) for the first time to estimate the amount of soil wind erosion of cultivated land in the Great Plains [5]. The wind erosion prediction system (WEPS), developed from the wind erosion equation (WEQ), is a process-based high-resolution model designed to simulate weather, field conditions, and erosion dynamics [6]. The Bocharov soil wind erosion model [7], established by former Soviet Union soil scientists in 1984, takes human activity factors into account and provides an innovative idea for erosion prediction. Corresponding works in China began in the 1970s, which were accompanied by continuous and in-depth research on desertification [8]. Representative research focused on the impact of soil mechanical composition [9], water content [10,11], ploughing [12], livestock trampling [13], and other factors on soil wind erosion and the characteristics of erosion related to specific surfaces [14–16]. Based on these basic studies, Chinese scientists have developed mechanistic [17–19] and empirical models for different underlying surfaces (farmland, grassland, shrubbery, and sandy land) [20–22].

With the development of remote sensing technology, it is possible to monitor soil wind erosion in a large scale [23,24]. On this basis, a quantitative method has been carried out [25]. Such a method can be summarized into three categories. The first one is based on field surveys and a soil wind erosion model [26]. The second one uses remote sensing to retrieve soil wind erosion characteristics across a wide geographic range [27–29]. The third one estimates the spatial and temporal distribution of soil wind erosion by combining field surveys, remote sensing, erosion simulation, and land use reclassification [30–32].

Currently, the soil wind erosion models established by scientists have become instrumental in simulating, monitoring, and mitigating soil erosion in arid areas [33]. However, it is worth noting that these models have constraints that limit their application. For example, WEPS requires detailed input data about soil properties, crop systems, climate, and management practices. In areas where these data are not readily available or are of poor quality, the predictions made by WEPS may be less accurate [33]. In addition, WEPS is designed for field-scale predictions and might not be as effective for larger or smaller scales. It is also more suited for short to medium-term predictions and may not be as reliable for long-term forecasting [34]. A comparative study of the WEPP and SWAT models found that using soil wind erosion climate factors to represent soil moisture factors is not theoretically applicable to arid, semi-arid, and semi humid regions [35]. Hence, investigating soil wind erosion models tailored for typical arid and semi-arid regions holds significant importance for the sustainable development of these areas.

The Tarim River stands as the largest inland river within China. The arid climate prevailing across the Tarim River Basin renders it one of the nation's most susceptible regions [36]. Various natural adversities, notably spring droughts, soil desiccation, wind erosion, and sandstorms, significantly impede the socio-economic advancement of this locale. Among these challenges, the deterioration of soil and air quality, exacerbated by wind-driven erosion, poses severe threats to both socio-economic prosperity and natural ecosystems, rendering them particularly vulnerable. Hence, attaining a comprehensive, systematic, and precise comprehension of soil wind erosion within the Tarim River Basin is imperative for elucidating the dynamics of water and soil loss in this region. Furthermore, such an understanding forms a critical foundation for devising overarching strategies for ecological restoration within the Tarim River Basin.

Against this backdrop, the present study endeavors to achieve the following objectives: (1) systematically investigate the spatiotemporal dynamics of soil wind erosion within the Tarim River Basin based on observed data from 2010, 2015, and 2018; (2) analyze the evolving multitemporal variations in soil wind erosion within the mainstem region of the Tarim River and its nine subbasins; and (3) delve into the driving forces and underlying mechanisms governing soil wind erosion in the Tarim River Basin. In summary, our aim is to delineate priority areas necessitating the implementation of soil and water conservation initiatives within the Tarim River Basin. This research endeavor not only enriches our understanding of soil wind erosion characteristics in quintessential inland river basins situated within the arid domains of central Asia but also furnishes crucial insights indispensable for formulating macro-policies aimed at mitigating soil and water losses.

## 2. Materials and Methods

### 2.1. Study Area

The Tarim River Basin is located in the south of southern Xinjiang (northwest part of China), with the Tianshan Mountains in the north, the Pamir Mountains Plateau in the west, the Kunlun Mountains and Arguin Mountains in the south, the Kuruktag Mountains in the east, and the Taklamakan Desert in the middle of Tarim River Basin. The study area includes nine subbasins (Aksu River Basin, Kashgar River Basin, Yarkand River Basin, Hotan River Basin, Kaidu Kongque River Basin, Dina River Basin, Weigan Kuqa River Basin, Keriya River Basin, and Qiemo River Basin) with an area of about 550,000 km$^2$ (Figure 1).

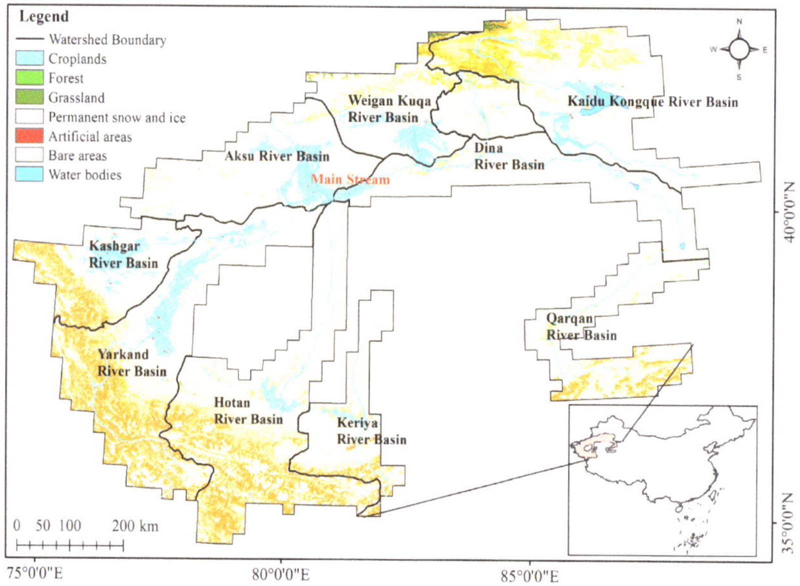


**Figure 1.** Study area.

The Tarim River Basin features a unique climate, positioning it among the most arid regions globally. Annually, it receives scant precipitation, frequently less than 50 mm, primarily because it lies in the rain shadow of adjacent mountain ranges, including the Tianshan Mountains to the north and the Kunlun Mountains to the south. The region experiences extreme temperatures due to its continental climate [36]. Summers can be very hot, with temperatures often exceeding 40 °C, while winters can be extremely cold, with temperatures dropping well below freezing. Humidity levels in the Tarim River Basin are generally low throughout the year, contributing to the overall arid conditions. The basin is known for its strong winds and frequent dust storms, particularly in the spring [37]. These can be attributed to the dry, barren landscape and the temperature gradients between the hot desert and the cooler mountains. While generally dry, the basin does receive some precipitation, which is highly variable and largely dependent on the season and specific location within the basin. The mountainous regions, in particular, receive more precipitation, often as snow, significantly contributing to the river's flow [38].

The vegetation of the Tarim River Basin consists of mountain vegetation and plain vegetation. In the mainstream region of the Tarim River, the tree species are mainly *Populus euphratica*, the shrub species include *Tamarix chinensis* Lour and *Halostachys caspica (M. Bieb.) C. A. Mey.*, and there are also *Haloxylon ammodendron (C. A. Mey.) Bunge, Halimodendron halodendron* (Pall.) *Voss*, etc. The herbs mainly include *Phragmites australis, Apocynum venetum* L., *Glycyrrhiza uralensis Fisch, Karelinia caspia* (Pall.) Less., *Alhagi sparsifolia*, etc.

### 2.2. Data Sources

Some of the data that are important in calculating soil wind erosion are shown in Table 1. Specifically, the fractional vegetation cover (FVC; the resolution is 30 m) was obtained by using the conversion model from the normalized difference vegetation index (NDVI, extracted via the Google Earth Engine platform) to the FVC. The land cover (at 0.5 m resolution) was based on a national basic survey program, which was based on the Land Use Status Classification Standard (GB/T 21010-2007) [39].

**Table 1.** Data sources.

| Type | The Data Sources |
|---|---|
| Wind force factor | Based on the observation data in the monitoring area from 1991 to 2018 |
| Topsoil moisture factor | Based on AMSR-E Level2A brightness temperature data inversion in 2010, 2015, and 2018 |
| Roughness factor | Field survey data in 2018 and 2019 as well as existing data and the literature in the previous period (from 2006 to 2016) |
| Vegetation coverage | Remote-sensing-derived monthly mean FVC data from 2008–2010, 2013–2015, and 2016–2018 |
| Land cover | National basic survey program |
| Topography | The Shuttle Radar Topography Mission |

We also integrated a suite of environmental datasets to investigate the interplay between vegetation health, soil properties, human impact, and climatic variables on soil wind erosion. The NDVI data, serving as a proxy for vegetation health and productivity, were sourced from NASA's EarthData Search (https://www.earthdata.nasa.gov/, accessed on 8 May 2023), which provides access to MODIS and Landsat satellite imagery. The soil water content and sand content were obtained from ISRIC's SoilGrids (https://soilgrids.org/, accessed on 12 June 2023), offering high-resolution global soil information. The human footprint index, reflecting the cumulative impact of human activities on terrestrial systems, was derived from datasets available through the Socioeconomic Data and Applications Center (SEDAC, https://sedac.ciesin.columbia.edu/, accessed on 16 July 2023). Climate variables, including the mean annual temperature and mean annual precipitation, were sourced from WorldClim (https://worldclim.org/, accessed on 16 July 2023), which offers global climate data suitable for ecological and environmental modeling. The integration of these datasets allowed for a comprehensive analysis of the environmental factors influencing soil wind erosion, highlighting the significance of multidimensional environmental data in understanding and managing ecosystems.

### 2.3. Soil Wind Erosion Model

In this study, we used models for different land use types (cultivated land, grass land, and sand model) adopted in the first national water conservancy survey adopted by the Chinese State Council. On this basis, according to the standards of Soil Erosion Classification (SL 190-2007) [40], the area, spatial distribution, and soil wind erosion area of different erosion intensities were determined.

The cultivated land model is:

$$Q_{fa} = 0.018 \cdot (1 - W) \cdot \sum_{j=1} T_j \cdot exp\left\{a_1 + \frac{b_1}{Z_0} + c_1 \cdot \left[(A \cdot U_j)^{0.5}\right]\right\} \tag{1}$$

where $Q_{fa}$ is the soil wind erosion modulus of cultivated land $t/(hm^2 \cdot a)$, $W$ is the topsoil moisture factor, $T_j$ is the cumulative time of each wind speed grade during the occurrence of soil wind erosion in a year, min, $Z_0$ is the surface roughness, $A$ is the wind speed revision factor related to tillage practices, $U_j$ is the wind factor, and $a_1$, $b_1$, and $c_1$ are constants related to soil type, and their values are shown in the Technical Regulations for Dynamic Monitoring of Regional Soil and Water Loss trial.

The grass (shrub) model is:

$$Q_{fg} = 0.018 \cdot (1 - W) \cdot \sum_{j=1} T_j \cdot exp\left[a_2 + b_2 V^2 + c_2 / \left[(A \cdot U_j)^{0.5}\right]\right] \quad (2)$$

where $Q_{fg}$ is the soil wind erosion modulus of grassland, $V$ is the vegetation coverage, and $a_2$, $b_2$, and $c_2$ are constants related to soil type. The other parameters have the same meaning as Equation (1).

The desert model is delineated by:

$$Q_{fs} = 0.018 \cdot (1 - W) \cdot \sum_{j=1} T_j \cdot exp\left[a_3 + b_3 V + c_3 \ln(A \cdot U_j) / (A \cdot U_j)\right] \quad (3)$$

where $Q_{fs}$ is the soil wind erosion modulus of sandy land, and $a_3$, $b_3$, and $c_3$ are constants related to soil type. The other parameters have the same meanings as Equations (1) and (2).

In refining the calculation of the erosion modulus for the Tarim River Basin, the categorization of soil types linked to erosion was enhanced. For example, on the basis of the existing land use classification, four typical land types, including saline–alkali land, sand barrier, industrial facilities, and impervious surfaces, were extracted. These categories were incorporated to refine the calculation of the soil wind erosion in typical areas of the Tarim River Basin. Furthermore, for gravel-covered surfaces, the modeled result was adjusted according to the proportion of gravel. The soil wind erosion of undeveloped saline–alkali land was not included in the calculation. However, for those that have been developed, their erosion was calculated in accordance with desert conditions. For surfaces covered by sand barriers, the field observation was used to adjust the corresponding results.

### 2.4. *Classification of Soil Wind Erosion Intensity*

After obtaining the biweekly soil wind erosion modulus of cultivated land, grassland, and desert, we firstly added up these biweekly data during the monitoring period in order to gain the total erosion modulus. Then, according to the standards of Soil Erosion Classification (SL 190-2007) [40], the soil erosion intensity level of each calculating grid and the entire study area was determined.

### 2.5. *Statistical Analysis*

We employed correlation analysis to examine the strength and direction of the association between soil wind erosion and environmental variables. The Pearson correlation coefficient (r) was calculated to quantify the linear relationship between the variables. A correlation coefficient close to 1 indicates a strong positive relationship, while a coefficient close to -1 indicates a strong negative relationship. A coefficient around 0 suggests no linear correlation. Regression analysis was conducted to model the relationship between the dependent variable, soil wind erosion, and six independent variables, namely, the NDVI, soil water content, sand content, human footprint index, mean annual temperature, and mean annual precipitation. We utilized a linear regression model to predict the soil wind erosion based on the independent variables. The model's coefficients were estimated using the ordinary least-squares method, providing insights into the impact of each independent variable on the dependent variable. The goodness-of-fit of the model was evaluated through the R-squared method.

## 3. Results

### 3.1. Spatial Variation in Soil Wind Erosion

In general, the soil wind erosion exhibited an increasing trend from the west (upper reach of the Tarim River Basin) to the east (middle and lower reach of the Tarim River Basin) part of the study area. In the west part, the erosion modulus was relatively lower; in comparison, the erosion modulus was relatively higher. The status of the soil wind erosion of each subbasin of the Tarim River Basin was different (Figure 2). The area of the soil wind erosion of the subbasins was sorted as follows: Kaidu Kongque River Basin > Qarqan River Basin > Hotan River Basin > Mainstream of Tarim River > Yarkand River Basin > Keriya River Basin > Kashgar River Basin > Aksu River Basin > Dina River Basin > Weigan Kuqa River Basin. The areas with severe soil wind erosion were mainly distributed in Kaidu Kongque River Basin, Qarqan River Basin, Hotan River Basin, and the mainstream of the Tarim River. The Kaidu Kongque River Basin, Qarqan River Basin, Hotan River Basin, and the mainstream of Tarim River are also the four regions with the largest desert area in the study area. At the same time, the Qarqan River Basin, Hotan River Basin, and the mainstream of Tarim River are the regions with the smallest grassland (or shrub) vegetation coverage in the study area. The proportion of erosion area (to the total area of basin) was higher in the Kaidu Kongque River Basin, Dina River Basin, and Qarqan River Basin compared with the other river basins (Figure 2).

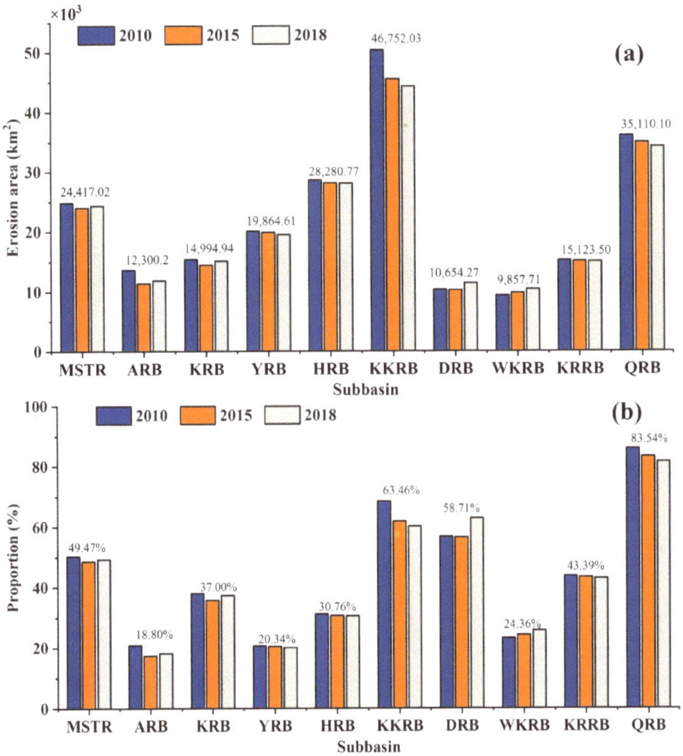


**Figure 2.** Erosion area and its proportion to total area in each subbasin of Tarim River Basin ((**a**), Erosion area, (**b**), the proportion of erosion area to total area). MSTR, ARB, KRB, YRB, HRB, KKRB, DRB, WKRB, KRRB, and QRB indicate mainstream of Tarim River Basin, Aksu River Basin, Kashgar River Basin, Yarkand River Basin, Hotan River Basin, Kaidu Kongque River Basin, Dina River Basin, Weigan Kuqa River Basin, Keriya River Basin, and Qiemo River Basin, respectively.

In 2018, the average soil wind erosion modulus in the Tarim River Basin was 1539.33 $t{\cdot}km^{-2}{\cdot}a^{-1}$, with a soil wind erosion area of 214,441.39 $km^2$. This accounted for approximately 38.79% of the Tarim River Basin's total area. Mild erosion accounted for 67.84% of the total erosion area, followed by moderate erosion (accounting for 13.16%), and strong and above erosion was less than 10.00%. There were significant differences in the soil wind erosion modulus among the subbasins of the Tarim River Basin, with an average soil wind erosion modulus ranging from 274.83 to 4537.20 $t{\cdot}km^{-2}{\cdot}a^{-1}$ (Figure 3). The average value of the soil wind erosion modulus, from large to small, was in the following order: Qarqan River Basin, Kaidu Kongque River Basin, mainstream of Tarim River, Dina River Basin, Kashgar River Basin, Kriya River Basin, Hotan River Basin, Weigan Kuche River Basin, Yarkand River Basin, and Aksu River Basin (Figure 4).

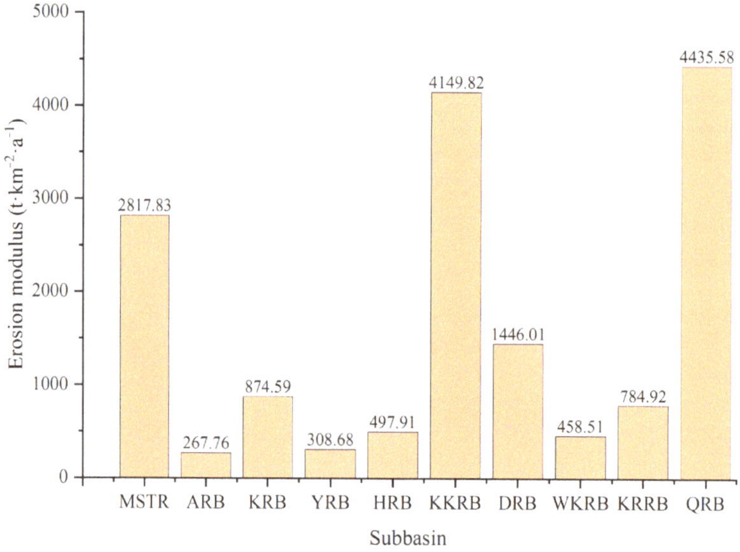


**Figure 3.** Soil wind erosion intensity statistics for various subbasins in the Tarim River Basin in 2018. MSTR, ARB, KRB, YRB, HRB, KKRB, DRB, WKRB, KRRB, and QRB indicate mainstream of Tarim River Basin, Aksu River Basin, Kashgar River Basin, Yarkand River Basin, Hotan River Basin, Kaidu Kongque River Basin, Dina River Basin, Weigan Kuqa River Basin, Keriya River Basin and Qiemo River Basin, respectively.

Significant variations were observed in the intensity of soil wind erosion across the different subbasins. In 2018, the proportion of the light erosion area to the total erosion area in the mainstream region of the Tarim River and the Qarqan River basin was 34.64% and 38.53%, respectively, while in the other basins, it ranged from 70.70% to 91.78%, with an average of 81.31% (Figure 4). The proportions of the medium erosion area in the mainstream region of the Tarim River, Keriya River, Weigan Kuche River, Qarqan River, and Hotan River basins to the total erosion area (from 10.88% to 36.45%) were relatively higher compare with the other river basins (with an average of 6.15%). The proportions of the strong erosion area in the Qarqan River and Dina River basins to the total erosion area ranged from 12.34% to 20.28%, with an average of 3.49% in the other basins. The proportions of the extremely strong and severe erosion areas in the Qarqan River Basin, the mainstream region of the Tarim River, and the Kaidu Kongque River Basin were relatively higher than the other regions (Figure 4).

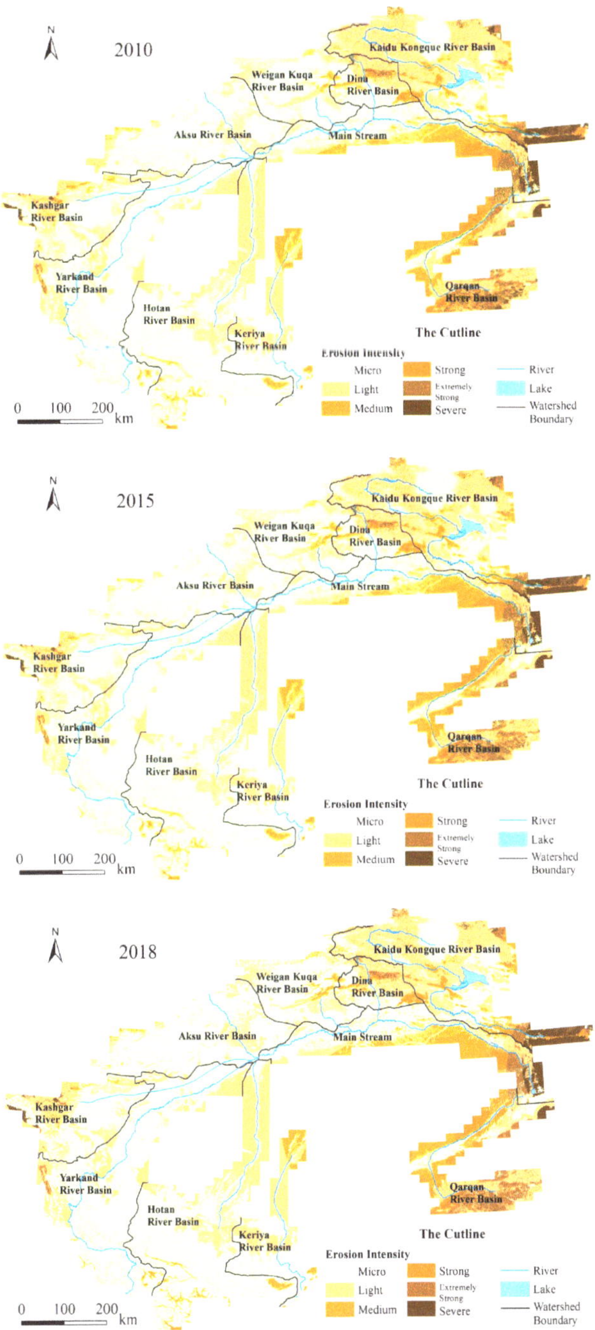


**Figure 4.** Soil wind erosion intensity in Tarim River Basin in 2010, 2015, and 2018.

### 3.2. Multitemporal Variation of Soil Wind Erosion

The average soil wind erosion area (average value of 2010, 2015 and 2018) of the study area was 217,355.14 km$^2$, which accounted for about 38.79% of the area in the monitoring area of the Tarim River Basin. Compared to 2010, the total area of soil wind erosion decreased by 9581.78 km$^2$ in 2018, which accounted for 4.28% of the total erosion area in 2010 (Table 2). Among these different kinds of erosion intensities (light, medium, strong, extremely strong, and severe), the light-intensity erosion decreased from 2010 to 2015 and then increased slightly to 2018, exhibiting a decreasing trend in the study area from 2010 to 2018. The medium-intensity erosion was exactly the opposite of the light-intensity erosion, and it obtained an increasing–decreasing trend during the study period. The strong-intensity erosion continuously increased from 2010 to 2018 (Table 2), whereas the extremely strong-intensity erosion continuously decreased. The severe erosion demonstrated a trend in line with that of the light-intensity erosion (Table 2).

**Table 2.** Area and percentage of different intensities of soil wind erosion in Tarim River Basin.

| Erosion Intensity | Erosion Area (km$^2$) | | | Percentage of Total Erosion Area (%) | | |
|---|---|---|---|---|---|---|
| | 2010 | 2015 | 2018 | 2010 | 2015 | 2018 |
| Light (and micro) | 154,015.92 | 144,778.67 | 144,962.37 | 68.75 | 67.78 | 67.60 |
| Medium | 28,137.31 | 28,494.36 | 28,477.82 | 12.56 | 13.34 | 13.28 |
| Strong | 15,457.60 | 15,998.71 | 16,983.76 | 6.90 | 7.49 | 7.92 |
| Extremely strong | 16,532.91 | 15,656.94 | 15,161.01 | 7.38 | 7.33 | 7.07 |
| Severe | 9879.42 | 8672.20 | 8856.43 | 4.41 | 4.06 | 4.13 |
| Total erosion area | 224,023.16 | 213,600.87 | 214,441.38 | 100 | 100 | 100 |

We calculated the average area of wind erosion for the nine subbasins and the mainstream region of the Tarim River. We found that the Kaidu Kongque River Basin obtained the highest average area of wind erosion (46,752.03 km$^2$) among the nine subbasins and the mainstream region of Tarim River, although the area decreased from 50,372.96 km$^2$ in 2010 to 45,562.77 km$^2$ in 2015 and to 44,320.37 km$^2$ in 2018 (Figure 2). The Qarqan River Basin also had a relatively higher area of erosion, with an average area of 35,110.10 km$^2$ during the study period. In comparison, the erosion area of the Weigan Kuqa River Basin and the Dina River Basin were relatively lower among the nine subbasins and the mainstream region of Tarim River, with an average area of 9857.71 km$^2$ and 10,654.27 km$^2$, respectively, from 2010 to 2018 (Figure 2). Among these nine subbasins and the mainstream region of the Tarim River, the erosion area of the Yarkand River Basin, Hotan River Basin, Kaidu Kongque River Basin, Keriya River Basin, and Qarqan River Basin showed a decreasing trend from 2010 to 2018. In comparison, the mainstream region of the Tarim River, the Aksu River Basin, the Kashgar River Basin, and the Dina River Basin displayed a decreasing–increasing trend during the study period. In addition, the Weigan Kuqa River Basin showed an increasing trend from 2010 to 2018. The average proportion of the erosion area to the total area in these nine subbasins and the mainstream region of Tarim River was also worth investigating. We found that the Qarqan River Basin obtained the highest average proportion, with a value of 83.54%, which indicates that the soil wind erosion was widespread in the basin (Figure 2). Additionally, the Kaidu Kongque River Basin and Dina River Basin possessed a relatively higher proportion of erosion area to the total basin's area, with values of 63.46% and 58.71%, respectively. In comparison, the proportion was relatively lower in the mainstream region of the Tarim River, the Aksu River Basin, the Kashgar River Basin, the Yarkand River Basin, the Hotan River Basin, the Weigan Kuqa River Basin, and the Keriya River Basin, especially the Aksu River Basin, where the average proportion during the study period was only 18.80%, indicating a relatively scarce level of soil wind erosion in the basin (Figure 2).

### 3.3. *The Impact of Environmental Variables on Wind Erosion*

In our detailed investigation into the various factors affecting soil wind erosion, we meticulously evaluated the roles of the NDVI, soil water content, sand content, human footprint index, mean annual temperature, and mean annual precipitation. Through rigorous statistical methodologies, our research unveiled nuanced insights into their interactions with soil wind erosion. Notably, the analysis identified the NDVI and human footprint index as significant negative influencers on soil wind erosion, with correlation coefficients ($r^2$) of 0.12 and 0.18, respectively, and with statistical significance ($p < 0.05$, Figure 5). The mean annual precipitation was negatively correlated with soil wind erosion, presenting a correlation coefficient ($r^2$) of 0.31 with statistical significance ($p < 0.05$). This suggests that increased precipitation levels are associated with reduced soil wind erosion, possibly due to enhanced vegetation growth and soil cohesion from moisture, which in turn protect the soil surface from wind detachment and transportation. Interestingly, this study did not find significant correlations between soil wind erosion and the other factors examined, namely, the soil water content, sand content, and mean annual temperature.

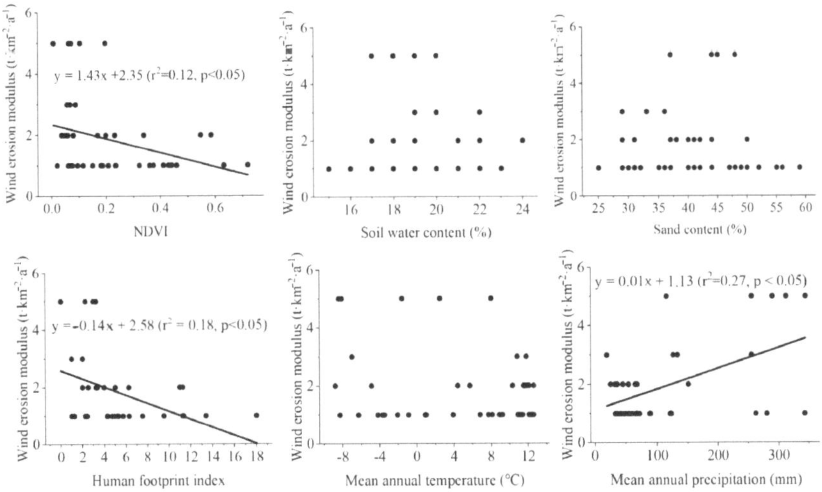


**Figure 5.** Impacts of environmental and anthropogenic factors on soil wind erosion (Each dot represents an observation point, with the x-axis displaying the factor being analyzed and the y-axis showing the corresponding wind erosion modulus).

## 4. Discussions

### *4.1. The Reliability of Our Method*

In this study, we employed models that were previously utilized in the inaugural national water conservancy survey, endorsed by the Chinese State Council, to simulate the spatiotemporal variations in the soil wind erosion within our designated study area. The models were specified to different land use types (cultivated land, grass land, and sand model). Our findings indicate that the soil wind erosion intensity was generally higher in the eastern part of the study area, situated in the middle and lower reaches of the Tarim River Basin. It was reported that the WEPS-model-simulated wind erosion is relatively higher in Wuqia County, at the Tieganlike station, in Ruoqiang County, in Wensu County, in Tacheng County, in Kelamayi City, in Mulei county, in Akedala County, at the Shisanjianfang station, at the Hongliuhe station, and at the Kumishi station among the 64 stations in Xinjiang Province [41]. It should be noted that Wuqia County, the Tieganlike station, and Wensu County were within our study area and are located in the middle reach of the Tarim River Rasin, and Ruoqiang County is located in the lower reach of the basin, indicating the reasonability of our method. In addition, another study on soil erosion observation and simulation in the Taklimakan Desert revealed that the soil wind erosion intensity is 10 ug·m$^{-2}$·s$^{-1}$ at the upper reach of the Keriya River Basin, which is equivalent to 315.36 t·km$^{-2}$·a$^{-1}$ [42]. In the present study, we also found that the soil wind erosion at the upper reach of the Keriya River Basin belongs to the light-intensity type, which corresponds to from 200 to 2500 t·km$^{-2}$·a$^{-1}$ in terms of the erosion modulus according to the national standards of Soil Erosion Classification (SL 190-2007) [40]. The similarity of the corresponding research further proved the reliability of our method.

*4.2. The Drivers of Soil Wind Erosion in the Tarim River Basin*

Compared with 2010, the area of soil wind erosion in the Tarim River Basin in 2015 decreased by 10,422.29 $km^2$; among the different erosion types, the areas of light, extremely strong, and severe erosion decreased by 9237.25 $km^2$, 875.97 $km^2$, and 1207.22 $km^2$, respectively. However, medium and strong erosion increased by 357.05 $km^2$ and 541.11 $km^2$, respectively. At the same time, the increase in the area of erosion does not mean that strong erosion was transformed into light or medium erosion but that the original moderate and strong erosion was transformed into the erosion type with a lower intensity, and the original extremely strong and severe erosion was transformed into moderate and strong erosion, which has been reported elsewhere [43] and is the result of the joint action of human activities and natural factors [44].

The impact of human activities is mainly reflected in the following two aspects: Firstly, some projects implemented in the research area (such as sand barriers, soil stubble, farmland film covering, and the construction of photovoltaic facilities) have increased the surface roughness, increased the critical starting wind speed for soil wind erosion, changed the structure of the near-surface airflow, avoided direct wind action on surface soil particles, and reduced the force of airflow on surface soil particles, which in turn has led to a decrease in the modulus of surface soil wind erosion [45]. Secondly, changes in the surface vegetation cover have led to changes in the soil wind erosion. Compared to 2010, in 2015, there was a total increase of 766.79 $km^2$ in arable land, forest land, and grassland, and the sand and gravel surface decreased by 2588.03 $km^2$; therefore, the changes in the area of these land use/cover types have led to corresponding changes in the erosion area involved in the calculation.

As presented previously, the vegetation delineated by the NDVI is an important variable that influences soil wind erosion. Plants and their roots stabilize the soil, making it harder for the wind to pick up and transport soil particles. When vegetation is sparse or absent, such as in deserts or areas affected by deforestation, soil becomes more vulnerable to wind erosion. In our study area, the vegetation coverage underwent profound changes; in 2015, the vegetation coverage increased by 47.52% compared to 2010, thus effectively reducing the soil wind erosion on exposed surfaces, as mentioned in a previous study [46]. In areas with higher precipitation, we also observed a negative correlation between precipitation and wind erosion. The observation that areas with higher precipitation levels exhibit more robust vegetation cover led to the identification of a negative correlation between precipitation and wind erosion. This relationship can be understood through several key mechanisms that underscore the protective role of vegetation against soil degradation processes. Firstly, vegetation anchors soil through complex root systems, which consolidate soil particles and enhance soil stability. This root-induced cohesion significantly diminishes the susceptibility of soil to wind erosion by preventing the dislodgment of soil particles. Secondly, the physical presence of vegetation above ground acts as a barrier to wind, effectively reducing the wind velocity at the soil surface. This reduction in wind speed directly influences the wind's capacity to mobilize and transport soil particles, thus mitigating wind erosion. In addition, increased precipitation contributes to higher soil moisture levels, which, in turn, make soil particles heavier and more resistant to aeolian transport. But, interestingly, we did not find any significant relationship between the soil water content and erosion modulus, which demonstrates that the complex interplay between the soil water content and erosion processes is influenced by multiple soil properties, environmental conditions, and external factors. A comprehensive analysis considering these multifaceted interactions is essential to elucidate the nuanced effects of soil moisture on erosion dynamics.

Beyond factors like the NDVI, human activity, and precipitation, wind emerges as a pivotal natural force in soil wind erosion, which is especially pronounced in arid and semi-arid regions. The phenomenon occurs as strong winds lift and transport loose soil particles, significantly eroding topsoil and degrading land. Among the influences, the wind speed is paramount; stronger winds have the energy to mobilize larger soil particles over greater distances, exacerbating erosion [47]. We investigated the linear relationship between the annual average wind speed and erosion intensity at 46 points selected randomly within the study area and found that there exists a significant linear relationship between them (Figure 6). In addition, this process underscores the necessity of integrating wind speed with other variables in erosion studies to accurately assess and mitigate the impact [48].

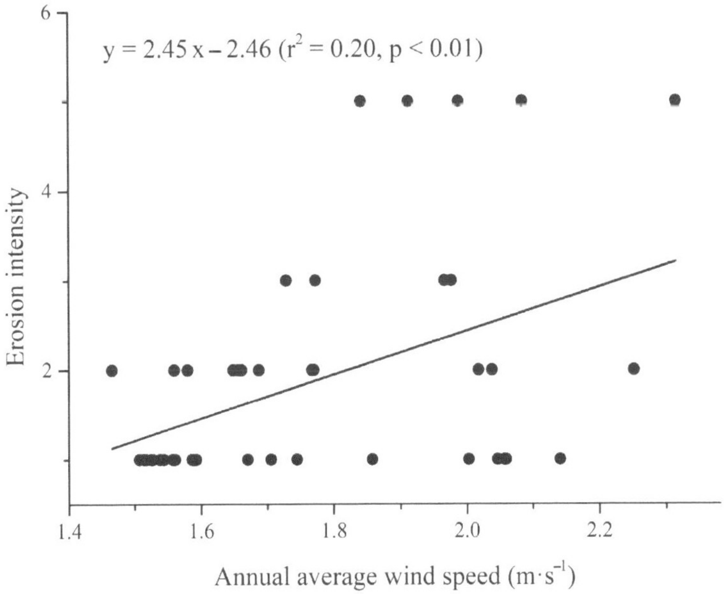


**Figure 6.** The linear relationship between annual average wind speed and erosion intensity (Each dot represents an observation point).

## 5. Conclusions

In summary, our study delved into the spatiotemporal dynamics of soil wind erosion within the Tarim River Basin, employing a specialized soil wind erosion model. Several key conclusions emerge from our analysis: Firstly, the soil wind erosion area encompassed approximately 40% of the basin's total area, primarily characterized by mild erosion. Although the proportion of the soil wind erosion area in the Tarim River Basin fell below the average for the Xinjiang Uygur Autonomous Region, it surpassed the average for northern China. Secondly, our findings reveal that, as of 2018, soil wind erosion hotspots are concentrated in the Kaidu Kongque River Basin, Qarqan River Basin, Hotan River Basin, and the mainstream region of the Tarim River. Thirdly, a positive trend was observed, as the total area affected by soil wind erosion has decreased since 2010, indicating a gradual improvement in the natural environment of the Tarim River Basin. Lastly, our study underscores the imperative for prioritizing soil and water conservation efforts, particularly in the downstream regions of the Tarim River Basin, to safeguard against further environmental degradation. These findings collectively provide valuable insights for informing targeted conservation strategies aimed at preserving the ecological integrity of this vital basin.

**Author Contributions:** Conceptualization, Q.Z. and Z.X.; methodology, Q.Z. and F.G.; software, Q.Z. and F.G.; validation, Q.Z. and S.Z.; formal analysis, Q.Z. and S.Z.; investigation, Q.Z., F.G., S.Z. and X.C.; resources, Q.Z.; data curation, Q.Z.; writing—original draft preparation, Q.Z., X.D. and Z.X.; writing—review and editing, Q.Z. and Z.X.; visualization, Q.Z., F.G., S.Z. and X.C.; supervision, Z.X.; project administration, Z.X.; funding acquisition, Z.X. All authors have read and agreed to the published version of the manuscript.

**Funding:** This research was funded by the Natural Science Foundation of Xinjiang Uygur Autonomous Region (No. 2022D01C399 and No. 2022E01052).

**Data Availability Statement:** The original contributions presented in the study are included in the article, further inquiries can be directed to the corresponding author.

**Acknowledgments:** The authors would like to thank the associate editor and the anonymous reviewers for their valuable comments and constructive suggestions. We also acknowledge the support of Xu Ma.

**Conflicts of Interest:** The authors declare no conflicts of interest.

## References

1. Chepil, W. Dynamics of wind erosion: I. Nature of movement of soil by wind. *Soil Sci.* **1945**, *60*, 305–320. [CrossRef]
2. Lyles, L. Possible effects of wind erosion on soil productivity. *J. Soil Water Conserv.* **1975**, *30*, 279–283.
3. Wu, Z.; Wang, M.; Zhang, H.; Du, Z. Vegetation and soil wind erosion dynamics of sandstorm control programs in the agro-pastoral transitional zone of northern China. *Front. Earth Sci.* **2019**, *13*, 430–443. [CrossRef]
4. Bartkowski, B.; Schepanski, K.; Bredenbeck, S.; Müller, B. Wind erosion in european agricultural landscapes: More than physics. *People Nat.* **2023**, *5*, 34–44. [CrossRef]
5. Woodruff, N.P.; Siddoway, F. A wind erosion equation. *Soil Sci. Soc. Am. J.* **1965**, *29*, 602–608. [CrossRef]
6. Wagner, L.E. A history of wind erosion prediction models in the united states department of agriculture: The wind erosion prediction system (weps). *Aeolian Res.* **2013**, *10*, 9–24. [CrossRef]
7. Bocharov, A. *A Description of Devices Used in the Study of Wind Erosion of Soils*; AA Balkema: Rotterdam, The Netherlands, 1984.
8. Wang, F.; Pan, X.; Wang, D.; Shen, C.; Lu, Q. Combating desertification in China: Past, present and future. *Land Use Policy* **2013**, *31*, 311–313. [CrossRef]
9. Yang, X.; Zhu, Z.; Jaekel, D.; Owen, L.; Han, J. Late quaternary palaeoenvironment change and landscape evolution along the keriya river, xinjiang, China: The relationship between high mountain glaciation and landscape evolution in foreland desert regions. *Quat. Int.* **2002**, *97*, 155–166. [CrossRef]
10. Dong, Z.; Li, S.; Zhao, Y.; Lei, J.; Wang, Y.; Li, C. Stable oxygen-hydrogen isotopes reveal water use strategies of tamarix taklamakanensis in the Taklimakan desert, China. *J. Arid Land* **2020**, *12*, 115–129. [CrossRef]
11. Hao, X.-M.; Li, C.; Guo, B.; Ma, J.-X.; Ayup, M.; Chen, Z.-S. Dew formation and its long-term trend in a desert riparian forest ecosystem on the eastern edge of the taklimakan desert in China. *J. Hydrol.* **2012**, *472*, 90–98. [CrossRef]
12. Chen, Z.; Cui, H.; Wu, P.; Zhao, Y.; Sun, Y. Study on the optimal intercropping width to control wind erosion in north China. *Soil Tillage Res.* **2010**, *110*, 230–235. [CrossRef]
13. Du, H.; Zuo, X.; Li, S.; Wang, T.; Xue, X. Wind erosion changes induced by different grazing intensities in the desert steppe, northern China. *Agric. Ecosyst. Environ.* **2019**, *274*, 1–13. [CrossRef]
14. Li, D.; Xu, E.; Zhang, H. Influence of ecological land change on wind erosion prevention service in arid area of northwest China from 1990 to 2015. *Ecol. Indic.* **2020**, *117*, 106686. [CrossRef]
15. Chi, W.; Zhao, Y.; Kuang, W.; He, H. Impacts of anthropogenic land use/cover changes on soil wind erosion in China. *Sci. Total Environ.* **2019**, *668*, 204–215. [CrossRef] [PubMed]
16. Du, H.; Wang, T.; Xue, X. Potential wind erosion rate response to climate and land-use changes in the watershed of the ningxia–inner mongolia reach of the yellow river, China, 1986–2013. *Earth Surf. Process. Landf.* **2017**, *42*, 1923–1937. [CrossRef]
17. Chen, C.; Zhao, G.; Zhang, Y.; Bai, Y.; Tian, P.; Mu, X.; Tian, X. Linkages between soil erosion and long-term changes of landscape pattern in a small watershed on the chinese loess plateau. *CATENA* **2023**, *220*, 106659. [CrossRef]
18. Dai, T.; Wang, L.; Li, T.; Qiu, P.; Wang, J. Study on the characteristics of soil erosion in the black soil area of northeast China under natural rainfall conditions: The case of sunjiagou small watershed. *Sustainability* **2022**, *14*, 8284. [CrossRef]
19. Shi, Z.; Cai, C.; Ding, S.; Wang, T.; Chow, T. Soil conservation planning at the small watershed level using rusle with gis: A case study in the three gorge area of China. *CATENA* **2004**, *55*, 33–48. [CrossRef]
20. Chen, J.; Xiao, H.; Li, Z.; Liu, C.; Wang, D.; Wang, L.; Tang, C. Threshold effects of vegetation coverage on soil erosion control in small watersheds of the red soil hilly region in China. *Ecol. Eng.* **2019**, *132*, 109–114. [CrossRef]
21. Zhang, Z.; Sheng, L.; Yang, J.; Chen, X.-A.; Kong, L.; Wagan, B. Effects of land use and slope gradient on soil erosion in a red soil hilly watershed of southern China. *Sustainability* **2015**, *7*, 14309–14325. [CrossRef]
22. Wei, J.-B.; Xiao, D.-N.; Zeng, H.; Fu, Y.-K. Spatial variability of soil properties in relation to land use and topography in a typical small watershed of the black soil region, northeastern China. *Environ. Geol.* **2008**, *53*, 1663–1672. [CrossRef]

23. Sepuru, T.K.; Dube, T. An appraisal on the progress of remote sensing applications in soil erosion mapping and monitoring. *Remote Sens. Appl. Soc. Environ.* **2018**, *9*, 1–9. [CrossRef]
24. King, C.; Baghdadi, N.; Lecomte, V.; Cerdan, O. The application of remote-sensing data to monitoring and modelling of soil erosion. *CATENA* **2005**, *62*, 79–93. [CrossRef]
25. García-Ruiz, J.M.; Beguería, S.; Nadal-Romero, E.; González-Hidalgo, J.C.; Lana-Renault, N.; Sanjuán, Y. A meta-analysis of soil erosion rates across the world. *Geomorphology* **2015**, *239*, 160–173. [CrossRef]
26. Lal, R. *Soil Erosion Research Methods*; St. Lucie Press: Delray Beach, FL, USA, 1994.
27. Krishna Bahadur, K. Mapping soil erosion susceptibility using remote sensing and gis: A case of the upper nam wa watershed, nan province, thailand. *Environ. Geol.* **2009**, *57*, 695–705. [CrossRef]
28. Chen, T.; Niu, R.-q.; Li, P.-x.; Zhang, L.-p.; Du, B. Regional soil erosion risk mapping using rusle, gis, and remote sensing: A case study in miyun watershed, north China. *Environ. Earth Sci.* **2011**, *63*, 533–541. [CrossRef]
29. Weifeng, Z.; Bingfang, W. Assessment of soil erosion and sediment delivery ratio using remote sensing and gis: A case study of upstream chaobaihe river catchment, north China. *Int. J. Sediment Res.* **2008**, *23*, 167–173.
30. Teng, H.-F.; Jie, H.; Yue, Z.; Zhou, L.-Q.; Zhou, S. Modelling and mapping soil erosion potential in China. *J. Integr. Agric.* **2019**, *18*, 251–264. [CrossRef]
31. Wang, X.; Zhao, X.; Zhang, Z.; Yi, L.; Zuo, L.; Wen, Q.; Liu, F.; Xu, J.; Hu, S.; Liu, B. Assessment of soil erosion change and its relationships with land use/cover change in China from the end of the 1980s to 2010. *CATENA* **2016**, *137*, 256–268. [CrossRef]
32. Yang, X.; Zhang, K.; Jia, B.; Ci, L. Desertification assessment in China: An overview. *J. Arid Environ.* **2005**, *63*, 517–531. [CrossRef]
33. Batista, P.V.; Davies, J.; Silva, M.L.; Quinton, J.N. On the evaluation of soil erosion models: Are we doing enough? *Earth-Sci. Rev.* **2019**, *197*, 102898. [CrossRef]
34. Funk, R.; Skidmore, E.L.; Hagen, L.J. Comparison of wind erosion measurements in germany with simulated soil losses by weps. *Environ. Model. Softw.* **2004**, *19*, 177–183. [CrossRef]
35. Shen, Z.; Gong, Y.; Li, Y.; Hong, Q.; Xu, L.; Liu, R. A comparison of wepp and swat for modeling soil erosion of the zhangjiachong watershed in the three gorges reservoir area. *Agric. Water Manag.* **2009**, *96*, 1435–1442. [CrossRef]
36. Tao, H.; Gemmer, M.; Bai, Y.; Su, B.; Mao, W. Trends of streamflow in the tarim river basin during the past 50 years: Human impact or climate change? *J. Hydrol.* **2011**, *400*, 1–9. [CrossRef]
37. Xu, Z.; Chen, Y.; Li, J. Impact of climate change on water resources in the tarim river basin. *Water Resour. Manage.* **2004**, *18*, 439–458. [CrossRef]
38. Xu, Z.; Liu, Z.; Fu, G.; Chen, Y. Trends of major hydroclimatic variables in the tarim river basin during the past 50 years. *J. Arid Environ.* **2010**, *74*, 256–267. [CrossRef]
39. *GB/T 21010-2007*; Current Land Use Classification. China Standard Press: Beijing, China, 2007.
40. *SL190-2007*; Standards for Classification and Gradation of SOIL Erosion. Ministry of Water Resources of the People's Republic of China: Beijing, China, 2008.
41. Pi, H.; Sharratt, B.; Lei, J. Wind erosion and dust emissions in central asia: Spatiotemporal simulations in a typical dust year. *Earth Surf. Process. Landf.* **2019**, *44*, 521–534. [CrossRef]
42. Kang, Y. Study on Soil Erosion Observation and Simulation in Taklimakan Desert. Master's Thesis, Xinjiang Normal University, Urumqi, China, 2018.
43. Chi, W.; Wang, Y.; Lou, Y.; Na, Y.; Luo, Q. Effect of land use/cover change on soil wind erosion in the yellow river basin since the 1990s. *Sustainability* **2022**, *14*, 12930. [CrossRef]
44. Ali, K.F.; De Boer, D.H. Spatially distributed erosion and sediment yield modeling in the upper indus river basin. *Water Resour. Res.* **2010**, *46*, W08504. [CrossRef]
45. Uddin, K.; Murthy, M.; Wahid, S.M.; Matin, M.A. Estimation of soil erosion dynamics in the koshi basin using gis and remote sensing to assess priority areas for conservation. *PLoS ONE* **2016**, *11*, e0150494. [CrossRef] [PubMed]
46. Biswas, S.S.; Pani, P. Estimation of soil erosion using rusle and gis techniques: A case study of barakar river basin, jharkhand, india. *Model. Earth Syst. Environ.* **2015**, *1*, 1–13. [CrossRef]
47. Wu, G.L.; Liu, Y.F.; Cui, Z.; Liu, Y.; Shi, Z.H.; Yin, R.; Kardol, P. Trade-off between vegetation type, soil erosion control and surface water in global semi-arid regions: A meta-analysis. *J. Appl. Ecol.* **2020**, *57*, 875–885. [CrossRef]
48. Abdullah, M.; Feagin, R.; Musawi, L. The use of spatial empirical models to estimate soil erosion in arid ecosystems. *Environ. Monit. Assess.* **2017**, *189*, 1–17. [CrossRef] [PubMed]

*land*

MDPI

*Article*

# Risk Assessment of World Corn Salinization Hazard Factors Based on EPIC Model and Information Diffusion

**Degen Lin [1,*], Chuanqi Hu [2], Fang Lian [3], Jing'ai Wang [4,5,6,7,*], Xingli Gu [8] and Yingxian Yu [1]**

1 School of Business, Wenzhou University, Wenzhou 325000, China; 21450101023@stu.wzu.edu.cn
2 Northwest Institute of Historical Environment and Socio-Economic Development, Shaanxi Normal University, Xi'an 710119, China; hcq0917@snnu.edu.cn
3 Integrated Research on Disaster Risk (IRDR) International Programme Office, Beijing 100094, China; fang.lian@irdrinternational.org
4 Faculty of Geographical Science, Beijing Normal University, Beijing 100875, China
5 Key Laboratory of Environmental Change and Natural Disaster, Ministry of Education, Beijing Normal University, Beijing 100875, China
6 State Key Laboratory of Earth Surface Processes and Resource Ecology, Beijing Normal University, Beijing 100875, China
7 Academy of Plateau Science Sustainability, The People's Government of Qinghai Province and Beijing Normal University, Xining 810016, China
8 College of Geography and Environmental Sciences, Zhejiang Normal University, Jinhua 321004, China; xinligu@zjnu.edu.cn
* Correspondence: 20200661@wzu.edu.cn (D.L.); jwang@bnu.edu.cn (J.W.)

**Abstract:** Salinization is a serious land degradation phenomenon. This study identified the salinity stress threshold as a causal factor for salinization, focusing on global maize fields as the study area. By excluding environmental stressors and setting salinization scenarios, the EPIC model was used to simulate the daily salinity stress threshold during the corn growth process. The global intensity and risk of salinization-induced disaster for maize were evaluated. Based on the principle of information diffusion, the intensity of salinization-induced disaster was calculated for different return periods. The main conclusions were as follows: (1) By excluding environmental stress factors and setting salinization scenarios, algorithms for the salinization index during the growing season and the intensity of salinization-induced disaster were proposed. (2) The salinity hazard factor is highly risky and concentrated in arid and semi-arid regions, while it is relatively low in humid regions. (3) As the recurrence period increases, the risk of salinization-induced hazard becomes higher, the affected area expands, and the risk level increases. (4) The salinization intensity results of this study are consistent with the research results of HWSD ($R^2$ = 0.9546) and GLASOD ($R^2$ = 0.9162).

**Keywords:** information diffusion; risk of hazard factor; salt stress; EPIC model

## 1. Introduction

With the advent of the "Anthropocene", land systems increasingly face complex challenges [1,2], salinization being a prime example requiring human attention. Soil salinization, a severe land degradation phenomenon, arises from the buildup of soluble salts in both the soil cultivation and surface layers [3]. Salinization undermines land productivity, diminishing agricultural output in irrigation areas and impeding agricultural development and food production [4]. Over a hundred countries and regions around the globe deal with varying levels of saline soils. According to data from the United Nations Educational, Scientific and Cultural Organization (UNESCO, Paris, France) and the Food and Agriculture Organization (FAO, Rome, Italy) of the United Nations, the total area of saline soils is around 954.38 million hectares. Salinization poses a significant agricultural risk (risk = hazard factors × exposure × vulnerability) [5]. In some timeframes and regions, exposure remains relatively steady, with the hazard severity typically determining the risk magnitude. Consequently, several scholars concentrate on the salinization hazards [6–8].

Despite decades of research, comprehensive, large-scale land salinization assessments remain an enduring challenge [9]. Numerous studies concur that obtaining accurate measures of salinization at large scales is a complex task [10–13]. When assessing hazards contributing to salinization, scholars typically select salinization-specific indicators for a comprehensive evaluation, such as cation exchange rate (ECe), average annual evapotranspiration volume, groundwater characteristics, and dry humidity [14]. Initially, the soil's conductivity is usually identified by the ECe of the saturated paste extraction [15–18]. Most studies employ ECe as an index to gauge soil salinization, considering soils with ECe exceeding 4 $dsm^{-1}$ as saline. Secondly, elevated evapotranspiration intensifies water-salt movements, increasing the salt concentration in soil water and on the surface [19,20]. Consequently, the average annual evapotranspiration volume serves as a significant index for assessing salinization risk. Thirdly, groundwater impacts on salinization manifest in two ways: i the depth of the water table, where shallow, semi-closed aquifers can trigger salinization by impeding surface soil drainage [21]; and ii groundwater mineralization, linked to irrigation, as the quality of groundwater used for irrigation influences soil texture. Long-term unsuitable irrigation increases salinization risk [22]. Data acquisition limitations and model parameter validity concerns lead some investigators to select alternate indicators for assessing salinization risk. Some base their evaluation on the depth of the saline soil and the groundwater table [23], others incorporate soil and climate attributes, irrigation water properties, conductivity, cation exchange rate, and dry humidity [14].

Additionally, many scholars have used models to assess the hazards of salinization factors. Currently, the models primarily used to evaluate salinization risk include UNSATCHEM [24], SALTMED [25], BUDGET [26], Pla [27], and Riverside [18], among others. The first three are complex models, requiring large amounts of data, suitable for assessing small-scale areas, such as farmland and small watersheds, while Pla [19] and Riverside [18] are simpler models, suitable for larger-scale areas, such as farms and large river basins. It is noteworthy that many new methods have emerged in recent years. For instance, Hassani and others utilized machine learning algorithms to make long-term predictions on global salinization issues based on EC′e″ [28]; FAO constructed The Global Map of Salt-Affected Soils (GSASmap) platform based on indices like EC′e, ESP, PH [29]; Kaya and others adopted remote sensing methods to assess soil salinization conditions in the western part of Turkey based on EC′e″ [30].

Contemporary research frequently employs external factors such as soil characteristics, climate variability, and groundwater dynamics as metrics to assess the hazards of soil salinization to agricultural productivity. Nonetheless, these indicators may inadvertently sidestep the direct effects of salinization on the crops themselves, potentially obscuring the isolated impact of saline conditions on plant growth. Addressing this gap, our study proposes the utilization of normalized salt stress values, correlated to the crop growth cycle, as a more precise measure of salinization impact, thereby elucidating the independent extent of salt-induced stress on crop vitality. The EPIC crop growth model, known for its robustness, is employed to establish salinization scenarios, which delineates the day-to-day susceptibility of crops to saline disturbances. The selection of crop species is pivotal for the applicability of such growth models. Maize, with its global significance as a staple crop [31], wide cultivation range [32], and intermediate salinity tolerance [33], represents an ideal candidate for this analysis.This research focuses on global cornfields, uses the EPIC0509 model to simulate the salinity stress value [34–37] during corn growth with days as steps, attempts to evaluate the risk of salinization hazard intensity about corn on a global scale, and calculates the salinization hazard intensity under different recurrence periods based on the principle of information diffusion [38].

## 2. Materials and Methods

### 2.1. Basic Concepts and Research Framework

#### 2.1.1. The EPIC Crop Growth Model

The erosion-productivity impact calculator (EPIC) model was developed in 1981 by Williams et al. to study the relationship between soil erosion and soil productivity [39–41]. Initially, the value of the EPIC model in simulating crop growth was not noticed. It was not until 1989 that the EPIC model began to be used as a crop growth model [42]. In 1996, Williams et al. incorporated environmental factors such as water quality, carbon cycle, and climate change into the EPIC model, subsequently renaming it the "Environmental Policy Impact Climate" model [43]. Due to the EPIC model's ability to simulate crop productivity over hundreds or even thousands of years in various climate scenarios, environmental conditions, and management systems, assessing the impact of multiple agricultural disasters on crop yield and land productivity, it has become one of the most popular crop growth models [44–46]. Some studies even suggest that, in terms of model calibration and crop yield evaluation, the EPIC model performs better than the CSM-CERES-Maize [47].

#### 2.1.2. Mechanism of Salt Stress in Maize

Salinity stress affects almost all crucial metabolic processes of corn growth [48]. Based on previous research, the damaging mechanisms of salinity stress on corn mainly include the following aspects: (1) Salinity stress leads to difficulties in corn water absorption. Due to the high salt content in saline soils, the soil solution water potential significantly decreases, causing difficulties in water absorption of corn roots, or they cannot absorb water at all. In severe cases, it even leads to the outward discharge of water, causing osmotic dehydration of the tissue and harming corn [49,50]. (2) Salinity stress has detrimental effects on the corn biomembrane. Studies indicate that the membrane plays a crucial role in generating primary and secondary stress responses. Salt stress influences various aspects of membrane function, including ion selective permeability, transport of inorganic and organic matter, membrane secretion function, membrane lipid composition, and ultrastructure [51–53]. (3) Salinity stress causes physiological disorders in corn. Salinity stress affects the physiological activities of corn, for instance, salt stress causes a decline in the net photosynthesis rate of corn [54–56], and salinity stress disrupts normal respiratory metabolism and protein synthesis of corn [57–59].

#### 2.1.3. Hazard Intensity Assessment and Information Diffusion Method

The core of the disaster factor hazard assessment is to establish a relationship between the disaster intensity and frequency. The hazard assessment of disaster-inducing factors can be divided into the average expected disaster intensity and the probability that the intensity of the disaster factor exceeds a certain value (the frequency of the disaster factor) [60]. Calculating the frequency of disaster factors requires a large amount of sample data. However, when the sample information is incomplete, the information diffusion method can be used to calculate the frequency of disaster factors, and a fuzzy set is obtained after diffusing the original incomplete information. This is a method of treating samples collectively using fuzzy mathematical methods to compensate for information deficiencies [61]. Specifically, using the diffusion function to convert sample data into sample sets, the simplest model is the normal diffusion model [38]. This paper calculates the intensity of disaster factors with different probabilities of occurrence (recurrence periods of 10, 20, 50, 100 years) by diffusing the salinization hazard factors of the 30-year samples through information diffusion.

#### 2.1.4. Research Framework

This study is conducted in four steps. The first step is to establish the database. The second step is to determine the hazard-inducing factor-salt stress and eliminate other environmental stress factors. The third step is to simulate the growth process of corn using the EPIC model and, based on the daily salt stress value, calculate the index of salinization hazard intensity. The results are then compared with existing salinization results (HWSD and GLASOD). The fourth step, based on the principle of information diffusion, is to calculate the hazard intensity of corn salinization under the recurrent periods of 10 years, 20 years, 50 years, and 100 years (Figure 1).

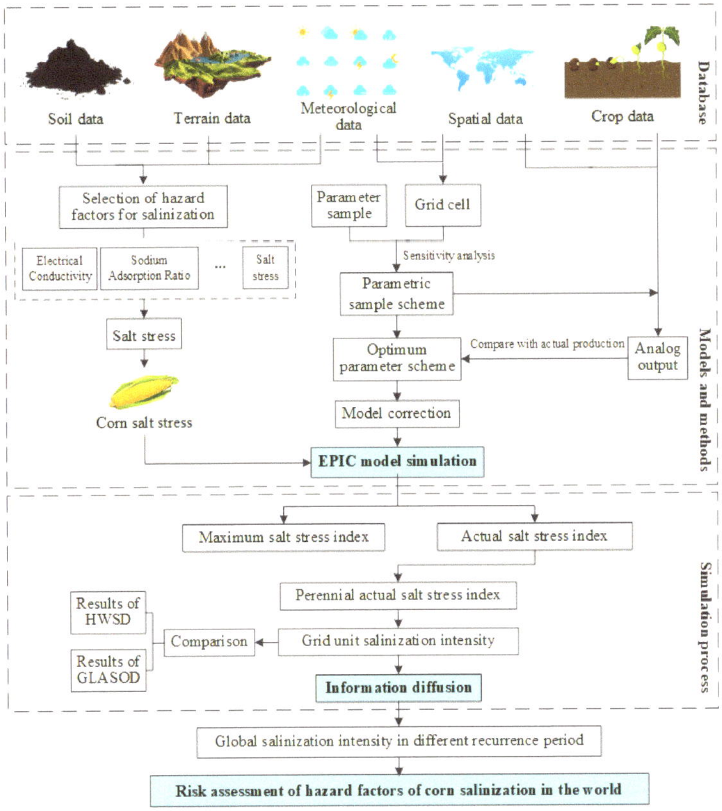


**Figure 1.** The research framework.

*2.2. Data*

Based on the research approach, this paper uses the EPIC0509 model to calculate the hazard intensity of salinization. In the research, a data list required was constructed, as shown in Table 1.

**Table 1.** Datasets for world corn salinization.

| Data Name | Data Content | Spatial Resolution | Temporal Resolution | Data Sources |
|---|---|---|---|---|
| DEM | Global elevation | 0.0833° × 0.0833° | 1997 | USGS [62] |
| Slope | Global slope | 0.0833° × 0.0833° | 1997 | GAEZ [63] |
| Soil Properties | Global soil distribution raster image and soil physical and chemical properties such as PH, soil depth, conductivity, etc. | 0.0833° × 0.0833° | 1995 | ISRIC |

**Table 1.** *Cont.*

| Data Name | Data Content | Spatial Resolution | Temporal Resolution | Data Sources |
|---|---|---|---|---|
| Meteorological | Global precipitation, temperature, solar radiation, and other information | 0.5° × 0.5° | 1971–2099 | Cross-sector impact model comparison projectRCP2.6 [64] |
| Planting Area | Global cultivation crop region | 5 min × 5 min | 1992 | Sustainability and the Global Environment, University of Wisconsin-Madison [65] |
| Corn Parameter Data | Corn EPIC Model Reference (US) | Site | - | Texas A&M University College of Agriculture and Life Sciences |
| Growth Period | Corn planting time and growth period length | 0.5° × 0.5° | 2000–2015 | Nelson Institute for Environmental Studies at the University of Wisconsin-Madison [66] |
| Irrigation | Global annual irrigation water of agriculture(mm) | 0.5° × 0.5° | 1995 | Institute of Industrial Science, University of Tokyo [67] |
| Fertilizer | Global annual fertilizer application for maize | 0.5° × 0.5° | 2012 | Earth stat [68] |
| Corn production | Production data for global country units | Vector unit | 1995–2004 | FAO |
| | China provincial unit production data | Vector unit | 1995–2004 | Department of Plantation Management, Ministry of Agriculture, China |
| | US state unit production data | Vector unit | 1995–2004 | United States Department Of Agriculture |
| | Australian state unit production data | Vector unit | 1995–2004 | Australian Bureau of Statistics |
| | India state unit production data | Vector unit | 1995–2004 | Department of Agriculture and Cooperation |
| Evaluation unit | World administrative divisions, rivers, lakes, etc. | Vector unit | 1995–2004 | ESRI, China Surveying and Mapping Geographic Information Bureau, CRU TS2.1, DIVA-GIS |
| Aridity Index | Global Map of Aridity | 10 arc minutes | 1961–1990 | FAO [69] |
| Other salinization research results | Excess salts | 0.5° × 0.5° | 1971–1981 | Harmonized World Soil Database [70] |
| | Cs | Vector unit | 1991 | GLASOD [71] |

### 2.3. Method

#### 2.3.1. Features and Simulation Process of EPIC0509

The version of the EPIC model used in this study is EPIC0509, which is a classic version officially developed by EPIC and released in 2006 [34]. This version has the following features: It operates on a daily timestep, capable of simulating crop growth conditions for 1–4000 years; the model provides basic soil, weather, tillage, and crop parameters; the soil can be divided into 10 layers; there is an optional weather generator; and the site-specific model has been improved to a field-scale model. It is widely used in crop growth simulation [28,72–74]. The main simulation process of the EPIC0509 model can be divided into 4 steps (Figure 2): (1) based on temperature (temperature and heat required by crops) to simulate the phenological development of crops, with heat unit index as the characterization index, this process affects the calculation of leaf area and root weight; (2) the growth of potential biomass of crops was simulated based on light energy (solar radiation, sunshine hours and light energy conversion rate), with leaf area index as the core index; (3) simulated the environmental stress (water, salt, temperature and nutrients) during the growth of crops. Among them, the maximum stress value was involved in the simulation calculation of leaf area and aboveground biomass; and (4) simulated crop yield based on aboveground biomass and harvest index, in which aboveground biomass was comprehensively affected by root weight, potential biomass, and environmental stress, and harvest index was affected by environmental stress and potential harvest index.

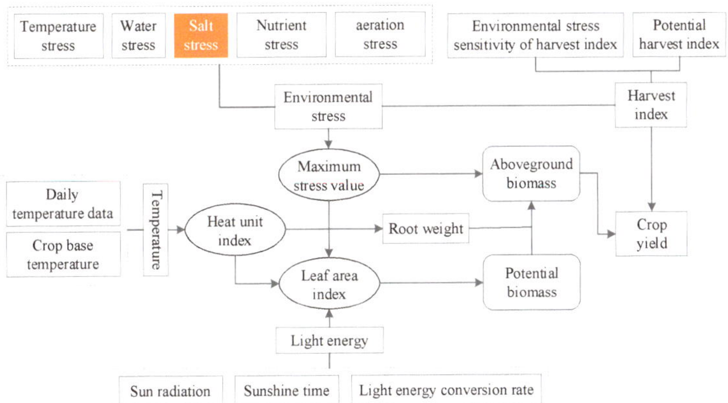


**Figure 2.** EPIC crop growth model simulation process.

### 2.3.2. Identifying Hazard Factors for Corn Salinization

In the previous experimental studies on the effects of salinization on crops, the evaluation was generally based on soil salinity combined with certain environmental conditions to construct hazard indicators, which cannot accurately reflect the independent impact of salinity on crops. The EPIC model simulates various environmental stress values in the crop growth process on a daily basis, which reflect the independent severity of different stresses on crop growth. The salinity stress value is calculated based on the crop's irrigation conditions and original soil conditions, and it quantifies the yield loss caused by the final soil salinity concentration.

Soil salinity is closely related to irrigation, and its calculation formula is as follows [42]:

$$WSLT_{i+1} = WSLT_i + 0.01 \times AIR_i \times CSLT_i \quad (1)$$

$WSLT$: Soil salinity content; $AIR$: maximum irrigation amount per time; $CSLT$: salt concentration in irrigation water.

Soil moisture content has a significant impact on soil salinity stress, and its calculation formula is as follows [42]:

$$ST_{i+1} = ST_i + AIR_i \quad (2)$$

$ST$: Soil moisture content; $AIR$: maximum irrigation amount per time.

The daily salt stress value is the hazard factor of this study, and its calculation formula is as follows [42]:

$$SWRZ_{i+1} = SWRZ_i + ST_i \times (RZ - Z_{i-1})/(Z_{i-1} - Z_i) \quad (3)$$

$$TSRZ_{i+1} = TSRZ_i + WSLT_i \times (RZ - Z_{i-1})/(Z_{i-1} - Z_i) \quad (4)$$

$$SS = a \times (0.15625 \times TSRZ/SWRZ - b) \quad (5)$$

$SS$: The daily salinity stress value. In addition, $SWRZ$ and $TSRZ$ are the intermediate variable, the initial value is 0.

The calculation formula for salinity stress during the maize growth period is as follows [75]:

$$SS_{Total} = \sum_{i=1}^{total} SS_i \quad (6)$$

$SS_{Total}$ represents the total salinity stress value during the maize growth period, and $SS_i$ represents the salinity stress value for the $i$-th day.

2.3.3. Intensity of Hazard Caused by Corn Salinization

Constructing an index for the intensity of salinization hazard based on daily salinity stress values [75].

$$SI = \frac{SS^i_{total}}{max(SS^i_{total})} \quad (7)$$

Among them, $SS^i_{total}$: total salt stress in the scenario $i$, $max(SS^i_{total})$: the largest total salt stress value of scenarios in the current year.

The maximum value of salinity stress in a given year is obtained through the simulation of salinization scenarios. In the EPIC model simulation, there are three main aspects that can cause yield reduction in crops, and corresponding measures are taken to mitigate them. First, field management measures, including diseases, pests, and management errors, are automatically excluded in the simulation. Second, soil erosion conditions, such as water erosion and wind erosion, are eliminated by disabling the water erosion and wind erosion modules in the model, thus not considering their influence during the simulation. Third, environmental stress factors, including temperature stress, nutrient stress (nitrogen, phosphorus, potassium), water stress, and ventilation stress, are addressed by setting appropriate parameters in the model to exclude these environmental stress factors (Table 2).

**Table 2.** Elimination of environmental stress elements.

| Elimination of Coercion Type | Elimination Method |
|---|---|
| Temperature stress | Management measures automatic fertilization |
| Nutrient stress | Management measures automatic fertilization |
| Water stress | Set up automatic irrigation to meet crop water requirements |
| Ventilation stress | Pre-experiment setting the maximum water supply so that no ventilation stress is generated |

According to the calculation formula of the salinity stress value, it can be inferred that salinity is involved in the hazard process of salinization through the initial soil salinity and irrigation water salinity, while the final soil salinity content affects crop yield. Therefore, in the model simulation process of the scenarios with the highest total salt stress, we set the initial soil salt concentration, namely conductivity, as 0, and established the highest salt concentration of irrigation water to obtain the extreme yield loss scenario.

### 2.3.4. Calculation of Crop Yield

The salt tolerance of crops to salinization is closely related to their species and variety [76]. The classification criteria for determining the salt tolerance of crops are based on soil salinity, evapotranspiration loss, and water daily stress index. The calculation formulas for crop yield under salt stress caused by different dominant factors vary [77]. In cases where the dominant factor of salt stress is soil salinity, the crop yield formula is as follows:

$$Y = 100 - b(EC_e - a) \tag{8}$$

$Y$ is relative yield, $EC_e$ is the electrical conductivity of soil solution (dS/m), $a$ is the threshold electrical conductivity tolerance of the crop (dS/m), and $b$ is the slope, which represents the reduction rate of yield per unit electrical conductivity.

### 2.3.5. Salinization Hazard Intensity Recurrence Algorithm

Information diffusion is a method based on fuzzy set theory for comprehensive evaluation of regional environmental risks [78]. Its specific application in this study is as follows:

The indices of salinization hazard intensity are statistically analyzed, with these hazard intensity indices denoted as $x_1, x_2, \cdots, x_m$, then

$$X = \{x_1, x_2, \cdots, x_m\} \tag{9}$$

$X$ is the observation sample set, $x_i$ is the index of the hazard intensity caused by salinization of a cornfield in the $i$th year ($i = 1, 2, \ldots, m$; $m = 24$);

Let the domain of the hazard intensity index be $U$:

$$U = \{u_1, u_2, \cdots, u_n\} \tag{10}$$

Each individual data sample, denoted as $x_i$, can diffuse the information it carries to all points in the domain of hazard intensity index $U$ using Formula (10):

$$f_i(u_j) = \frac{1}{h\sqrt{2\pi}} exp\left[-\frac{(x_i - u_j)^2}{2h^2}\right] \tag{11}$$

The variable $h$ is referred to as the diffusion coefficient, and its value can be determined based on the maximum and minimum values of the hazard intensity index in the sample set, as well as the number of samples $m$. The calculation formula for $h$ is as follows:

$$h = 2.6851(b - a)/(m - 1) \tag{12}$$

where $a = \underset{1 \le i \le m}{min}\{x_i\}, b = \underset{1 \le i \le m}{max}\{x_i\}$.

Based on this value, an estimation of the hazard intensity beyond the probability can be obtained [2].

## 3. Results

### 3.1. Mean Expected Hazard Intensity of Global Corn Salinisation

In this study, meteorological station data from 1971 to 2004 were selected. The actual conductivity content in the soil was used as a substitute for irrigation water salinity to calculate the salinization index (SI) for each year and grid in the 0.5 × 0.5 grid unit. The expected results of salinization hazard intensity were obtained and presented in Figure 3.

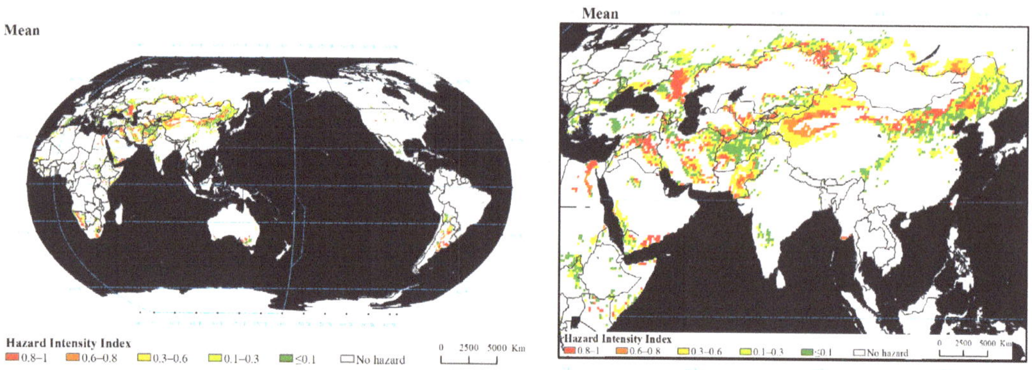


**Figure 3.** Mean expected hazard intensity of global corn salinization.

According to Figure 3, the red areas in the graph (with hazard index > 0.8) indicate the highest salinization hazard intensity for maize. These areas are primarily distributed in Central Asia, northwestern China, southern South America, and the western coast of southern Africa. Asia is identified as the region facing the most severe salinization threat. Oman, the southern part of the high mountains and basins in northwestern China, the plains and hills in central-western Kazakhstan, and the mountainous regions in the northeastern plains all have an average salinization hazard index above 0.5. Additionally, the Nile River Delta and the western side of the South African plateau, as well as the central-northern plateau in Algeria, are also high-risk areas for salinization.

The salinization hazard intensity is related to aridity levels [79]. Based on the aridity classification data from the FAO Global Map of Aridity, the salinization hazard intensity is correlated with the aridity level. Subsequently, the salinization data are categorized based on the aridity level to study the distribution of salinization grid points in each category (Figure 4).

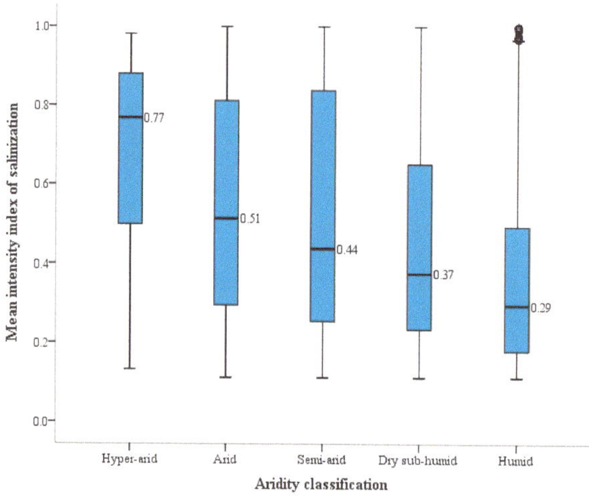


**Figure 4.** Boxplots of salinization index for different aridity types.

According to Figure 4, the salinization hazard intensity index is highest in hyper-arid regions, with a median value of 0.77 and an interquartile range of approximately 0.4, indicating a relatively high and concentrated salinization intensity in that area. Arid and semi-arid regions also exhibit high salinization hazard intensity, with median values of

0.51 and 0.44, respectively, and interquartile ranges of 0.5 and 0.6, suggesting relatively high salinization intensity but with more dispersed values in these regions. Dry sub-humid regions, as transitional zones between arid and humid areas, show a median salinization hazard intensity index of 0.37, lower than the previous three types, indicating a decrease in salinization intensity. Humid regions have the lowest salinization hazard intensity index, with a median value of 0.29 and an interquartile range of approximately 0.3. The values are concentrated between 0.2 and 0.5, indicating a relatively low and concentrated salinization intensity in that area. Additionally, the boxplot reveals some outliers with higher salinization hazard intensity index in the humid region. This suggests that the salinity index is generally low in Humid, but there are some areas of high salinity along the coast of the sea and inland lakes. In summary, arid and semi-arid regions exhibit higher salinization levels, while humid regions experience relatively lower salinization intensity.

### *3.2. Global Risk Assessment of Salinization Hazard Factors with Different Return Periods*

When sample information is incomplete in the assessment of hazard factor risk, the information diffusion method can be used to calculate the exceedance probability of a certain hazard intensity. In this study, based on the information diffusion model, the salinization hazard factors of a 30-year sample were calculated for different occurrence probabilities (return periods of 10, 20, 50, and 100 years). Figure 5 illustrates the results of salinization hazard intensity calculation. This study also conducted a statistical analysis of the top ten countries in terms of average salinization intensity under different return periods, as shown in Table 3. Additionally, we calculated and ranked the average salinization intensity under different return periods for the top ten largest countries in the world by land area (Table 4).

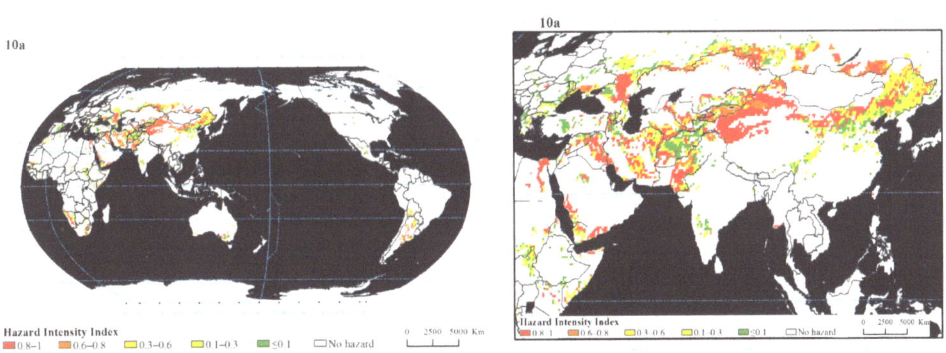


(a)

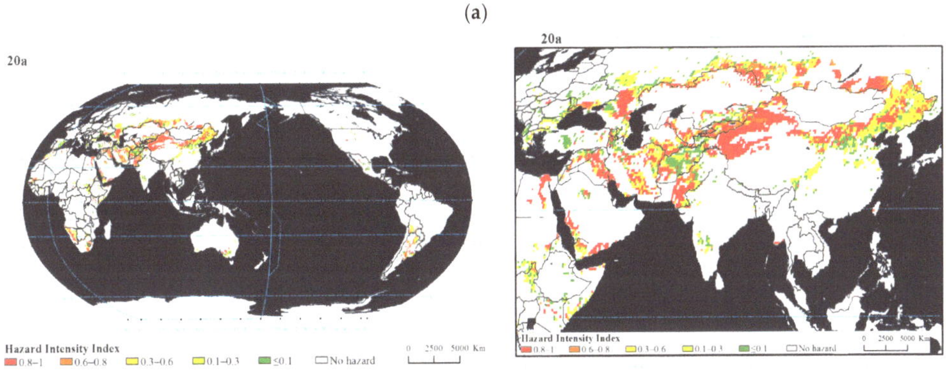


(b)

**Figure 5.** *Cont.*

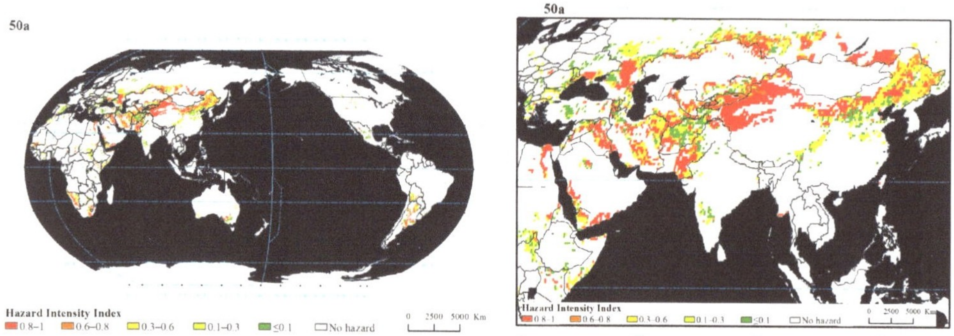


(c)

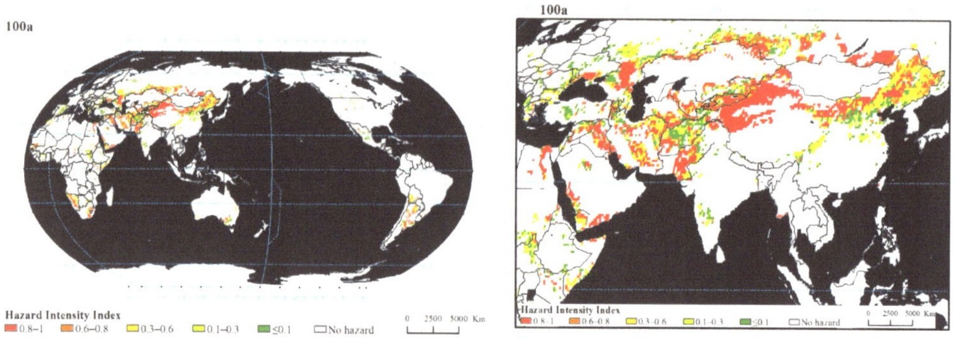


(d)

**Figure 5.** Global salinization hazard intensity under different return periods: (**a**) 10-year return period; (**b**) 20-year return period; (**c**) 50-year return period; (**d**) 100-year return period.

**Table 3.** Top ten countries and their average salinization intensity values under different return periods.

| Rank | 10-Year-Return-Period | | 20-Year-Return-Period | | 50-Year-Return-Period | | 100-Year-Return-Period | |
|---|---|---|---|---|---|---|---|---|
| | Country | Mean | Country | Mean | Country | Mean | Country | Mean |
| 1 | Oman | 0.99 | Oman | 0.99 | Oman | 1.00 | Oman | 1.00 |
| 2 | Egypt | 0.90 | Egypt | 0.91 | Egypt | 0.92 | Egypt | 0.92 |
| 3 | Mongolia | 0.73 | Mongolia | 0.78 | Mongolia | 0.82 | Mongolia | 0.84 |
| 4 | Kuwait | 0.65 | Kuwait | 0.66 | Kuwait | 0.66 | Kyrgyzstan | 0.67 |
| 5 | Turkmenistan | 0.62 | Turkmenistan | 0.64 | Turkmenistan | 0.65 | Kuwait | 0.66 |
| 6 | Yemen | 0.58 | Kyrgyzstan | 0.61 | Kyrgyzstan | 0.65 | Turkmenistan | 0.66 |
| 7 | Uzbekistan | 0.57 | Yemen | 0.60 | Yemen | 0.61 | Yemen | 0.62 |
| 8 | Kyrgyzstan | 0.56 | Uzbekistan | 0.58 | Uzbekistan | 0.59 | Uzbekistan | 0.60 |
| 9 | Algeria | 0.56 | Algeria | 0.57 | Algeria | 0.58 | Saudi Arabia | 0.59 |
| 10 | Iraq | 0.54 | Saudi Arabia | 0.56 | Saudi Arabia | 0.58 | Algeria | 0.58 |

**Table 4.** Ranking of the 10 largest countries in terms of average intensity of salinization for different return periods and their average salinization intensity values.

| Country | 10-Year-Return-Period | | 20-Year-Return-Period | | 50-Year-Return-Period | | 100-Year-Return-Period | |
|---|---|---|---|---|---|---|---|---|
| | Rank | Mean | Rank | Mean | Rank | Mean | Rank | Mean |
| Russia | 28 | 0.19 | 29 | 0.20 | 29 | 0.20 | 29 | 0.21 |
| Canada | 68 | 0.00 | 68 | 0.00 | 69 | 0.01 | 69 | 0.01 |
| China | 21 | 0.30 | 21 | 0.31 | 21 | 0.32 | 21 | 0.33 |
| United States | 69 | 0.00 | 69 | 0.00 | 70 | 0.00 | 70 | 0.00 |
| Brazil | 76 | 0.00 | 76 | 0.00 | 76 | 0.00 | 76 | 0.00 |
| Australia | 35 | 0.09 | 36 | 0.09 | 36 | 0.10 | 36 | 0.10 |
| India | 63 | 0.01 | 64 | 0.01 | 63 | 0.01 | 63 | 0.01 |
| Argentina | 27 | 0.20 | 27 | 0.20 | 28 | 0.21 | 28 | 0.21 |
| Kazakhstan | 14 | 0.45 | 14 | 0.46 | 14 | 0.47 | 14 | 0.48 |
| Algeria | 9 | 0.56 | 9 | 0.57 | 9 | 0.58 | 10 | 0.58 |

The influence range of high salinization-induced disaster intensity expands with an increase in the return period, as evidenced by Figure 5, Tables 3 and 4. Conversely, the influence range of low salinization-induced disaster intensity decreases. The ranking of total salinization intensity remains relatively stable across different recurrence periods. Specifically, during return periods of 10, 20, 50, and 100 years, Oman, Egypt, and Mongolia consistently ranked in the top three countries worldwide, with an average disaster intensity of over 0.7 (Table 3). However, the rankings of the last seven countries within the top ten for average salinization intensity have shown slight fluctuations. For instance, Kuwait ranked fourth during the 10-year and 20-year return periods, but dropped to fifth during the 100-year return period. Similarly, Kyrgyzstan ranked eighth during the 10-year return period but improved to fourth during the 100-year return period. In terms of the recurrence periods of 10, 20, 50, and 100 years, the rank of mean salinity intensity among the ten countries with the largest areas remained stable (Table 4). Among the countries mentioned, Algeria, Kazakhstan, China, and Russia exhibit relatively high average salinization intensity, ranking 9th, 14th, 21st, and 28th, respectively, worldwide. Remarkably, despite being the smallest among them, Algeria achieves the highest ranking. Algeria consistently experiences an average intensity of salinization greater than 0.5 during the four recurrence periods, placing it around 9th globally. This phenomenon is closely linked to the predominantly dry climate in Algeria's savanna and tropical desert climate zones. In contrast, Brazil, ranking fifth in terms of area, faces comparatively low salinization, with a ranking of 76. This occurrence can be attributed to the majority of Brazil's geographical location in a humid area.

## 4. Discussion

### *4.1. Comparison of Salinization Results*

#### 4.1.1. Model Validation

We conducted parameter sensitivity analysis, parameter adjustment, and validation on the model using corn yield to ensure the simulation accuracy of the EPIC0509 model. The specific process can be found in the Yin's paper [80–82].

#### 4.1.2. Compared to the Excess Salts Data

In order to evaluate the stability of the research results, we conducted a non-parametric correlation test, specifically the Spearman's rank correlation test, between the results of this study and the excess salts data from the Harmonized World Soil Database v 1.2 (HWSD), at the national and comparable geographical unit scales [60]. The results showed a significant correlation between the two at the 0.01 level. Additionally, the comparison results between the national and comparable geographical units were depicted as scatter plots and data distribution graphs for the salinization levels (as shown in Figure 6).

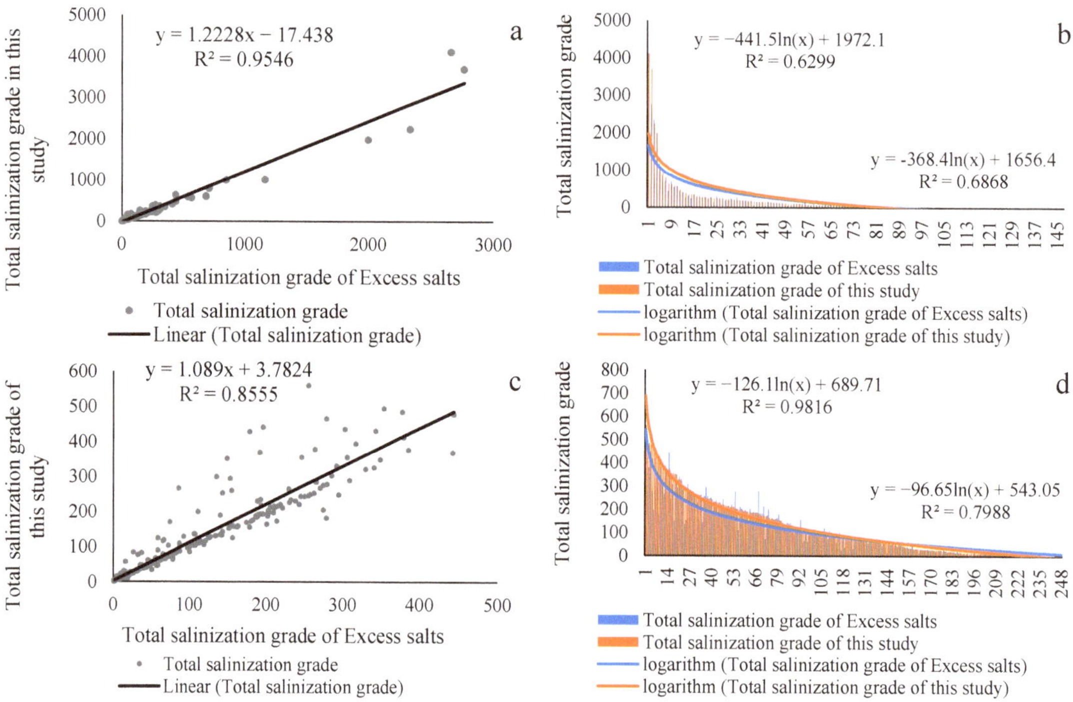


**Figure 6.** Comparison with Excess Salts at national unit and comparable geographic unit scales. ((**a**): scatter plot of salinity classes at the national unit scale; (**b**): Data distribution of salinity levels at the national unit scale; (**c**): Scatter plot of salinity levels at the comparable geographical unit scale; (**d**): Data distribution of salinity levels at the comparable geographical unit scale).

The scatter plot (a) and the data distribution graph (b) at the national unit scale show a strong consistency between the salinization intensity map of this study and the excess salts data from HWSD, with a high $R^2$ value of 0.95. However, due to the disparate areas between countries, the distribution of salinization total levels is highly concentrated, with a majority falling below 1000. From the scatter plot of salinization total levels at the comparable geographical unit scale, it can be observed that there is good consistency between the salinization intensity map of this study and the excess salts data from HWSD, with an $R^2$ value of 0.86. According to the data distribution graph of salinization total levels at the comparable geographical unit scale, the salinization levels of this study are slightly higher overall compared to those of HWSD's "Excess salts".

#### 4.1.3. Compared to the Salinization Data from GLASOD

The Cs index in the World map of the status of human-induced soil degradation, created by GLASOD, is used to measure the degree of salinization. To compare the results of this study with the salinization data from GLASOD, we conducted a non-parametric correlation test, specifically the Spearman's rank correlation test, at the national and comparable geographical unit scales [60]. The results showed a significant correlation between the two at the 0.01 level. Additionally, the comparison results between the national and comparable geographical units were depicted as scatter plots and data distribution graphs for salinization levels (Figure 7).

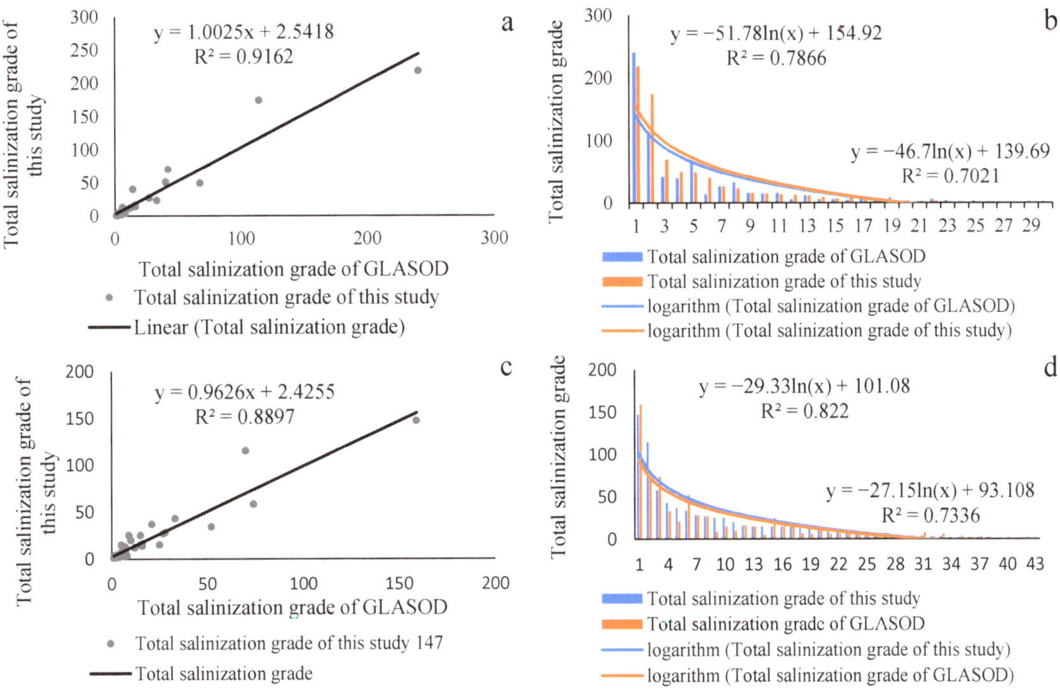


**Figure 7.** Comparison with GLASOD at national unit and comparable geographic unit scales. ((**a**): scatter plot of salinity classes at the national unit scale; (**b**): Data distribution of salinity levels at the national unit scale; (**c**): Scatter plot of salinity levels at the comparable geographical unit scale; (**d**): Data distribution of salinity levels at the comparable geographical unit scale).

Although there are not many studies that have the same research scope as this study and GLASOD at the national and comparable geographical unit scales, they still have a strong comparability. As shown in Figure 7, this study and GLASOD had high consistency in their research results at the national unit scale, with an $R^2$ of up to 0.92. The consistency at the comparable geographical unit scale was slightly lower, with an $R^2$ of 0.89. Overall, both at the national unit scale and the comparable geographical unit scale, the salinization intensity in this study is slightly higher than the salinization grade in GLASOD.

### *4.2. Research Value and Policy Recommendations*

Salinization significantly affects crop yields, particularly for irrigated crops. Examining the risk of salinization as a hazard can provide vital information for mitigating agricultural losses caused by salinization and promoting sustainable agricultural development. This study utilized salt stress as an indicator to assess hazardous factors, surpassing the limitations associated with using external environmental conditions to establish hazard indicators. The EPIC0509 model was employed to simulate the growth of corn, obtaining results on the global distribution and quantification of corn salinization intensity. This research has made valuable contributions to quantifying regional land degradation and addressing the gap in global salinity risk assessment mapping.

The risk assessment of corn salinization hazard factors under different return periods was conducted based on the principle of information diffusion. This paper introduces a novel approach to studying hazard factors of large-scale salinization, which sets the groundwork for salinization risk assessment. Using salt stress as an indicator for assessing the risk of salinization hazard factors may offer a potential direction for future research. It serves as a warning against irrational land cultivation practices and the use of transitional groundwater irrigation in ecological transition zones.

Based on the findings of this study, we propose the following recommendations:

(1) Adapt to climate change by adjusting agricultural production structure and selecting crop varieties with strong resilience to climate impacts.
(2) Optimize land use and resource allocation by planning agricultural layout based on differences in land productivity potential, improving resource efficiency, and enhancing land protection and improvement measures to prevent salinization.
(3) Promote carbon-neutral agriculture by reducing greenhouse gas emissions from farming, promoting low-carbon agricultural technologies such as organic farming and precision fertilization, and increasing farmland carbon sequestration through afforestation and wetland conservation.
(4) Strengthen salinization prevention and control efforts by enhancing monitoring and assessment, developing scientific prevention and control measures such as rational irrigation, drainage infrastructure construction, and soil improvement, and providing training and technical guidance to farmers to enhance their ability to cope with salinization.

### 4.3. The Outlook and Shortcomings

While this study introduces a new approach to assessing the risk of salinization factors, it overlooks the evaluation of salinization risk in conjunction with corn yield loss rates. The salinization scenario in this study only considers salt stress while disregarding the impact of temperature, precipitation stress, and other forms of land degradation. Additionally, the optimal scenario is applied to all other stress factors without considering their relationship with salinization. In future studies, the following steps will be taken: (1) Integrate disaster risk and production loss rates to construct a global corn salinization vulnerability curve and evaluate the risk of corn salinization worldwide. (2) Further examine the relationship between salinization, temperature stress, and precipitation stress. (3) Investigate the connection between salinization and other forms of land degradation, such as soil erosion and desertification, to achieve a comprehensive evaluation of land degradation. These efforts aim to provide a more comprehensive assessment of salinization and its impact on land degradation.

## 5. Conclusions

In this study, the salinization hazard factor was determined by the salt stress value. The global cornfield served as the research area, and salinization scenarios were established by eliminating environmental stress factors. Using the EPIC model, the day-step-length salt stress value during the corn growth process was simulated. The risk of corn salinization intensity was evaluated on a global scale, and the intensity of salinization under different return periods was calculated based on the principle of information diffusion. The main conclusions are as follows: (1) Environmental stress factors were eliminated, salinization scenarios were established, and algorithms for the growth season salinization index and disaster intensity index were proposed. (2) High-risk areas (with disruption index > 0.8) for corn salinization were primarily located in Central Asia, northwestern China, southern South America, and the southern coast of Africa. Hazard factors in arid and semi-arid regions posed a high risk, while wet areas had relatively lower risk. (3) The risk of salinization hazard increased with longer return periods (i.e., 10, 20, 50, and 100 years). The impact scope expanded, and the level of danger increased. Oman, Egypt, and Mongolia had an average salinization intensity greater than 0.7, ranking as the top three countries for all return periods. (4) The salinization intensity map produced in this study exhibited high consistency with the Excess salts of HWSD and Cs of GLASOD. The $R^2$ values between the two results at the country and regional units exceeded 0.9, while the $R^2$ values at the comparable geographic unit exceeded 0.8. However, the salinization grades in this study were slightly higher overall than those of excess salts of HWSD and Cs of GLASOD.

**Author Contributions:** Conceptualization, D.L. and F.L.; methodology, D.L. and F.L.; software, D.L.; validation, C.H. and Y.Y.; formal analysis, D.L. and C.H.; investigation, F.L., C.H. and X.G.; resources, J.W.; data curation, D.L. and F.L.; writing—original draft preparation, C.H. and D.L.; writing—review and editing, C.H. and D.L.; visualization, C.H.; supervision, J.W.; project administration, D.L. and J.W.; funding acquisition, J.W. All authors have read and agreed to the published version of the manuscript.

**Funding:** This study was funded by the 2023 National Social Science Foundation of China (NSSFC) Annual Program (23BSH006), and the Natural Science Foundation of Zhejiang Province, China (LQ21D010009), Wenzhou Philosophy and Social Science Planning Project (No. 22wsk033), and the National Natural Science Foundation of China (Project No. 41271286).

**Data Availability Statement:** The data used to support the findings of this study can be made available by the corresponding author upon request.

**Conflicts of Interest:** The authors declare no conflict of interest.

## References

1. Wu, W.B.; Verburg, P.H.; Tang, H.J. Understanding land system dynamics and its consequences. *J. Geogr. Sci.* **2018**, *28*, 1563–1566. [CrossRef]
2. Shi, P.J.; Song, C.Q.; Cheng, C.X. Geographical synergetics: Fromunderstanding human-environment relationship to designing human-environment synergy. *Acta Geograpaica Sin.* **2019**, *74*, 3–15.
3. FAO. *Learning about Soil Salinization*; Food and Agriculture Organization: Rome, Italy, 2020.
4. Martinez-Beltran, J. Overview of Salinity Problems in the World and FAO Strategies to Address the Problem. In *Managing Saline Soils and Water: Science, Technology and Social Issues*; Proceedings of the International Salinity Forum; Water Science and Policy Center: Riverside, CA, USA; p. 2005.
5. Davidson, R.A.; Lambert, K.B. Comparing the Hurricane Disaster Risk of U.S. Coastal Counties. *Nat. Hazards Rev.* **2001**, *2*, 132–142. [CrossRef]
6. Huang, J.; Prochazka, M.J.; Triantafilis, J. Irrigation salinity hazard assessment and risk mapping in the lower Macintyre Valley, Australia. *Sci. Total Environ.* **2016**, *551*, 460–473. [CrossRef] [PubMed]
7. Stofberg, S.F.; Essink, G.H.O.; Pauw, P.S.; De Louw, P.G.; Leijnse, A.; van der Zee, S.E. Fresh water lens persistence and root zone salinization hazard under temperate climate. *Water Resour. Manag.* **2017**, *31*, 689–702. [CrossRef]
8. Bessaim, M.M.; Missoum, H.; Bendani, K.; Bekkouche, M.S. Mitigation of Salinity Hazard from Low Permeable Soil by Electrochemical Treatment: A Laboratory Based Investigation. In *Recent Advances in Geo-Environmental Engineering, Geomechanics and Geotechnics, and Geohazards*; Springer: Berlin/Heidelberg, Germany, 2019; pp. 93–95. [CrossRef]
9. Lobell, D.; Lesch, S.; Corwin, D.; Ulmer, M.; Anderson, K.; Potts, D.; Doolittle, J.; Matos, M.; Baltes, M. Regional-scale assessment of soil salinity in the Red River Valley using multi-year MODIS EVI and NDVI. *J. Environ. Qual.* **2010**, *39*, 35–41. [CrossRef]
10. Furby, S.; Caccetta, P.; Wallace, J. Salinity monitoring in Western Australia using remotely sensed and other spatial data. *J. Environ. Qual.* **2010**, *39*, 16–25. [CrossRef] [PubMed]
11. Caccetta, P.; Dunne, R.; George, R.; McFarlane, D. A methodology to estimate the future extent of dryland salinity in the southwest of Western Australia. *J. Environ. Qual.* **2010**, *39*, 26–34. [CrossRef]
12. Spies, B.; Woodgate, P. *Salinity Mapping Methods in the Australian Context*; Department of the Environment and Heritage: Canberra, Australia, 2005; pp. 23–24.
13. Metternicht, G.I.; Zinck, J. Remote sensing of soil salinity: Potentials and constraints. *Remote Sens. Environ.* **2003**, *85*, 1–20. [CrossRef]
14. De Paz, J.; Visconti, F.; Zapata, R.; Sánchez, J. Integration of two simple models in a geographical information system to evaluate salinization risk in irrigated land of the Valencian Community, Spain. *Soil Use Manag.* **2004**, *20*, 333–342. [CrossRef]
15. Amezketa, E. An integrated methodology for assessing soil salinization, a pre-condition for land desertification. *J. Arid. Environ.* **2006**, *67*, 594–606. [CrossRef]
16. Bouksila, F.; Bahri, A.; Berndtsson, R.; Persson, M.; Rozema, J.; Van der Zee, S.E. Assessment of soil salinization risks under irrigation with brackish water in semiarid Tunisia. *Environ. Exp. Bot.* **2013**, *92*, 176–185. [CrossRef]
17. Rengasamy, P. World salinization with emphasis on Australia. *J. Exp. Bot.* **2006**, *57*, 1017–1023. [CrossRef] [PubMed]
18. Richards, L.A. *Diagnosis and Improvement of Saline and Alkali Soils*; US Government Printing Office: Washington, DC, USA, 1954; Volume 78, p. 154.
19. Sentis, I. Soil Salinization and Land Desertification. *Soil Degrad. Desertif. Mediterr. Environ.* 1996, pp. 105–129. Available online: https://indico.ictp.it/event/a0114/material/2/14.pdf (accessed on 10 September 2023).
20. Castrignanò, A.; Buttafuoco, G.; Puddu, R. Multi-Scale Assessment of the Risk of Soil Salinization in an Area of South-Eastern Sardinia (Italy). *Precis. Agric.* **2008**, *9*, 17–31. Available online: https://link.springer.com/content/pdf/10.1007/s11119-008-9054-4.pdf (accessed on 10 September 2023). [CrossRef]

21. Mirlas, V. Assessing soil salinity hazard in cultivated areas using MODFLOW model and GIS tools: A case study from the Jezre'el Valley, Israel. *Agric. Water Manag.* **2012**, *109*, 144–154. [CrossRef]
22. Datta, K.; Sharma, V.; Sharma, D. Estimation of a production function for wheat under saline conditions. *Agric. Water Manag.* **1998**, *36*, 85–94. [CrossRef]
23. Bui, E.N.; Smettem, K.R.; Moran, C.J.; Williams, J. Use of soil survey information to assess regional salinization risk using geographical information systems. *J. Environ. Qual.* **1996**, *25*, 433–439. [CrossRef]
24. Šimůnek, J.; Suarez, D.L. Two-dimensional transport model for variably saturated porous media with major chemistry. *Water Resour. Res.* **1994**, *30*, 1115–1133. [CrossRef]
25. Ragab, R. An integrated modelling approach for irrigation water management using saline and non-saline water: The SALTMED model. In Proceedings of the International Symposium on Techniques to Control Salination for Horticultural Productivity, Antalya, Turkey, 7–10 November 2000; Volume 573, pp. 129–138.
26. Raes, D.; Van Goidsenhoven, B.; Goris, K.; Samain, B.; De Pauw, E.; El Baba, M.; Tubail, K.; Ismael, J.; De Nys, E. BUDGET, a management tool for assessing salt accumulation in the root zone under irrigation. In Proceedings of the Inter Regional Conference on Environment-Water, ICID, Fortaleza, Brazil, 27–30 August 2001.
27. Rubio, J.; Calvo, A. *Solil Degradation and Desertification in Mediterranean Envoronments*; Geoforma Ediciones: Logrono, Barcelona, 1996.
28. Arunrat, N.; Pumijumnong, N.; Hatano, R. Predicting local_scale impact of climate change on rice yield and soil organic carbon sequestration: A case study in Roi Et Province, Northeast Thailand. *Agric. Syst.* **2018**, *164*, 58–70. [CrossRef]
29. FAO. Global Map of Salt-Affected Soils. Available online: https://www.fao.org/3/cb7247en/cb7247en.pdf (accessed on 10 September 2023).
30. Kaya, F.; Schillaci, C.; Keshavarzi, A.; Başayiğit, L. Predictive Mapping of Electrical Conductivity and Assessment of Soil Salinity in a Western Türkiye Alluvial Plain. *Land* **2022**, *11*, 2148. [CrossRef]
31. Shiferaw, B.; Prasanna, B.M.; Hellin, J.; Bänziger, M. Crops that feed the world 6. Past successes and future challenges to the role played by maize in global food security. *Food Secur.* **2011**, *3*, 307–327. [CrossRef]
32. Yang, R.; Cao, R.; Gong, X.; Feng, J. Cultivation has selected for a wider niche and large range shifts in maize. *PeerJ* **2022**, *10*, e14019. [CrossRef] [PubMed]
33. Farooq, M.; Hussain, M.; Wakeel, A.; Siddique, K.H. Salt stress in maize: Effects, resistance mechanisms, and management: A review. *Agron. Sustain. Dev.* **2015**, *35*, 461–481. [CrossRef]
34. Williams, J.R.; Wang, E.; Meinardus, A.; Harman, W.; Siemers, M.; Atwood, J.D. *EPIC Users Guide V. 0509*; Blackland Research and Extension Center: Temple, TX, USA, 2006.
35. Williams, J.R. *EPIC: The Erosion-Productivity Impact Calculator*; United States Department of Agriculture: Washington, DC, USA, 1989.
36. Singh, V.P.; Woolhiser, D.A. Mathematical modeling of watershed hydrology. *J. Hydrol. Eng.* **2002**, *7*, 270–292. [CrossRef]
37. Gassman, P.W.; Reyes, M.R.; Green, C.H.; Arnold, J.G. SWAT peer-reviewed literature: A review. In Proceedings of the 3rd International SWAT Conference, Zurich, Switzerland, 11–15 July 2005. Available online: https://swat.tamu.edu/docs/swat/conferences/2005/PDF/Session_I/Gassman.pdf (accessed on 10 September 2023).
38. Huang, C.F. *Optimality Processing to the Sample Knowledge of Non-Completeness*; Department of Mathematics, Beijing Normal University: Beijing, China, 1992; Volume 28, pp. 129–148.
39. National Soil Erosion-Soil Productivity Research Planning Committee. Soil erosion effects on soil productivity: A research perspective. *J. Soil Water Conserv.* **1981**, *36*, 82–90.
40. Williams, J.R. The erosion-productivity impact calculator (EPIC) model: A case history. *Philos. Trans. R. Soc. Lond. Ser. B Biol. Sci.* **1990**, *329*, 421–428.
41. Williams, J.R.; Jones, C.A.; Dyke, P.T. A modeling approach to determining the relationship between erosion and soil productivity. *Trans. ASAE* **1984**, *27*, 129–144. [CrossRef]
42. Williams, J.R.; Jones, C.A.; Kiniry, J.R.; Spanel, D.A. The EPIC crop growth model. *Trans. ASAE* **1989**, *32*, 497–511. [CrossRef]
43. Williams, J.; Nearing, M.; Nicks, A.; Skidmore, E.; Valentin, C.; King, K.; Savabi, R. Using soil erosion models for global change studies. *J. Soil Water Conserv.* **1996**, *51*, 381–385.
44. Stockle, C.O.; Williams, J.R.; Rosenberg, N.J.; Jones, C.A. A method for estimating the direct and climatic effects of rising atmospheric carbon dioxide on growth and yield of crops: Part I—Modification of the EPIC model for climate change analysis. *Agric. Syst.* **1992**, *38*, 225–238. [CrossRef]
45. Zhi-Qiang, W.; Wei-Hua, F.; Fei, H.E.; Hong, X.U. Effect of climate change on wheat yield in northern China: A research based on EPIC model. *J. Nat. Disasters* **2008**, *17*, 109–114.
46. Anderson, C.J.; Babcock, B.A.; Peng, Y.; Gassman, P.W.; Campbell, T.D. Placing bounds on extreme temperature response of maize. *Environ. Res. Lett.* **2015**, *10*, 124001. [CrossRef]
47. Bao, Y.; Hoogenboom, G.; McClendon, R.; Vellidis, G. A comparison of the performance of the CSM-CERES-Maize and EPIC models using maize variety trial data. *Agric. Syst.* **2017**, *150*, 109–119. [CrossRef]
48. Maas, E.V.; Hoffman, G.J.; Chaba, G.D.; Poss, J.A.; Shannon, M.C. Salt sensitivity of corn at various growth stages. *Irrig. Sci.* **1983**, *4*, 45–57. [CrossRef]

49. Brugnoli, E.; Björkman, O. Growth of cotton under continuous salinity stress: Influence on allocation pattern, stomatal and non-stomatal components of photosynthesis and dissipation of excess light energy. *Planta* **1992**, *187*, 335–347. [CrossRef] [PubMed]
50. Pessarakli, M.; Huber, J.; Tucker, T. Dry matter yield, nitrogen absorption, and water uptake by sweet corn under salt stress. *J. Plant Nutr.* **1989**, *12*, 279–290. [CrossRef]
51. BLISS, R.D.; Platt-Aloia, K.; Thomson, W. Changes in plasmalemma organization in cowpea radicle during imbibition in water and NaCl solutions. *Plant Cell Environ.* **1984**, *7*, 601–606. [CrossRef]
52. Lauchli, A.; Schubert, S. The Role of Calcium in the Regulation of Membrane and Cellular Growth Processes under Salt Stress. In *Environmental Stress in Plants*; Springer: Berlin/Heidelberg, Germany, 1989; pp. 131–138.
53. Lynch, J.; Cramer, G.R.; Läuchli, A. Salinity reduces membrane-associated calcium in corn root protoplasts. *Plant Physiol.* **1987**, *83*, 390–394. [CrossRef] [PubMed]
54. Hichem, H.; El Naceur, A.; Mounir, D. Effects of salt stress on photosynthesis, PSII photochemistry and thermal energy dissipation in leaves of two corn (*Zea mays* L.) varieties. *Photosynthetica* **2009**, *47*, 517–526. [CrossRef]
55. Cha-Um, S.; Kirdmanee, C. Effect of salt stress on proline accumulation, photosynthetic ability and growth characters in two maize cultivars. *Pak. J. Bot* **2009**, *41*, 87–98.
56. Khodary, S. Effect of salicylic acid on the growth, photosynthesis and carbohydrate metabolism in salt-stressed maize plants. *Int. J. Agric. Biol* **2004**, *6*, 5–8.
57. Zhang, Y.F.; Yin, B. Advances in Study of Salt-stress Tolerance in Maize. *J. Maize Sci.* **2008**, *16*, 83–85.
58. Gunes, A.; Inal, A.; Alpaslan, M.; Eraslan, F.; Bagci, E.G.; Cicek, N. Salicylic acid induced changes on some physiological parameters symptomatic for oxidative stress and mineral nutrition in maize (*Zea mays* L.) grown under salinity. *J. Plant Physiol.* **2007**, *164*, 728–736. [CrossRef] [PubMed]
59. Volkmar, K.; Hu, Y.; Steppuhn, H. Physiological responses of plants to salinity: A review. *Can. J. Plant Sci.* **1998**, *78*, 19–27. [CrossRef]
60. Yin, Y.; Zhang, X.; Yu, H.; Lin, D.; Wu, Y.; Wang, J.A. Mapping Drought Risk (Maize) of the World. In *World Atlas of Natural Disaster Risk*; Springer: Berlin/Heidelberg, Germany, 2015; pp. 211–226. [CrossRef]
61. Huang, C.F.; Liu, X.L.; Zhou, G.X.; Li, X.J. Aricultural natural disater risk assessment method according to the historic disaster data. *J. Nat. Disasters* **1998**, *7*, 1–9.
62. Miliaresis, G.C.; Argialas, D. Segmentation of physiographic features from the global digital elevation model/GTOPO30. *Comput. Geosci.* **1999**, *25*, 715–728. [CrossRef]
63. IIASA; FAO. *Global Agro-Ecological Zones–Model Documentation (GAEZ V. 3.0)*; International Institute of Applied Systems Analysis: Vienna, Austria; Food and Agricultural Organization: Rome, Italy, 2012. Available online: https://pure.iiasa.ac.at/id/eprint/13290/1/GAEZ_Model_Documentation.pdf (accessed on 1 September 2022).
64. Warszawski, L.; Frieler, K.; Huber, V.; Piontek, F.; Serdeczny, O.; Schewe, J. The inter-sectoral impact model intercomparison project (ISI–MIP): Project framework. *Proc. Natl. Acad. Sci. USA* **2014**, *111*, 3228–3232. [CrossRef]
65. Vitousek, P.M.; Mooney, H.A.; Lubchenco, J.; Melillo, J.M. Human domination of Earth's ecosystems. *Science* **1997**, *277*, 494–499. [CrossRef]
66. Sacks, W.J.; Deryng, D.; Foley, J.A.; Ramankutty, N. Crop planting dates: An analysis of global patterns. *Glob. Ecol. Biogeogr.* **2010**, *19*, 607–620. [CrossRef]
67. Tan, G.; Shibasaki, R. Global estimation of crop productivity and the impacts of global warming by GIS and EPIC integration. *Ecol. Model.* **2003**, *168*, 357–370. [CrossRef]
68. Mueller, N.D.; Gerber, J.S.; Johnston, M.; Ray, D.K.; Ramankutty, N.; Foley, J.A. Closing yield gaps through nutrient and water management. *Nature* **2012**, *490*, 254. [CrossRef] [PubMed]
69. ESRI.; GRID. Map of Aridity. Available online: https://data.apps.fao.org/catalog/iso/221072ae-2090-48a1-be6f-5a88f061431a (accessed on 1 September 2022).
70. Portal, F.S. Harmonized World Soil Database. Available online: https://www.fao.org/soils-portal/soil-survey/soil-maps-and-databases/harmonized-world-soil-database-v12/zh/ (accessed on 1 September 2022).
71. ISRIC. Global Assessment of Human-Induced Soil Degradation (GLASOD). Available online: https://www.isric.org/projects/global-assessment-human-induced-soil-degradation-glasod (accessed on 1 September 2022).
72. Zhang, X.; Zhao, W.; Liu, Y.; Fang, X.; Feng, Q. The relationships between grasslands and soil moisture on the Loess Plateau of China: A review. *Catena* **2016**, *145*, 56–67. [CrossRef]
73. Wang, X.; Wang, S.; Chen, J.; Cui, H.; Wu, Y.; Ravindranath, N.H.; Rahman, A. Simulating potential yields of Chinese super hybrid rice in Bangladesh, India and Myanmar with EPIC model. *Acta Geogr. Sin.* **2018**, *28*, 1020–1036. [CrossRef]
74. Guo, H.; Wang, R.; Garfin, G.M.; Zhang, A.; Lin, D.; Liang, Q.; Wang, J. Rice drought risk assessment under climate change: Based on physical vulnerability a quantitative assessment method. *Sci. Total Environ.* **2021**, *751*, 141481. [CrossRef] [PubMed]
75. Lian, F. Crop Salinity Vulnerability Curve Construction Based on Epic Model—A Case Study of Global Mazie. Doctoral Dissertation, Beijing Normal University, Beijing, China, 2016; pp. 24–25.
76. Tanji, K.K.; Kielen, N.C. *Agricultural Drainage Water Management in Arid and Semi-Arid Areas*; FAO: Rome, Italy, 2002.
77. Katerji, N.; Van Hoorn, J.; Hamdy, A.; Mastrorilli, M. Salinity effect on crop development and yield, analysis of salt tolerance according to several classification methods. *Agric. Water Manag.* **2003**, *62*, 37–66. [CrossRef]

78. Huang, C.F. Informationmatrix method for risk analysis of natural disaster. *J. Nat. Disasters* **2006**, *15*, 1–10.
79. Schofield, R.V.; Kirkby, M.J. Application of salinization indicators and initial development of potential global soil salinization scenario under climatic change. *Glob. Biogeochem. Cycles* **2003**, *17*, 4-1–4-13. [CrossRef]
80. Yin, Y.; Zhang, X.; Lin, D.; Yu, H.; Wang, J.; Shi, P. GEPIC-V-R model: A GIS-based tool for regional crop drought risk assessment. *Agric. Water Manag.* **2014**, *144*, 107–119. [CrossRef]
81. Guo, H.; Zhang, X.; Lian, F.; Gao, Y.; Lin, D.; Wang, J. Drought Risk Assessment Based on Vulnerability Surfaces: A Case Study of Maize. *Sustainability* **2016**, *8*, 813. [CrossRef]
82. Zhang, X.; Guo, H.; Wang, R.; Lin, D.; Gao, Y.; Lian, F.; Wang, J. Identification of the Most Sensitive Parameters of Winter Wheat on a Global Scale for Use in the EPIC Model. *Agron. J.* **2017**, *109*, 58–70. [CrossRef]

land

Article

# Impact of the Grain for Green Project on the Well-Being of Farmer Households: A Case Study of the Mountainous Areas of Northern Hebei Province, China

Kun Wang [1], Piling Sun [1,2,3,*], Xin Wang [1], Junxiong Mo [1], Nan Li [1] and Jinye Zhang [1]

1 School of Geography and Tourism, Qufu Normal University, Rizhao 276826, China
2 Rizhao Key Laboratory of Territory Spatial Planning and Ecological Construction, Rizhao 276962, China
3 College of Land Science and Technology, China Agriculture University, Beijing 100193, China
* Correspondence: spling86@qfnu.edu.cn; Tel.: +86-187-6332-1108

**Abstract:** There are close dynamic relationships among the livelihood, well-being, and ecological environment of farmer households. It is of great significance to scientifically clarify the impact of the Grain for Green policy on the livelihoods and well-being of farmer households in mountainous areas. Based on data from a survey of 392 farmer households in Zhangbei County, the system of indicators for livelihood assets and well-being of farmer households were constructed using the sustainable livelihood framework (SLF). The livelihood assets and well-being levels of different types of farmer households were measured, and a multiple linear regression model was used to analyze the impact of the Grain for Green policy implementation on the well-being levels of farmer households. The results showed that (1) the Grain for Green project caused changes in the livelihood of farmer households. The average livelihood diversity of farmer households was 3.008, and the returned farmland households (3.022) were higher than the nonreturned farmland households (2.975) in Zhangbei County. The level of natural assets among the total average livelihood assets of farmer households was the highest at 0.374, while the level of physical assets was the lowest at 0.018. The level of livelihood assets of returned farmland households (0.948) was lower than that of nonreturned farmland households (1.117). (2) The Grain for Green policy had an improving effect on the level of well-being of farmer households, but the effect was not significant. The level of well-being of all farmer households in Zhangbei County was 0.517, with the level of wealth contributing the most to the well-being of farmer households at 40.20% and the quality of the ecological environment contributing the least at 11.99%. The level of well-being of returned farmland households (0.518) was slightly higher than that of nonreturned farmland households (0.514). (3) The influencing degree of each factor on the level of well-being varied significantly. There are three main paths through which the Grain for Green policy affects the well-being of farmer households: by reallocating human assets, optimizing natural assets, and enhancing financial assets. The factor of household size had the highest degree, at 0.366, while educational attainment of household members, household labor capacity, annual household expenditure, livelihood diversity, number of large production tools, and total value of livestock were also important drivers of household well-being, and area of arable land was negatively associated with household well-being. There were also differences in the factors influencing the level of well-being of different types of farmer households.

**Keywords:** Grain for Green project; livelihood assets; livelihood diversity; well-being of farmer households; mountainous areas of northern Hebei Province

## 1. Introduction

There are close dynamic relationships among the livelihood, well-being, and ecological environment of farmer households [1]. To relieve regional ecological pressure, improve the ecological environment, and enhance the well-being of farmer households, the Chinese government implemented the Grain for Green policy in 1999, and successively carried out ecological projects such as the construction of the Three Northern Protective Forests System (phases IV and V), the construction of the Beijing–Tianjin Wind and Sand Source Control, the construction of the Yangtze River Protective Forest System (phases II and III), and the treatment of rock desertification in Southwest China [2]. The Grain for Green project is a management mode of ecological restoration by stopping cultivation of sloping land and planting trees and grass to restore vegetation. Participants in the Grain for Green policy need to systematically stop cultivating sloping land that is prone to soil erosion. Participating farmer households need to plant trees and grasses on these lands to restore vegetation according to local conditions, and they can be subsidized by the government. In the past 20 years, China had implemented reforestation and grass restoration of up to $3.43 \times 10^5$ km$^2$ and contributed more than 4% to the global greening area in the same period, which had significantly improved the ecological environment and ecosystem services quality. The changes in ecosystems affect the survival, livelihood development and the well-being of farmer households, either directly or indirectly. Ecosystems protection, livelihood development, and the improvement of well-being of farmer households are the core components of achieving the United Nations (UN) 2030 Sustainable Development Goals [3]. In the vast fragile mountainous areas of China, the contradictions between ecological environmental protection, livelihood development of farmer households, and well-being of farmer households are prominent. How to protect the ecological environment to improve the livelihood and well-being of farmer households in ecologically fragile areas is an urgent issue. Therefore, it is important to scientifically clarify the impacts of Grain for Green policy on the livelihoods and well-being of farmer households in mountainous areas.

Studies outside of China paid less attention to ecological projects, especially in terms of their relationships with livelihoods and well-being of farmer households. However, methods for measuring and assessing ecosystem services, livelihoods, and well-being of farmer households are all relatively well established. Ecosystem services can be used to measure the effectiveness of ecological policies, and the value and physical quantity of ecosystem services can be estimated through value-equivalent scales and the Integrated Valuation of Ecosystem Services and Tradeoffs (InVEST) model [4,5]. The livelihoods of farmer households are measured by drawing on the sustainable livelihood framework, and the human development index; life satisfaction, and well-being index frameworks are used to assess well-being of farmer households [6–8]. Existing research had also been performed on the relationships between ecosystem services and well-being of farmer households, which had been measured with the coupled coordination model, elasticity coefficient, and bivariate spatial autocorrelation, with different results. Ciftcioglu et al. [9] believed that the impact of ecosystem services on the satisfaction of material needs was significant, and ecosystem services appeared to be a significant positive influence on the well-being of farmer households. Jones et al. [10] pointed out that ecosystem services had a negative influence on human health and physical well-being. Although there are few foreign related studies, they provide support for this study in terms of theory and methodology.

Many scholars have studied the relationships among ecological conservation, livelihoods of farmer households, and well-being of farmer households in China, focusing on the relationship between livelihoods of farmer households and ecosystem services [11–13], the relationship between ecosystem services and well-being of farmer households [3,14–17], and the relationship between livelihoods and well-being of farmer households [18]. Recently, ecological protection and high-quality development have become the focus of efforts in China. With the implementation of ecological civilization and rural revitalization strategies, research on the relationship among ecosystem services, livelihoods of farmer households, and well-being of farmer households has become a hot academic issue. Based on the ecosystem services availability assessment, Liu et al. [19] believe that ecosystem services have a significant positive impact on the well-being of farmer households. Based on the theory of ecosystem services value, scholars have identified spatial and temporal variations in per capita ecological well-being and economic efficiency in Chinese cities at the prefecture level and above and have elucidated patterns of different types of ecological well-being [20].

A study of the Yellow River headwaters showed that ecological compensation had an important impact on the well-being of farmer households, which could compensate for the loss of well-being to stimulate conservation and improve the well-being of farmer households [21]. Therefore, reasons, implications, and pathways of ecological compensation mechanisms were proposed from the perspective of resource opportunity–cost [22]. Moreover, coupling relationships between ecosystem service functions and well-being of farmer households were found in the areas with relatively poor economic development [23,24]. Concurrently, research on the impact of the Grain for Green policy on the well-being of farmer households is emerging. You et al. [25] examined the impact of the Grain for Green policy on the well-being of farmer households and the direction of optimizing ecological compensation. Liu et al. [26] explored the impact of the Grain for Green policy in the Loess Plateau region on changes in ecosystem services and well-being of farmer households. Through a comparative analysis of farm household panel data, Yao et al. [27] found that there were differences in the impact of the Grain for Green policy on the well-being of farmer households in the Yellow River Basin and Yangtze River Basin of China. Overall, the existing studies focused on the contribution of ecosystem services to human well-being, and the dependence of human well-being on ecosystem services, as well as the coupling relationship between ecosystem service and human well-being. However, the impact of Grain for Green project on the well-being of farmer households is still lacking.

The mountainous areas of northern Hebei Province are a typical ecologically fragile area and agro-pastoral ecotone in China, with a single type of livelihood and a low level of well-being for farmer households. Many ecological projects exist in this region, such as the Three Northern Protective Forests System, the Taihang Mountains Greening project, and the Beijing–Tianjin Wind and Sand Source Control. Grain for Green is an important means of implementing these ecological projects. The Grain for Green project has a significant impact on land holdings of farmer households and regional vegetation, which in turn affects their livelihoods and well-being. The objectives of this study include the following main aspects: (1) to measure the livelihood assets and well-being levels of farmer households in the mountainous areas of northern Hebei; (2) to explore the influencing factors on the well-being of farm households; and (3) to reveal the impact paths of the Grain for Green project on the well-being of farmer households.

## 2. Materials and Methods

### 2.1. Study Area

Zhangbei County is located in the northwestern part of Zhangjiakou City, at the transition zone between the Inner Mongolia Plateau and the North China Plain. The geographical location is 40°57′–41°34′ N and 114°10′–115°27′ E. It covers 4185 km$^2$, including 18 towns (7 towns and 11 townships). The elevation ranges between 1300 m and 2128 m, and the terrain is complex and diverse, including plateaus, hills, mountains, and basins (Figure 1). Agriculture and animal husbandry are intertwined within the county, and the ecological environment is sensitive and fragile. Due to the continental monsoon climate of the middle latitude temperate zone, the average annual temperature of the area is only 3.2 °C and the average annual precipitation is only 392.7 mm, which is suitable for staggered seasonal vegetable cultivation. By the end of 2021, the total population of Zhangbei County was 0.356 million. Zhangbei County is a component of an important green ecological barrier in northern China, responsible for building a strong ecological barrier in the Beijing–Tianjin–Hebei region. The Grain for Green project was piloted since 2000 and fully launched in 2002. The county completed the Grain for Green project for 625.07 km$^2$, of which 320.87 km$^2$ were returned to forest and 304.20 km$^2$ were afforested on barren hills. The policy involves 18 townships and more than 60,000 farmer households in the county. In the past 20 years, the vegetation coverage rate in Zhangbei County increased significantly, and the ecological environment continually improved, which affected the livelihoods and well-being of farmer households. Given the large-scale and significant Grain for Green policy achievements in Zhangbei County, a survey was conducted on the impacts of the policy on the livelihoods and well-being of farmer households in some townships of Zhangbei County.

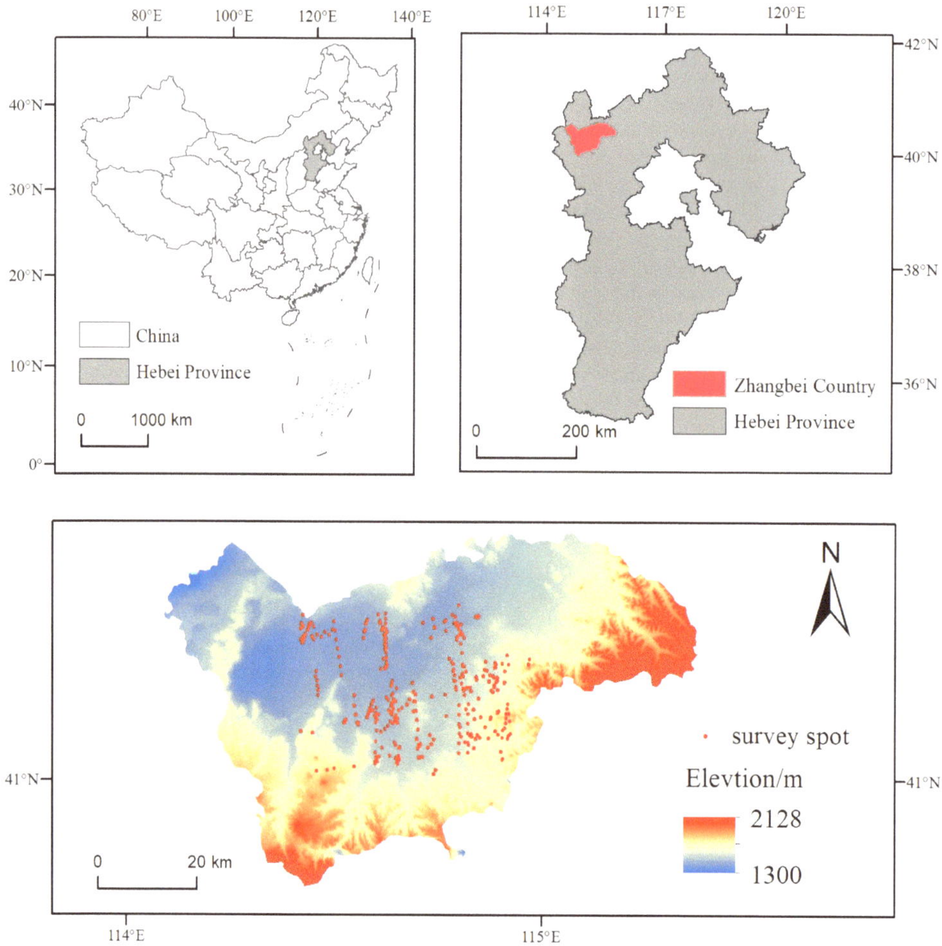


**Figure 1.** Location of the study area.

## 2.2. Data Sources

Based on participatory rural appraisal (PRA), the data used in this study were obtained through a questionnaire and semi-structured interviews with farmer households. Based on the physical geography, accessibility, and implementation of the Grain for Green policy in Zhangbei County, 392 farmer households in 26 villages in 8 townships were selected for research interviews (Figure 1). The survey used random sampling in the policy coverage area to ensure the randomness of the selected farmer households. Based on the farmers' own evaluation, analysis, implementation, supervision, and assessment of the Grain for Green policy status, an in-depth understanding of the policy and the living and production environment of the farmers in Zhangbei County was conducted. A total of 398 questionnaires were distributed in this survey, and 6 invalid questionnaires were eliminated resulting in a total of 392 valid questionnaires involving 1173 people. The questionnaire efficiency was 98.5%. The average survey time was between 25 and 30 min per farmer household. A random sampling method was used to select the survey sample, and the

392 sample households selected were to some extent able to reflect the basic characteristics of the farmer households in the study area. The survey results from these number of questionnaires also have a strong explanatory power.

Since the implementation of the Grain for Green policy, Zhangbei County successively returned farmland to forest and grass on sloping farmland above 25°, on sloping farmland with important water sources between 15° and 25°, and on sandy farmland. Farmer households without the above-mentioned types of arable land were not able to participate in the Grain for Green policy. There are also some farmer households whose households are dominated by women, who had no other source of livelihood and chose not to participate in the policy. Therefore, in this study, the surveyed households are divided into returned farmland households and nonreturned farmland households based on whether they participated in the policy. Among the surveyed farmer households, there are 273 returned farmland households, accounting for 69.6% of the total valid sample households and 119 nonreturned farmland households, accounting for 30.4% of the total valid sample households.

The survey covered the basic information of farmer households, the basic information of land, the status of livelihood assets, and the farmer households' response to the Grain for Green policy. The basic information of farmer households included number of family members, age, education level, nature of employment, working hours, and salary income. The basic land information included land type, crop planting situation, the status of farmer households' land returned to forest and grass, and land input status. The status of livelihood assets included human assets, natural assets, physical assets, social assets, and financial assets. The farmer households' response to the Grain for Green policy included knowledge and judgment of the policy, willingness to participate in the policy, behavioral choice concerning the policy, and expectations of the policy.

### 2.3. *Methods*

#### 2.3.1. Theoretical Analysis Framework

The implementation of the Grain for Green project inevitably leads to changes in the livelihoods and land-use behavior of farmer households, thus affecting their well-being. This study established a theoretical analysis framework along the lines of "Grain for Green → livelihood characteristics of farmer households → well-being of farmer households → analysis of factors influencing well-being of farmer households → policy recommendations" (Figure 2). Since livelihood assets do not fully reflect the livelihood characteristics of farm households, this study introduced livelihood diversity. This study quantitatively assessed the livelihood diversity of farmer households through survey results; based on SLF, a system of indicators for livelihood assets of farmer households was constructed from five aspects: human assets, natural assets, physical assets, social assets, and financial assets, to quantitatively assess the status of livelihood assets of farmer households after the implementation of the Grain for Green policy. Based on the conceptual framework of human well-being, a system of indicators for the well-being of farmer households was constructed to quantitatively assess the level of farmer households' well-being after fallowing from four aspects: labor force conditions, wealth level, ecological environment quality, and social conditions. A multiple linear regression stepwise analysis model was established to quantitatively assess the factors influencing the well-being of farmer households, and to explore how the Grain for Green project affects well-being of farmer households by acting on livelihood diversity and livelihood assets. Based on the relevant conclusions, relevant policy recommendations are put forward to provide a scientific basis for the government to formulate relevant policies.

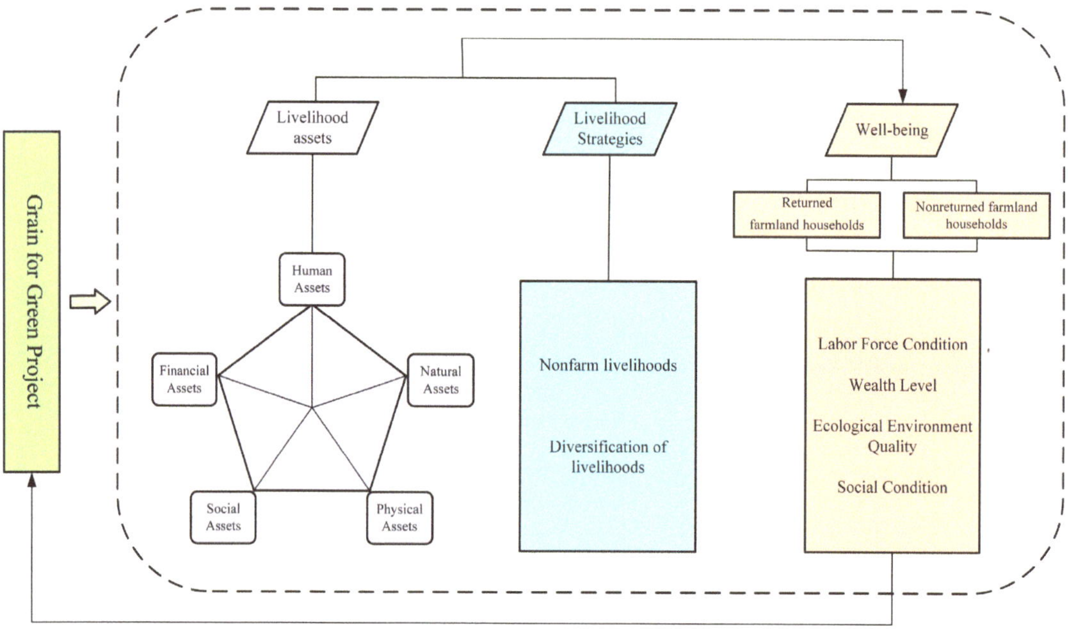


**Figure 2.** Analysis framework.

### 2.3.2. Measurement of Livelihood Diversity

According to the survey results, livelihood diversity is increased by one unit whenever a farmer household engages in a certain livelihood activity, until the total number of types of livelihood activities engaged in by the household is reached. The calculation formula is as follows:

$$N = \sum \beta_i \quad (1)$$

where $N$ is farmer household livelihood diversity and $\beta_i$ is farmer household livelihood type.

### 2.3.3. Measurement of Livelihood Assets and Well-Being

1. Indicator Selection

Referring to the sustainable livelihood framework of the United Kingdom Department for International Development [28,29], the evaluation system for livelihood assets of farmer households in Zhangbei County was constructed from five dimensions, including human, natural, physical, social, and financial assets (Table 1), with corresponding adjustments to the actual situation in the study area. The implementation of the Grain for Green policy may lead to changes in the local ecological environment, so the livelihood assets of farmer households need to include the ecological environment of the area they live in. The indicator of the area fallowed to farming was used to measure this livelihood asset. The farmers of China tend to go out to work during the agricultural leisure season, so the indicator of the number of channels for outworking was included in social assets. Zhangbei County is located in an agro-pastoral ecotone, and livestock farming is an important means of livelihood for farmer households in this area and an important way to reserve wealth, so the total value of livestock was included in financial assets. The well-being of farmer households involves production, living, health, and safety conditions of rural farmer households, concentrating on the improvement of living standards, production conditions, infrastructure, and social security. This method was based on the conceptual framework of human well-being and its multidimensional and hierarchical structure, following the principles of scientificity, validity, and hierarchy.

The system of indicators for well-being of farmer households was constructed from the four dimensions of household characteristics, wealth level, ecological environment quality, and social conditions (Table 2).

**Table 1.** Indicators of livelihood assets of farmer households.

| Asset Type | Indicator | Symbol | Indicator Meaning and Value | Nature | Weight |
|---|---|---|---|---|---|
| Human assets (HA) | Household size | H1 | Total household size | Positive | 0.010 |
| | Educational attainment of household members | H2 | No formal education (little literacy) = 0; elementary school = 1; junior high school = 2; high school, junior college = 3; college, senior college = 4; university undergraduate and above = 5 | Positive | 0.042 |
| | Household labor capacity | H3 | Assigned according to age, where 6 and below = 0; 7–18 = 1; 18–25 = 2; 26–45 = 5; 46–60 = 4; 60 and above (including military/students aged 19–60) = 3 | Positive | 0.016 |
| Natural assets (NA) | Area of arable land | N1 | Household arable land area | Positive | 0.045 |
| | Area of watered land | N2 | Area of household watered land | Positive | 0.328 |
| | Area fallowed to farming | N3 | Area of land fallowed to farming by household | Negative | 0.001 |
| Physical assets (PA) | Residential index | P1 | Residential index = $\alpha M + \beta C + \gamma N$, where $M$, $C$, and $N$ denote the number of houses, house structure, and residential age, respectively; $\alpha$, $\beta$, and $\gamma$ are the weights of the three, respectively, using the entropy value method to calculate $\alpha = 0.3949$, $\beta = 0.4023$, and $\gamma = 0.2028$. House structure: civil structure = 1, brick structure = 2, brick and mixed structure = 4, others: such as brick = 3, relief = 0. If the house consists of different structures, the weighted summation is calculated in proportion to the number of rooms | Positive | 0.005 |
| | Number of large production tools | P2 | Number of asset types owned by farmer households | Positive | 0.013 |
| Social assets (SA) | Number of public officials among relatives | S1 | Number of public officials among relatives | Positive | 0.290 |
| | Number of channels for outworking | S2 | Number of types of channels for outworking | Positive | 0.024 |
| | Number of social contacts | S3 | Assigned according to the number of cell phone contacts, no cell phone = 0, 0–20 people = 1, 21–50 people = 2, 51–100 people = 3, 101 and above = 4 | Positive | 0.023 |
| Financial assets (FA) | Total value of livestock | F1 | Total value of livestock = number of livestock × unit price of livestock (differentiate between young and adult livestock, unit price obtained from research data) | Positive | 0.138 |
| | Annual household income | F2 | The sum of farm income, wage income, financial income (direct grain subsidy, old-age insurance, low-income insurance, social security), subsidies for retired farming and other income (medicine collection, etc.) in one year | Positive | 0.063 |
| | Annual household expenditure | F3 | Sum of children's school fees, food consumption, gift expenditure, health expenditure, consumption of durable goods, and consumption of daily necessities, etc., in one year | Negative | 0.001 |

**Table 2.** Indicators of well-being levels of farmer households.

| Well-Being Dimension | Indicator | Symbol | Indicator Assignment | Nature | Weight |
|---|---|---|---|---|---|
| Labor force condition (FL) | Household size | F1 | Number of farmer household members | Positive | 0.052 |
| | Household labor capacity | F2 | Assignment according to age, where 6 years and below = 0; 7–18 years = 1; 18–25 years = 2; 26–45 years = 5; 46–60 years = 4; 60 years and above (including military/students aged between 19 and 60) = 3 | Positive | 0.059 |
| | Educational attainment of labor force | F3 | No formal education (little literacy) = 0; elementary school = 1; junior high school = 2; high school, junior college = 3; college or senior college = 4; university undergraduate and above = 5 | Positive | 0.092 |
| | Number of people engaged in non-cultivation agriculture | F4 | Assigned according to the nature of employment, where farming = 0; farming + other employment = 1; other employment = 2 | Positive | 0.112 |
| Wealth level (WL) | Annual household income | T1 | Sum of agricultural income, wage income, financial income (direct food subsidy, old-age insurance, low-income insurance, social security), fallowing subsidy, and other income (medicine picking, etc.) in one year | Positive | 0.152 |
| | Annual household expenditure | T2 | Sum of school fees of children, food consumption, gift expenditure, health expenditure, consumption of durable goods, consumption of daily necessities, etc., in one year | Negative | 0.102 |
| | Arable land per capita | T3 | Arable land/total population | Positive | 0.109 |
| | Housing area per capita | T4 | Average housing area per capita | Positive | 0.028 |
| Ecological environment quality (EQ) | Water safety | Z1 | Assigned according to farmers' perception of changes in water pollution after fallowing, strong = 0; constant = 1; diminished = 2 | Positive | 0.039 |
| | Air safety | Z2 | Assigned according to farmers' perception of whether air quality has improved after fallowing, worse = 0; no change = 1; better = 2 | Positive | 0.020 |
| | Soil and water conservation | Z3 | Values Assigned according to whether farmers' soil erosion has improved after fallowing, severe = 0; no change = 1; improved = 2 | Positive | 0.021 |
| | Soil safety | Z4 | Assigned according to whether the farmer is serious about soil contamination after fallowing, serious = 0; not serious = 1 | Positive | 0.011 |
| Social conditions (SC) | Traffic accessibility | S1 | Distance of farmer households to the nearest road | Negative | 0.141 |
| | Resource accessibility | S2 | Distance of farmer households to the nearest hospital or school | Negative | 0.054 |
| | Policy satisfaction | S3 | Assignment of values according to whether farmers are willing to participate in the Grain for Green policy, indifferent = 0; very unwilling = 1; not very willing = 2; average = 3; very willing = 4 | Positive | 0.011 |

2. Indicator Weight Determination

The nature of the indicators is divided according to their attributes of the selected indicators and the assigned values. Positive indicators are those that represent upward or forward progress, and the higher the value of these indicators the better the evaluation, while the opposite is true for negative indicators. To facilitate the calculation and make the indicators' trend the same, the negative indicators are converted to positive. Since the indicators have different dimensions, the study adopts the standardization of the mean value to process the indicators independent of the dimension. The entropy value method was used to determine the index weights [30]. The calculation formula is as follows:

$$Y_{ij} = \frac{y_{ij}}{\sum_{i=1}^{m} y_{ij}} \tag{2}$$

$$e_j = -\frac{1}{ln\ m} \sum_{i=1}^{m} Y_{ij} ln\ Y_{ij} \tag{3}$$

$$d_j = 1 - e_j \tag{4}$$

$$w_j = \frac{d_j}{\sum_{j=1}^{n} d_j} \tag{5}$$

where $y_{ij}$ is the value of the $j$th indicator of the $i$th statistical unit after dimensionless processing, $e_j$ is the entropy value of the $j$th indicator, $m$ is the number of samples, $d_j$ denotes the coefficient of variability of the $j$th indicator, and $w_j$ is the weight of the $j$th indicator.

3. Measurement of Indicators

The livelihood assets and well-being level of farmer households was calculated using the linear weighted summation. The calculation formula is as follows:

$$F = \sum W_j Y_{ij} \tag{6}$$

where $F$ is the total value of livelihood assets or the well-being level of farmer households, $W_j$ is the weight of the livelihood asset index of the $j$th item or the well-being index of the $j$th item, and $Y_{ij}$ is the standardized value of the $j$th item in the $i$th statistical unit.

#### 2.3.4. Multiple Linear Regression

The well-being of farmer households is often the result of multiple factors, rather than a single factor [31]. The well-being level of farmer households is influenced by numerous factors, such as livelihood diversity and livelihood assets of farmer households. This study attempted to find the optimal combination of each indicator of livelihood of farmer households to explain the level of well-being of farmer households. Therefore, we chose multiple linear regression to solve this problem. In this study, we constructed a multiple linear regression model with 14 indicators of livelihood assets of farmer households and livelihood diversity as independent variables and the level of well-being of farmer households as dependent variable, as a way to quantitatively identify specific factors affecting the well-being level of farmer households. The regression model is as follows:

$$Y = B_0 + B_1 X_1 + B_2 X_2 + \cdots + B_i X_i + \varepsilon \tag{7}$$

where $Y$ represents the welfare level of farmer households, $X_i$ ($i$ = 1, 2, 3, ... ) represents the relevant indicators, $B_0$ represents the regression constant, $B_i$ represents the regression coefficient, and $\varepsilon$ represents a random error term.

## 3. Results

### 3.1. Analysis of the Livelihood Characteristics of Farmer Households

Based on the interpretation of the sustainable livelihood framework, the livelihood status of farmer households depends on the comprehensive role of various livelihood assets, and their livelihood goals are also achieved on this basis [32,33]. However, livelihood assets cannot indicate the choice of farmer households of livelihood activities, so the characteristics of livelihoods of farmer households include two aspects: livelihood assets and livelihood diversity.

#### 3.1.1. Livelihood Diversity Characteristics of Farmer Households

Farmer households are rational economic agents that can choose different livelihood activities according to their available livelihood assets. Farmer households can maintain their status by diversifying their livelihood activities through different asset allocations to cope with risks and shocks [34]. After conducting a survey of 392 farmer households, three distinct types of farming livelihoods were identified: food crops, cash crops, and vegetable crops. In addition, nonfarming livelihoods were also present, including animal husbandry, forestry, part-time work, self-employment, permanent work, and freelance work.

The statistics show that the average livelihood diversity of the returned farmland households is higher than that of the nonreturned farmland households (Figure 3), which is because returned farmland households have shifted to other livelihood activities after retiring from farming. However, the difference in average livelihood diversity between the two types of farmer households is not significant, probably because most nonreturned households are engaged in more than one type of cultivation.

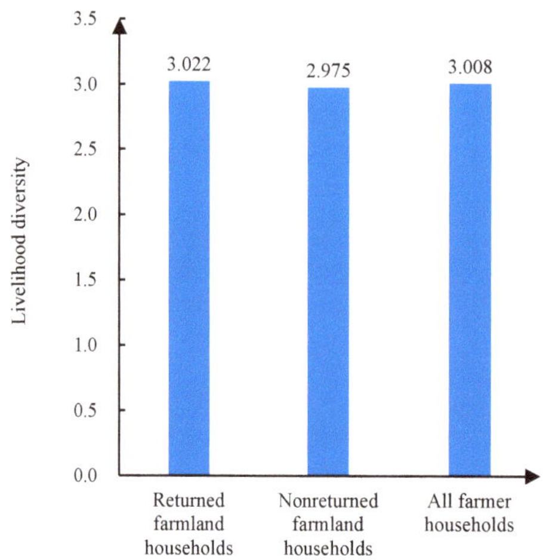


**Figure 3.** Average livelihood diversity of different types of farmer households.

#### 3.1.2. Livelihood Asset Characteristics of Farmer Households

Zhangbei County is located in the Bashang plateau area, and the landscape can be divided into the dam-head mountainous area, the hilly area, and the undulating plateau area [35]. The special topography and climatic conditions have led to ecological fragility and frequent natural disasters in the area. As a result, in general, Zhangbei farmer households have low levels of livelihood assets and live in poverty.

The order of average livelihood asset levels of farmer households in Zhangbei County is natural assets > social assets > financial assets > human assets > physical assets. A comparison of returned farmland and nonreturned farmland households reveals that the average total asset levels of returned farmland households are lower than those of nonreturned farmland households. Among them, only the financial assets of the returned farmland households are higher than those of the nonreturned farmland households, while the rest of the assets are lower than those of the nonreturned farmland households. The level of natural assets in Zhangbei County is 0.374 (Figure 4), contributing the most to the total assets, at 37.43% (Figure 5). In terms of the contribution of each indicator to natural assets, the area of watered land contributes the most, at 87.69% (Table 3). In comparing returned farmland and nonreturned farmland households, the level of natural assets of nonreturned farmland households is greater than that of returned farmland households. The human assets per capita of all farmer households in Zhangbei County are 0.068, contributing 6.77% to the total assets (Table 3). The education attainment of household members contributes the most to the level of human assets, at 61.69%, while the size of the household contributes the least, at 14.76%. The level of human assets of nonreturned farmland households is greater than that of returned farmland households. The level of physical assets in Zhangbei County is 0.018, contributing 1.83% to the total assets and accounting for the lowest proportion. The number of large production tools and the residential index contribute 72.96% and 27.04%, respectively. The physical assets of returned farmland households are slightly lower than those of nonreturned farmland households, with the contribution of each indicator following the trend of total farmer households. The level of social assets is 0.337, which is the second highest contribution to total assets, at 33.75%. The number of public officials among relatives contributes the most to the social assets, at 86.09%. The social assets of returned farmland households (0.328) are lower than those of nonreturned farmland households (0.359). The average level of the financial assets of farmer households in Zhangbei County is 0.202, contributing 20.21% to the total assets. The value of livestock makes the largest contribution to the financial assets, exceeding 2/3 of the total financial assets.

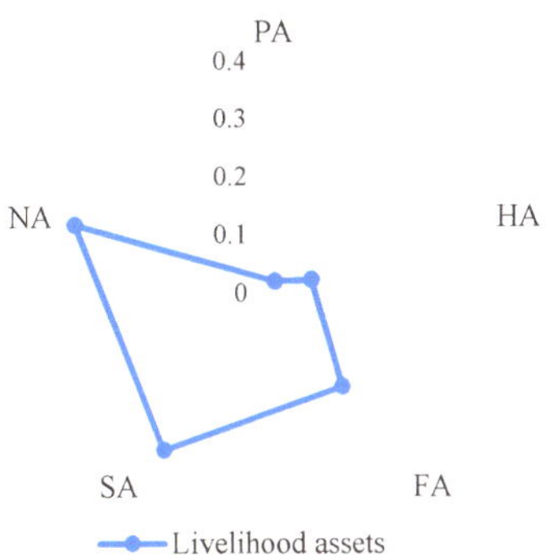


**Figure 4.** Livelihood assets of farmer households.

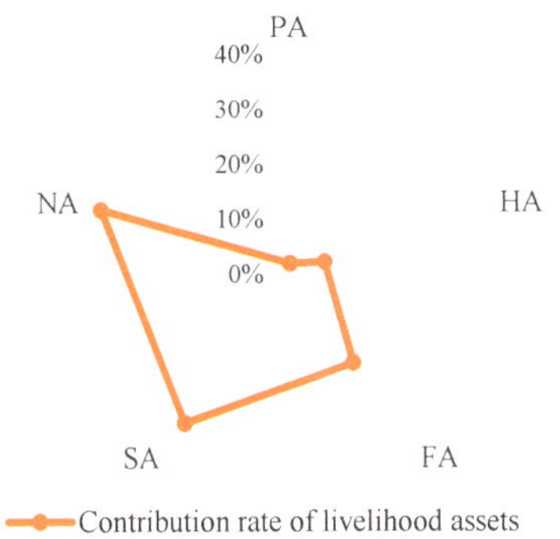


**Figure 5.** Contribution of livelihood assets.

**Table 3.** Level and contribution of average livelihood assets of farmer households.

| Livelihood Assets | | Returned Farmland Households | | Nonreturned Farmland Households | | All Farmer Households | |
|---|---|---|---|---|---|---|---|
| | | Asset Level | Contribution Rate % | Asset level | Contribution Rate % | Asset Level | Contribution Rate % |
| Human assets | Household size | 0.010 | 14.97 | 0.010 | 14.30 | 0.010 | 14.76 |
| | Educational attainment of household members | 0.041 | 61.43 | 0.044 | 62.26 | 0.042 | 61.69 |
| | Household labor capacity | 0.016 | 23.60 | 0.017 | 23.43 | 0.016 | 23.55 |
| Natural assets | Area of arable land | 0.044 | 13.38 | 0.046 | 9.61 | 0.045 | 11.92 |
| | Area of watered land | 0.285 | 86.21 | 0.428 | 90.06 | 0.328 | 87.69 |
| | Area fallowed to farming | 0.001 | 0.41 | 0.002 | 0.34 | 0.001 | 0.38 |
| Physical assets | Residential index | 0.005 | 26.54 | 0.005 | 28.11 | 0.005 | 27.04 |
| | Number of large production tools | 0.013 | 73.46 | 0.014 | 71.90 | 0.013 | 72.96 |
| Social assets | Number of public officials among relatives | 0.280 | 85.30 | 0.315 | 87.74 | 0.290 | 86.09 |
| | Number of channels for outworking | 0.025 | 7.48 | 0.022 | 5.99 | 0.024 | 7.00 |
| | Number of social contacts | 0.024 | 7.22 | 0.023 | 6.27 | 0.023 | 6.91 |
| Financial assets | Total value of livestock | 0.143 | 69.31 | 0.127 | 65.69 | 0.138 | 68.26 |
| | Annual household income | 0.062 | 30.30 | 0.066 | 33.90 | 0.063 | 31.34 |
| | Annual household expenditure | 0.001 | 0.39 | 0.001 | 0.41 | 0.001 | 0.39 |

### *3.2. Analysis of the Level of Well-Being of Farmer Households*

The level of well-being of all farmer households in Zhangbei County is 0.517; that of returned farmland households is 0.518 and that of nonreturned farmland households is 0.514. The contribution of each indicator to the well-being of farmer households shows that the wealth level contributes the most, at 40.20%, while the ecological environment quality contributes the least, at 11.99% (Figure 6). A comparison of returned farmland households with nonreturned farmland households reveals that the labor force condition of nonreturned farmland households is greater than those of returned farmland households, because older people in general are less able to work and are more inclined to return farmland.

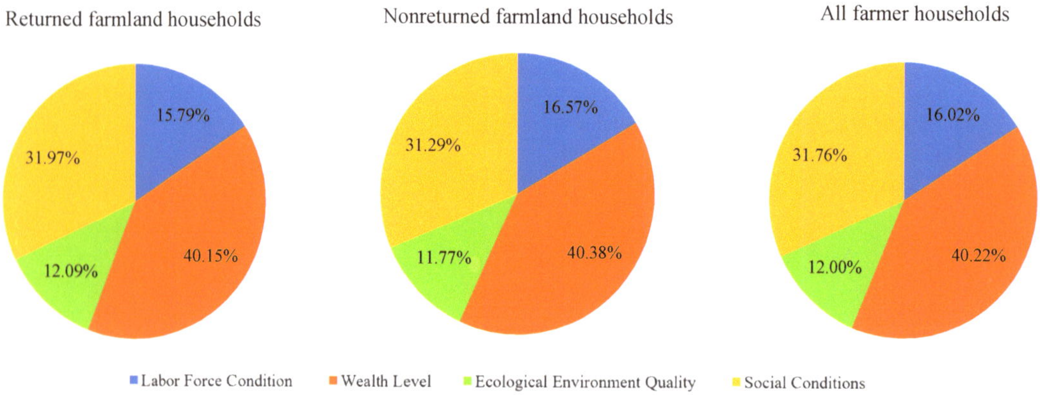


**Figure 6.** Contribution rates to well-being of farmer households.

As can be understood from Table 4, the number of people engaged in non-cultivation agriculture plays a vital role in labor force conditions, contributing the most, while the household size contributes the least. Comparing returned farmland households with nonreturned farmland households, the labor force conditions in nonreturned farmland households contribute more to the level of well-being than in returned farmland households. In terms of the contribution of each indicator to the wealth level, the arable land per capita contributes the most, with a contribution of 48.65%, followed by annual household expenditure. The wealth level of nonreturned farmland households is greater than that of returned farmland households. Comparing the contribution of each factor to the ecological environment quality, in descending order these are water safety > air safety > soil and water conservation > soil safety. The overall contribution of the ecological environment quality in returned farmland households is slightly higher than that in nonreturned farmland households. Social conditions make the second highest contribution to the level of well-being of farmer households. Traffic accessibility makes the largest contribution to social conditions at 70.73%, significantly higher than the other two indicators. The level of social conditions is higher for returned farmland households than for nonreturned farmland households.

**Table 4.** Levels and contributions of each indicator of well-being per farmer household.

| | Indicator | Returned Farmland Households | | Nonreturned Farmland Households | | All Farmer Households | |
|---|---|---|---|---|---|---|---|
| | | **Well-Being Level** | **Contribution Rate %** | **Well-Being Level** | **Contribution Rate %** | **Well-Being Level** | **Contribution Rate %** |
| Labor force condition | Household size | 0.014 | 17.51 | 0.016 | 18.36 | 0.015 | 17.78 |
| | Household labor capacity | 0.017 | 20.78 | 0.018 | 21.49 | 0.017 | 21.00 |
| | Educational attainment of labor force | 0.023 | 27.61 | 0.024 | 28.11 | 0.023 | 27.76 |
| | Number of people engaged in non-cultivation agriculture | 0.028 | 34.10 | 0.027 | 32.04 | 0.028 | 33.46 |
| Wealth level | Annual household income | 0.008 | 3.62 | 0.007 | 3.55 | 0.008 | 3.60 |
| | Annual household expenditure | 0.091 | 43.48 | 0.089 | 43.36 | 0.090 | 43.44 |
| | Arable land per capita | 0.102 | 48.78 | 0.100 | 48.36 | 0.101 | 48.65 |
| | Housing area per capita | 0.009 | 4.12 | 0.010 | 4.73 | 0.009 | 4.30 |
| Ecological environment quality | Water safety | 0.022 | 34.66 | 0.020 | 33.68 | 0.021 | 34.37 |
| | Air safety | 0.018 | 28.28 | 0.017 | 27.87 | 0.018 | 28.16 |
| | Soil and water conservation | 0.013 | 20.01 | 0.013 | 20.73 | 0.013 | 20.22 |
| | Soil safety | 0.011 | 17.04 | 0.011 | 17.72 | 0.011 | 17.24 |
| Social conditions | Traffic accessibility | 0.117 | 70.39 | 0.114 | 71.53 | 0.116 | 70.73 |
| | Resource accessibility | 0.039 | 23.71 | 0.036 | 22.40 | 0.038 | 23.33 |
| | Policy satisfaction | 0.010 | 5.89 | 0.010 | 6.07 | 0.010 | 5.95 |

### 3.3. Impact of Livelihoods Assets on the Well-Being of Farmer Households

The well-being level of farmer households is influenced by a variety of factors, of which the livelihood assets of farmer households are often the basis. Figure 7 shows that there is a significant positive relationship between livelihood assets of farmer households and the level of well-being. The negative value of livelihood assets of farmer households in the figure is caused by logarithmic data of livelihood assets of farmer households. The purpose of using logarithms is to make the normal relationship between the distribution of livelihood assets data of farmer households more apparent, as well as to reduce the absolute value of the data, and does not affect the scientific validity of the final conclusions. The linear regression analysis of livelihood assets and well-being levels passed the significance test at the 1% level, which further suggests that higher livelihood assets of farmer households are associated with higher levels of well-being of farmer households.

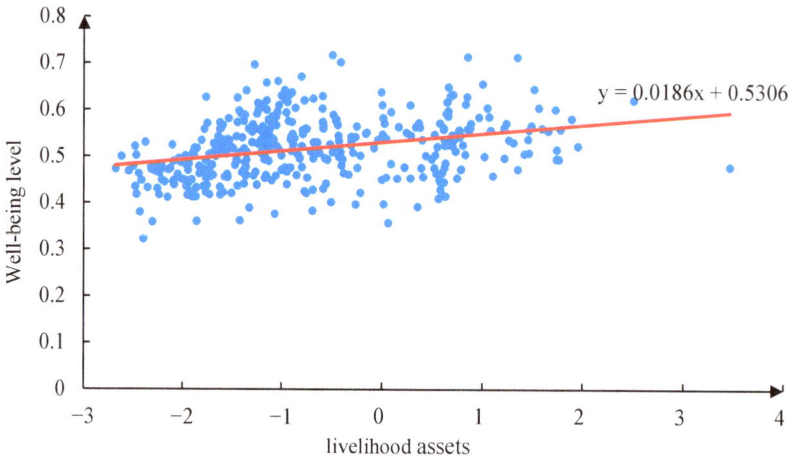


**Figure 7.** Scatter plot of correlation between livelihood assets and well-being levels of farmer households. Note: The purpose of using the logarithm of the dependent variable is to narrow its range of values and highlight the correlation between the variables.

Because there is a significant correlation between the level of farmer household well-being and livelihood assets, this study used the value of farmer household well-being in Zhangbei County as the dependent variable; the indicators of livelihood diversity and livelihood assets are the independent variables. Using SPSS 26, a multiple linear regression model was applied to analyze the impact of the livelihoods of farmer households on the well-being of farmer households under the effect of the Grain for Green policy, and to identify the factors influencing the well-being of farmer households (Table 5).

**Table 5.** Factors influencing well-being of farmer households.

| Impact Factor | All Farmer Households | | Returned Farmland Households | | Nonreturned Farmland Households | |
|---|---|---|---|---|---|---|
| | Standard Coefficient | Sig. | Standard Coefficient | Sig. | Standard Coefficient | Sig. |
| Household size | 0.366 *** | 0.000 | 0.345 *** | 0.000 | 0.392 *** | 0.000 |
| Educational attainment of household members | 0.323 *** | 0.000 | 0.305 *** | 0.000 | 0.415 *** | 0.000 |
| Household labor capacity | 0.166 *** | 0.010 | 0.203 ** | 0.015 | 0.156 | 0.128 |
| Area of arable land | −0.208 *** | 0.000 | −0.19 *** | 0.000 | −0.199 *** | 0.000 |
| Area of watered land | 0.034 | 0.343 | 0.042 | 0.380 | −0.012 | 0.813 |
| Area fallowed to farming | −0.048 | 0.097 | −0.035 | 0.317 | — | — |
| Residential index | 0.015 | 0.603 | 0.053 | 0.145 | −0.044 | 0.379 |
| Number of large production tools | 0.086 *** | 0.005 | 0.035 | 0.376 | 0.116 ** | 0.021 |
| Number of public officials among relatives | 0.026 | 0.395 | 0.017 | 0.659 | −0.008 | 0.867 |
| Number of channels for outworking | 0.009 | 0.767 | 0.014 | 0.688 | −0.030 | 0.540 |
| Number of social contacts | −0.005 | 0.862 | −0.056 | 0.146 | 0.067 | 0.180 |
| Total value of livestock | 0.064 ** | 0.034 | 0.076 ** | 0.034 | 0.081 | 0.106 |
| Annual household income | 0.055 | 0.091 | 0.081 ** | 0.034 | 0.056 | 0.344 |
| Annual household expenditure | 0.157 *** | 0.000 | 0.222 *** | 0.000 | 0.071 | 0.179 |
| Livelihood diversity | 0.107 *** | 0.003 | 0.091 ** | 0.047 | 0.149 *** | 0.009 |
| Constant | — | 0.000 | — | 0.000 | — | 0.000 |

Note: "***" and "**" indicate significance at the 1% and 5% levels, respectively.

#### 3.3.1. Analysis of Factors Influencing the Well-Being of all Farmer Households

The results show an adjusted $R^2 = 0.679$, indicating that the model has a high degree of model fit. Eight variables had significant effects on the level of well-being of farmer households in terms of livelihood diversity and livelihood assets of farmer households. Household size, education attainment of household members, area of arable land, annual household expenditure, number of large production tools, livelihood diversity, household labor capacity, and total value of livestock are the key factors influencing the well-being of farmer households in Zhangbei County because they all passed the significance test at the 1% or 5% level.

4. Livelihood diversity affects the well-being of farmer households

Table 5 shows that the coefficient of 0.107 for the effect of livelihood diversity on the well-being of farmer households passed the significance test at the 1% level, indicating that the higher the diversity of livelihoods of farmer households, the higher the level of well-being of farmer households. The Grain for Green policy has led farmer households to shift from farming to other livelihood activities, enriching their livelihood diversity, expanding their choice of livelihood strategies, and enhancing their defense and coping capacities against external risks, thus increasing their well-being.

5. Human assets drive the well-being of farmer households

Household size, education level of household members, and household labor capacity are important components of human assets, and their coefficients of influence on the well-being of farmer households are 0.366, 0.323, and 0.166, respectively, with their influences on the well-being of farmer households ranking first, second, and fourth, respectively. All three indicators passed the significance test at the 1% level, indicating that the higher the level of human assets, the higher the well-being level of farmer households. The Grain for Green policy has changed arable land holdings and increased other livelihood activities, such as forestry livelihoods. The Grain for Green policy released some human assets of farmer households from farming to other livelihoods, optimizing the allocation of human assets. Human assets are the basis for the participation of farmer household members in livelihood activities. Larger household sizes can contribute to an increased cumulative production time, leading to higher productivity. Higher levels of education mean that farmer households are better able to withstand external shocks. The external shocks faced by farmer households located in the agro-pastoral ecotone include natural disasters, declining crop and livestock prices, and epidemic diseases affecting livestock farming. Higher levels of education can help farmer households mitigate these shocks in two ways. Farmer households with a higher education level can use their knowledge to take effective measures against natural disasters that hit agriculture and against epidemics that harm livestock farming. In addition, higher education levels mean that farmer households can pursue more types of occupations, and they can choose to pursue other occupations to mitigate the negative effects of declining farm incomes on themselves. Higher levels of labor capacity mean that farmer households are more efficient in creating wealth. This suggests that human assets are a powerful driver of well-being of farmer households.

6. Natural assets affect the well-being of farmer households

The coefficient of the effect of arable land area on the well-being of farmer households is −0.208, which passed the significance test at the 1% level, indicating that arable land area is negatively correlated with the level of well-being of farmer households. The higher the arable land area of farmer households, the lower the level of well-being of farmer households. Zhangbei County has poor natural conditions and insufficient heat conditions. Moreover, a large amount of arable land is sloping and there are many stones on the surface of arable land. Farmer households engaged in grain production under such conditions have huge inputs and little returns, and even incur losses. Therefore, these reduce the level of well-being of farmer households. The arable land retired by farmer households is usually poor-quality farmland dominated by sloping fields. Therefore, the Grain for

Green policy improves the well-being of farmer households by reducing the quantity of poor-quality farmland.

7. Physical assets affect the well-being of farmer households

The coefficient of influence for the number of large production tools on the well-being of farmer households is 0.086, which passed the significance test at the 1% level, indicating that the higher the number of large production tools, the higher the level of well-being of farmer households. Most arable land retired by households is of inferior quality, such as sloping land, where the cultivation conditions are not conducive to the operation of large production tools. After the implementation of the Grain for Green policy, the quality of farmland held by farmer households increased, but the amount of holding decreased. The reduction in farmland holding led some farmer households to transfer their farmland to other farmers. A large amount of arable land was operated by local contractors. Retirement of poor-quality farmland and concentration of quality farmland make the use of large production tools much more efficient, which in turn increases the efficiency of the farmer households' use of farmland and provides the household with opportunities for other nonfarm part-time work, increasing household income. The physical assets of farmer households thus contribute to their level of well-being in terms of production efficiency and livelihood types.

8. Financial assets affect the well-being of farmer households

The coefficient of the effect of annual household expenditure on the well-being of farmer households is 0.157, which passed the significance test at the 1% level; the coefficient of the effect of the total value of livestock on the well-being of farmer households is 0.064, which passed the significance test at the 5% level, indicating that the higher the financial assets, the higher the well-being level of farmer households. The local farmer households pay for gifts to friends and relatives to maintain relationships. The large proportion of elderly people led to a large expenditure on health of farmer households. Gift expenditure and health expenditure affect the well-being of farmer households. Zhangbei County is located at an agro-pastoral ecotone, and rural livestock is more convenient and much more beneficial than farming for the same cost of investment. Before the implementation of the Grain for Green policy, farmer households put most of their time into agricultural cultivation and did not have much time to engage in livestock farming. The shift in labor to livestock farming after farmland has been retired has increased the level of well-being of farmer households by satisfying their dietary needs as well as reserving wealth.

#### 3.3.2. Analysis of Factors Influencing the Well-Being of Different Types of Farmer Households

The analysis of the impact of livelihoods of farmer households on the well-being of different types of farmer households reveals that the factors influencing the well-being differed between returned farmland households and nonreturned farmland households (Table 5). The results show adjusted $R^2$ values of 0.670 and 0.726, respectively. The goodness of fit of the two models reached 67.0% and 72.6%, respectively, indicating the good fit of the models. Household size, education attainment of household members, livelihood diversity, and area of arable land are common influencing factors for both returned farmland and nonreturned farmland households. Additionally, the level of well-being of returned farmland households is also influenced by annual household expenditure, household labor capacity, annual household income, and total value of livestock; that of nonreturned farmland households is also influenced by the number of large production tools.

Table 5 demonstrates that educational attainment of household members and household size remain the two most significant influencing factors for both returned farmland and nonreturned farmland households, and both passed the significance test at the 1% level. This indicates that human assets remain fundamental to the well-being of farmer households. Similarly, the area of arable land remains a significant influence on the well-being of both types of farmer households and is negatively correlated with the well-being of farmer households. The coefficient of impact of livelihood diversity on the well-being of returned farmland and nonreturned farmland households is 0.091 and 0.149, respectively, indicating that livelihood diversity has an impact on the well-being of different types of farmer households, but more so on nonreturned farmland households. Nonreturned farmland households have a lower livelihood diversity, and a richer livelihood diversity has a more significant effect on their well-being.

Annual household expenditure, household labor capacity, annual household income, and total value of livestock are significant influencing factors on the level of well-being of returned farmland households, with impact coefficients of 0.222, 0.203, 0.081, and 0.076, respectively. Annual household expenditure and household labor capacity are influencing factors for returned farmland households because these households have a higher proportion of elderly and young children, a higher dependency ratio, and higher expenditure on health care, which have greater impacts on their level of well-being. Annual household income is an influencing factor for returned farmland households because the average total assets of such households are lower than those of nonreturned farmland households, and to make up for the lack of total livelihood assets, household income needs to be increased. The total value of livestock becomes an impact factor for returned farmland households because they need to engage in other livelihood activities to earn an income after they have participated in the Grain for Green policy. As a typical agro-pastoral ecotone, Zhangbei County has a predominance of grassland, making it easier to raise livestock, and livestock farming has become the first choice for expanding household income.

The number of large production tools has a significant influence on the level of well-being of nonreturned farmland households, with an influence coefficient of 0.116, which passed the significance test at the 5% level. However, the number of large production tools is not an influential factor in the well-being of returned farmland households. This is because large production tools can be used by multiple farmer households and the demand for large production tools decreased after Grain for Green policy implementation.

## 4. Discussion

### 4.1. Geographical Differences and Similarities of Livelihoods of Farmer Households

The level of livelihood assets of farmer households in the mountainous areas of northern Hebei is low, because of the geographic limitations of such areas. However, the structural weight of livelihood assets in this study differs from other related studies. The high level of natural assets and lower level of human assets of farmer households in this study differs from the high level of human assets of farmer households in alpine ecologically fragile areas [36]. This may be related to geographical differences. The alpine ecological environment results in less available arable land, extremely low crop yields, and low livestock carrying capacity, which in turn leads to relatively low natural assets of farmer households and highlights the structural weight of human assets. This conjecture can be confirmed by the finding that rural reservoir migrants have the highest share of human assets and the second highest share of financial assets in their livelihood assets [37]. The loss of arable land by rural reservoir migrants due to reservoir construction led to an extremely low level of natural assets and increased the proportion of human assets; the large government compensation for migrant relocation raised the proportion of financial assets. Thus, livelihood assets of farmer households are related to the geographical environment and policy conditions of their regions, with significant geographical differences. Farmer households in alpine ecologically fragile areas addressed the problem of insufficient livelihood assets by tapping the potential of human assets and transforming the use of natural assets [36]. The present study's area is located in an agro-pastoral ecotone, which is also a kind of ecologically fragile area. Therefore, the government can learn from the above-mentioned practices and guide farmer households to explore the potential of their superior assets and change the way they use their inferior assets to improve their level of livelihood assets.

The comparison of the total livelihood assets of returned farmland and nonreturned farmland households revealed that the implementation of the Grain for Green policy in the mountainous region of northern Hebei did not improve the livelihood assets of farmer households. This is completely opposite to the impact on the livelihood assets of farmer households in the Qinba Mountain area of Gansu after the Grain for Green policy implementation [38]. The Qinba region experienced the vigorous development of an ecological and economic forest and improved infrastructure, which improved the livelihood assets of farmer households. In addition, the climatic conditions in the Qinba Mountains are more advantageous for forestry development than those in northern Hebei. The improvements in livelihood assets of farmer households in the Qinba region after the Grain for Green policy implementation are the result of a combination of human and natural factors. This result provides guidance for improving livelihood assets of farmer households after the implementation Grain for Green policy in our study area. The Grain for Green policy requires additional policies supporting and guaranteeing improvements in the livelihood assets of farmer households. Due to the inexperience of local farmer households in forestry development, the government needs to introduce relevant guiding policies for industrial development to support local forestry development and protect the livelihood assets of farmer households.

The livelihood diversity of returned farmland households was higher than that of nonreturned farmland households, but the difference was not significant. This finding is consistent with those of other related studies [38,39]. This suggests that the Grain for Green policy has a positive impact on the livelihood strategies of farmer households in different regions, but there is still much room for this positive impact to improve. Reforestation and grass restoration affect human assets by acting on natural assets, causing a portion of labor to be released from agricultural production. Theoretically, this labor force should have gone in other directions, and the livelihood diversity of farmer households should subsequently have increased, but the facts were not exactly as expected. The quality of the labor force is limited by education level, skills, the insufficient number of channels to work outside, and insufficient social contacts. Therefore, the government should increase cooperation with enterprises, build information exchange and promotion platforms, broaden access of farmers to outside information, encourage laborers to work seasonally, and guide farmers to nonfarm employment to solve the employment problem.

### 4.2. Grain for Green Policy Impacts on the Well-Being of Farmer Households

Human well-being is a comprehensive, multidimensional, and strongly subjective concept and is closely related to human living conditions and perceptions [40]. The mountainous areas of northern Hebei have harsh natural conditions and low levels of economic development. Farmer households living under such economic conditions have different perceptions of the benefits arising from the implementation of the Grain for Green policy. Related studies show that the subjective well-being of farmer households in poor mountainous areas is most influenced by the wealth factor, and that well-being enhancement depends on economic development [41,42]. The high contribution of the wealth level to the well-being of farmer households in the study area indicates that such households perceive the economic benefits of the Grain for Green policy most strongly. The analysis of the factors affecting the well-being of farmer households in this study also supports this assertion, and the economic benefits of the Grain for Green policy act on the well-being of farmer households in three main ways. First, the Grain for Green policy improves the well-being of farmer households through the reallocation of human assets. The retirement of farmland releases some of the labor force to engage in other more rewarding and less laborious livelihood activities. Second, the Grain for Green policy improves the well-being of farmer households through the optimization of natural assets. Farmer households choose to retire low-quality, sloping land, increasing their income and reducing their labor burden; the remaining higher-quality, irrigated land is conducive to more efficient use of large production tools, further increasing yields, household income, and the well-being of the farmer households. Third, the Grain for Green policy improves the well-being of farmer households by enhancing their financial assets. Farmer households who have retired farmland take advantage of livestock farming in the local agricultural–pastoral staggered zone, which can increase income and reserve wealth.

The primary objective of the Grain for Green policy is to enhance regional ecological quality. The policy has changed land-use types from cropland to woodland and grassland, driving changes in the ecosystem's service provisioning capacity [43]. The ecological benefits of the policy are reflected in the enhancement of ecosystem regulation and support services in the region [44], which are consistent with the results of our survey. A related study shows that ecosystem supply services have a greater impact on human well-being at lower levels of human economic development, but there is a lag in supply services [45]. This suggests that rural inhabitants in agro-pastoral ecotones have weaker perceptions of the regulating and supporting services, and stronger perceptions of supply services of the ecosystem. This is an important reason for the low contribution of ecosystem quality to the well-being of farmer households in the study area. Ecological improvement resulting from the Grain for Green policy is not the main path to improved well-being of farmer households. Therefore, the government should conduct various forms of publicity about the Grain for Green policy, to enhance recognition of farmers of the policy and their ecological perceptions.

The comparison with similar studies revealed that the Grain for Green policy did not significantly enhance the level of well-being of farm households [26,46], which corroborates the findings of our study. Liu et al. [46] believed that the ecological quality was improved faster than the socio-economic improvement in the relatively short period of time after the implementation of the Grain for Green policy. Therefore, the reason for this result may be that the Grain for Green policy has a certain lagging effect on promoting local economic development. You et al. [25] even pointed out that the implementation of the Grain for Green policy reduced the level of well-being of farm households, which is not consistent with the results of our study. This is due to the lack of government compensation to the farmers. The farmer households involved in our study all received policy subsidies from the government, including grain subsidies and monetary subsidies. Therefore, the government should do a good job in compensating the returned farmland households.

The mechanisms and pathways by which the implementation of ecological projects affects the well-being of farmer households vary across regions. The Grain for Green policy changes the ecosystem's service functions, which in turn can affect the well-being of farmer households. Therefore, the impact of ecological projects on the well-being of farmer households has been described in terms of changes in ecosystem service functions [26]. Due to weak perceptions of farmer households of the regulating and supporting functions of ecosystem services, the above research approach cannot well explain the mechanism of ecological projects effects on the well-being of farmer households from the perspectives of economic behavior and government security. Therefore, it is necessary to introduce livelihood characteristics into the study of such effects [47].

### 4.3. Limitations

This study used multiple linear regression to study the factors influencing the well-being of farmer households. This statistical model has been widely used in many fields and can effectively analyze combinations of factors, but it cannot analyze the interactions between the influencing factors. There is a certain spatial coupling relationship between farmer household livelihood characteristics and well-being levels, which cannot be measured using multiple linear regression.

Due to the unavailability of data related to farmer households before the Grain for Green policy implementation in Zhangbei County, this study can only illustrate the current situation of well-being of farmer households and cannot compare livelihoods and well-being levels of farmer households before and after policy implementation. Therefore, a comparison of livelihoods and well-being differences between returned farmland and nonreturned farmland households illustrates part of the impact of the policy on well-being of farmer households.

In the process of measuring livelihood assets of farmer households, some indicators are not included because they are difficult to quantify. For example, some ecological livelihood assets of farmer households cannot be effectively quantified and are not fully included in natural assets, which may affect the results of livelihood assets of farmer households.

## 5. Conclusions

This study systematically analyzed the impact of the Grain for Green policy on the well-being of farmer households in Zhangbei County, Hebei Province, China. The following three findings were obtained:

(1) The Grain for Green project caused changes in the livelihood of farmer households. The average livelihood diversity of farmer households was 3.008, and the returned farmland households (3.022) were higher than the nonreturned farmland households (2.975) in Zhangbei County. The average level of livelihood assets of farmer households in Zhangbei County was natural assets > social assets > financial assets > human assets > physical assets. The comparison between returned farmland households and nonreturned farmland households showed that the livelihood diversity of returned farmland households was higher than that of nonreturned farmland households, but the total livelihood assets were lower than those of nonreturned farmland households.

(2) The Grain for Green policy had an improving effect on the level of well-being of farmer households, but the effect was not significant. The well-being level of all farmer households in Zhangbei County was 0.517, with an overall low level of well-being. The well-being level of returned farmland households was slightly higher than that of nonreturned farmland households. The wealth level accounted for the largest share of the well-being structure of farmer households, and the contribution of ecological quality was the smallest.

(3) The degree of influence of the factors affecting the level of well-being of farmer households in Zhangbei County varied significantly. There were three main paths through which the Grain for Green policy affected the well-being of farmer households: by reallocating human assets, optimizing natural assets, and enhancing financial assets. The factor of household size had the highest degree, at 0.366, while educational attainment of household members, household labor capacity, annual household expenditure, livelihood diversity, number of large production tools, and total value of livestock were also important drivers of household well-being, and area of arable land was negatively associated with household well-being. There were also differences in the factors influencing the level of well-being of different types of farmer households.

**Author Contributions:** Conceptualization, K.W. and P.S.; methodology, K.W.; software, X.W.; validation, P.S., J.M. and N.L.; formal analysis, K.W., P.S. and J.Z.; investigation, K.W., P.S., X.W., J.M., N.L. and J.Z.; writing—original draft preparation, K.W. and J.M.; writing—review and editing, K.W. and P.S.; supervision, P.S. All authors have read and agreed to the published version of the manuscript.

**Funding:** This research was funded by the National Natural Science Foundation of China (No. 42101258) and the Natural Science Foundation of Shandong (No. ZR2019QD006).

**Institutional Review Board Statement:** Not applicable.

**Informed Consent Statement:** Not applicable.

**Data Availability Statement:** The datasets used in this article can be obtained by readers after the article is published online.

**Conflicts of Interest:** The authors declare no conflict of interest.

## References

1. Hao, H.G.; Gou, M.M.; Zhang, H.Y.; Zhang, Q.; Liu, Y.J. Assessing ecological compensation policies based on ecosystem services and human well-being: A review of recent progress. *Acta Ecol. Sin.* **2018**, *38*, 6810–6817.
2. Shao, Q.Q.; Liu, S.C.; Ning, J.; Liu, G.B.; Yang, F.; Zhang, X.Y.; Niu, L.N.; Huang, H.B.; Fan, J.W.; Liu, J.Y. Assessment of ecological benefits of key national ecological projects in China in 2000–2019 using remote sensing. *Sci. Geogr. Sin.* **2022**, *77*, 2133–2153.

3. Li, W.Q.; Zhao, X.Y.; Du, Y.X.; Ma, P.Y. Spatio-temporal changes of the coupling relationship between ecosystem services and residents' well-being in Qinba Mountains Area. *J. Nat. Resour.* **2021**, *36*, 2522–2540. [CrossRef]
4. Costanza, R.; d'Arge, R.; De Groot, R.; Farber, S.; Grasso, M.; Hannon, B.; Limburg, K.; Naeem, S.; O'Neill, R.V.; Paruelo, J.; et al. The value of the world's ecosystem services and natural capital. *Nature* **1997**, *387*, 253–260. [CrossRef]
5. Sanchez-Canales, M.; Benito, A.L.; Passuello, A.; Terrado, M.; Ziv, G.; Acuna, V.; Schuhmacher, M.; Elorza, F.J. Sensitivity analysis of ecosystem service valuation in a Mediterranean watershed. *Sci. Total Environ.* **2012**, *440*, 140–153. [CrossRef]
6. Otoiu, A.; Titan, E.; Dumitrescu, R. Are the variables used in building composite indicators of well-being relevant? Validating composite indexes of well-being. *Ecol. Indic.* **2014**, *46*, 575–585. [CrossRef]
7. Aguado, M.; González, J.A.; López-Santiago, C.; Montes, C. Exploring subjective well-being and ecosystem services perception along a rural–urban gradient in the high Andes of Ecuador. *Ecosyst. Serv.* **2018**, *34*, 1–10. [CrossRef]
8. Dube, L.C.; Chatterjee, S. Assessing livelihood impact of forest carbon projects using sustainable livelihood framework. *Mitig Adapt Strateg. Glob Chang.* **2022**, *27*, 49. [CrossRef]
9. Ciftcioglu, G.C.; Ebedi, S.; Abak, K. Evaluation of the relationship between ornamental plants-based ecosystem services and human wellbeing: A case study from Lefke Region of North Cyprus. *Ecol. Indic.* **2019**, *102*, 278–288. [CrossRef]
10. Jones, S.K.; Boundaogo, M.; DeClerck, F.A.; Estrada-Carmona, N.; Mirumachi, N.; Mulligan, M. Insights into the importance of ecosystem services to human well-being in reservoir landscapes. *Ecosyst. Serv.* **2019**, *39*, 100987. [CrossRef]
11. Fu, B.; Wang, Y.K.; Xu, P.; Yan, K. A Review of Coupled Relationship of Rural Livelihoods and Ecosystem Services. *Ecol. Econ.* **2017**, *33*, 142–145+151.
12. Guan, D.J.; Sun, L.L.; Zhou, L.L. Construction and Application of Coupling Model of Ecosystem Service and Farmers' Livelihood in Three Gorges Reservoir Area. *Res. Soil Water Conserv.* **2020**, *27*, 269–277. [CrossRef]
13. Wang, F.C.; Zheng, H.; Zhang, W.; Wang, H.; Peng, W.J. Regional differences and the driving mechanism of relationships between rural household livelihood and ecosystem services: A case study in upstream watershed of Miyun Reservoir, China. *Chin. J. Appl. Ecol.* **2021**, *32*, 3872–3882. [CrossRef]
14. Qiao, X.N.; Zhang, T.; Yang, Y.J.; Niu, H.P.; Yang, D.G. Spatial flow of ecosystem services and impacts on human well-being in the Weigan River Basin. *Resour. Sci.* **2017**, *39*, 533–544.
15. Zhang, J.F.; Deng, W.; Zhu, C.L.; Zhao, Y.L. Spatial Relationship and Its Dynamic Features of Ecosystem Services and Human Wellbeing in the Upper Reaches of Minjiang River. *J. Mount. Sci.* **2017**, *35*, 388–398. [CrossRef]
16. Huo, R.; Xu, X.Y.; Gao, J.L.; Zhu, T.T.; Lu, B. Ecosystem Services' Spatial Characteristics Patterns of Ecosystem Services and Their Relationships with Residents' Well-being of Coal Resource-based Cities. *Chin. J. Soil Sci.* **2020**, *51*, 31–39. [CrossRef]
17. Yang, X.T.; Qiu, X.T.; Xu, Y.; Zhu, F.B.; Liu, Y.W. Spatial heterogeneity and dynamic features of the ecosystem services influence on human wellbeing in the West Sichuan Mountain Areas. *Acta Ecol. Sin.* **2021**, *41*, 7555–7567.
18. Yuan, D.B.; Chen, M.Q.; Liao, C.R.; Xie, X.X.; Liao, X.B.; Yao, D.L. Analysis of Subjective Well-being of Farmers with Land Transfer and Its Influencing Factors: From the Perspective of Livelihood Capitals. *Chin. Land Sci.* **2019**, *33*, 25–33.
19. Liu, D.; Chen, H.; Zhang, H.; Shi, Q.Q.; Geng, T.W. The impact of ecosystem services on human well-being and its group differences in the loess hilly and gully region. *Geogr. Res.* **2022**, *41*, 1298–1310.
20. Zheng, D.F.; Wang, Y.Y.; Cao, Y.Q.; Wang, Y.H.; Hao, S.; Lv, L.T. Classification and spatiotemporal patterns of ecological well-being based on ecosystem services: Taking China's prefecture-level and above cities for example. *Resour. Sci.* **2020**, *42*, 1110–1122. [CrossRef]
21. Li, H.M.; Zhang, X.; Zhang, J.F.; Zhang, A.L.; Yang, H.Z. Herdsman's multidimensional well-being in response of natural resources protection in the source region of the Yellow River, China: Case study based on household investigation in Maduo County. *Acta Ecol. Sin.* **2014**, *34*, 6767–6777.
22. Yuan, W.T.; Qiao, D.; Ke, S.F.; Hou, Q.; Yan, R.H. How to Improve Ecological Compensation Mechanism from the Perspective of Resource Opportunity Cost? A Case Study Based on "Welfare Upside Down" of Forest Resources Compensation in State-owned Forest Area. *Chin. Rural Surv.* **2022**, *2022*, 59–78.
23. Li, N.J.; Cao, G.Y.; He, B.H.; Luo, G.L. On the Relationship Between the Change in Farmer Wellbeing and Ecosystem Services—A Case Study of Wuling-Qinba Contiguous Destitute Areas in Chongqing. *J. S. Univ. Nat. Sci. Ed.* **2017**, *39*, 136–142. [CrossRef]
24. Xiong, Y.; Hou, K.L.; Zheng, S.R.; Zhang, K.; Yang, T.H.; Zhao, D.D.; Sun, B.; Chen, L. Relationship between Farmer's Well-Being and Ecosystem Services in Hilly and Mountainous Areas of South China Based on Structural Equation Model: A Case Study of Lechang in Guangdong Province. *Trop. Geogr.* **2020**, *40*, 843–855. [CrossRef]
25. You, Z.; Wu, T.; Gong, M.Q.; Zhen, S.Q.; Cheng, J.H. The Impact of the Grain for Green Program on Farmers' Well-Being and Its Mechanism—Empirical Analysis Based on CLDS Data. *Front. Ecol. Evol.* **2022**, *10*, 22. [CrossRef]
26. Liu, X.L.; Zhang, B.; Zheng, Q.R.; He, X.Q.; Zhang, T.F.; Jia, Y.Q.; Luo, Z.M. Impacts of Converting Farmland into Forests on Farmer Well-Being in the Earth-Rock Mountain Areas of the Loess Plateau. *Resour. Sci.* **2014**, *36*, 397–405.
27. Ding, Y.H.; Yao, S.B. A comparative analysis of the impacts of the sloping land conversion program on the well-being of households in the Yellow and Yangtze river basins. *J. Arid Land Resour. Environ.* **2017**, *31*, 45–50. [CrossRef]
28. Guo, L.P.; He, M.Y.; Chen, S.Y. The impact of farmers' livehood capital on the adoption of different preference technologies——An analysis based on the sustainable livelihood framework. *Chin. J. Agric. Resour. Region. Plan.* **2022**, *43*, 197–205.
29. Liu, G.G.; Ge, Y.X.; Li, Y.; Zhang, H.N. Impact of ecological compensation on sustainable livelihood capacity of farmers in water source area—Based on the improved DFID livelihood analysis framework. *Chin. J. Agric. Resour. Region. Plan.* **2022**, *43*, 97–109.

30. Liu, W.; Yu, Q.Q. Impact of Poverty Alleviation Resettlement on Rural Household Livelihood Resilience in Southern Shaanxi. *Geogr. Geo.-Info. Sci.* **2023**, *39*, 105–110.
31. Bai, W.; Wu, B.G. An Analysis on Rural Residents' Subjective Well-being and the Determinants: An Empirical Survey based on Rural Household Data from Ten Counties in Five Provinces. *Chin. Rural Surv.* **2017**, *2017*, 41–51+141–142.
32. Xu, H.S.; Le, Z. Livelihood capital, livelihood risk and farmers' livelihood strategy. *Issue. Agric. Econ.* **2012**, *33*, 100–105. [CrossRef]
33. Wu, Y. Poor Mountain Farmers Livelihood Capital Impact on Livelihoods Strategy Research: Based on the Survey Data Pingwu and Nanjiang County of Sichuan Province. *Issue. Agric. Econ.* **2016**, *37*, 88–94+112. [CrossRef]
34. Yu, Z.L.; Yang, X.J.; Yang, T. Exploring conditions, determinants and mechanisms of rural households' adaptability to tourism development: A case study of Jinsixia in Qinling Mountains. *Sci. Geogr. Sin.* **2013**, *68*, 1143–1156.
35. Deng, Q.J.; Li, W.; Zhu, Q.J.; Ma, X.M.; He, J.; Ma, X.J.; Luo, X. An analysis of the characteristics of water storage structure and the practice of groundwater exploration in the basalt area of Zhangbei County, Bashang, Hebei Province. *Geol. Bul. Chin.* **2020**, *39*, 1899–1907.
36. Guo, X.L.; Li, W.P.; Sun, G.J.; Zhang, L.P. Analysis of Livelihood Capital and Its Coupling Coordination Degree of Farmers in Ecologically Vulnerable Alpine Areas—A Case of Xiahe County in Gannan Tibetan Autonomous Prefecture. *Res. Soil Water Conserv.* **2022**, *29*, 330–335+343. [CrossRef]
37. Wu, J.; Chen, S.; Zhou, K.; Chen, X. Influence of Livelihood Capital of Rural Reservoir Resettled Households on the Choice of Livelihood Strategies in China. *Water* **2022**, *14*, 4055. [CrossRef]
38. Sun, Y.G.; Wang, C.S. Project of Conversion of Cropland to Forest (Grass) Effect on Farmers' Livelihood: A Case Study of Qinba Mountain Area in Gansu Province. *Issue Forest Econ.* **2017**, *37*, 54–58+106. [CrossRef]
39. Wei, H.L.; Bai, X. The Performance and Mechanism of the Impact of the Sloping Land Conversion Program on Households' Livelihood Strategies. *Ecol. Econ.* **2019**, *35*, 121–127.
40. Liu, D.Q.; Zhang, J.X.; Gong, J.; Qian, C.Y. Spatial and Temporal Relations among Land-Use Intensity, Ecosystem Services and Human Well-Being in the Longzhong Loess Hilly Region: A Case Study of the Anding District, Gansu Province. *Acta Ecol. Sin.* **2019**, *39*, 637–648.
41. Shen, J.Y.; Han, X.; Hou, Y.L.; Wen, Y.L. Subjective Well-Being of Households in Rural Poverty Regions in Xiangxi, Hunan Province. *Ecol. Econ.* **2014**, *36*, 2174–2182.
42. Li, K.M.; Yao, L.L.; Shi, Y.; Zhang, D.; Lin, Y.M. Estimate Well-Being of Urban and Rural Residents in the Yellow River Basin and Its Spatial-Temporal Evolution. *Ecol. Econ.* **2022**, *38*, 222–229.
43. MA (Millennium Ecosystem Assessment). *Ecosystems and Human Well-Being*; Island Press: Washington, DC, USA, 2005.
44. Fu, B.J.; Zhang, L.W. Land-use change and ecosystem services: Concepts, methods and progress. *Prog. Geogr.* **2014**, *33*, 441–446.
45. Wang, D.S.; Zheng, H.; Ouyang, Z.Y. Ecosystem services supply and consumption and their relationships with human well-being. *Chin. J. Appl. Ecol.* **2013**, *24*, 1747–1753. [CrossRef]
46. Liu, X.L.; Zhang, B.; Wu, P.S.; Luo, Z.M.; Jia, Y.Q. The coupling effect of converting farmland into forestland ecological economic system based on the farmers well-being in the earth-rock mountains areas of loess plateau—A case study of Ningwu County. *J. Nw. Univ. Nat. Sci. Ed.* **2016**, *52*, 113–117. [CrossRef]
47. Wu, Z.L.; Cao, H.; Jin, M. Influence Mechanism of Wetland Conservation on the Livelihood Outcomes of Farmers': An Empirical Study Based on the Survey of Rural Households in Natural Reserves. *J. Ecol. Rural Environ.* **2022**. [CrossRef]

MDPI AG
Grosspeteranlage 5
4052 Basel
Switzerland
Tel.: +41 61 683 77 34

*Land* Editorial Office
E-mail: land@mdpi.com
www.mdpi.com/journal/land

www.ingramcontent.com/pod-product-compliance
Lightning Source LLC
Chambersburg PA
CBHW041400210926
51850CB00013B/3141
* 9 7 8 3 7 2 5 8 4 4 7 9 1 *